T0202433

Lecture Notes in Computer Science 13820

More information about this series at https://link.springer.com/bookseries/558

Ileana Buhan · Tobias Schneider (Eds.)

Smart Card Research and Advanced Applications

21st International Conference, CARDIS 2022
Birmingham, UK, November 7–9, 2022
Revised Selected Papers

Editors
Ileana Buhan (iD)
Radboud University
Nijmegen, The Netherlands

Tobias Schneider (iD)
NXP Semiconductors
Gratkorn, Austria

ISSN 0302-9743   ISSN 1611-3349 (electronic)
Lecture Notes in Computer Science
ISBN 978-3-031-25318-8   ISBN 978-3-031-25319-5 (eBook)
https://doi.org/10.1007/978-3-031-25319-5

This Springer imprint is published by the registered company Springer Nature Switzerland AG
The registered company address is: Gewerbestrasse 11, 6330 Cham, Switzerland

Preface

These are the proceedings of the 21st International Conference on Smart Card Research and Advanced Applications (CARDIS 2022). This year's CARDIS was held in Birmingham, UK, and took place from November 7 to November 9, 2022. The Centre for Cyber Security and Privacy of the University of Birmingham, UK, organized the conference this year. A one-day Fall School event accompanied CARDIS 2022.

CARDIS has been the venue for security experts from industry and academia to discuss developments in the security of smart cards and related applications since 1994. Smart cards play an increasingly important role in our daily lives through their use in banking cards, SIM cards, electronic passports, and devices of the Internet of Things (IoT). Thus, it is naturally important to understand their security features and develop sound protocols and countermeasures while maintaining reasonable performance. In this respect, CARDIS aims to gather security experts from industry, academia, and standardization bodies to make strides in the field of embedded security.

The present volume contains 15 papers that were selected from 29 submissions. The 30 members of the Program Committee evaluated the submissions, wrote 91 reviews, and engaged in extensive discussions on the merits of each article. Three invited talks completed the technical program. Maria Eichlseder discussed fault attacks from a cryptographer's perspective in her keynote *"Fault Attacks and Cryptanalytic Countermeasures"*. Dan Page discussed the fundamental role of the ISA in the design of cryptographic systems and its role in the emergent discipline of micro-architectural leakage in his keynote *"How do you solve a problem like the ISA?"*. Ivan Kinash and Mikhail Dudarev from Licel reflected on the evolution of smartcards in the past twenty years during their keynote talk entitled *"Open source and industry standards: Supporting eUICC and eSIM development with Java Card technology"*.

Organizing a conference can be challenging, but we are happy to report that this year's conference went surprisingly smoothly. We express our deepest gratitude to David Oswald and all the members of his team who enabled CARDIS 2022 to succeed. We thank the authors who submitted their work and the reviewers who volunteered to review and discuss the submitted articles. The authors greatly appreciate the amazing work of our invited speakers, who gave entertaining and insightful presentations. We thank Springer for publishing the accepted papers in the LNCS collection and the sponsors Licel, Infineon, hardwear.io, NewAE, NXP, and Rambus for their generous financial support. We are grateful to the CARDIS steering committee for allowing us to serve as the program chairs of such a well-recognized conference. Finally, we thank all presenters, participants, and session chairs, physically and online, for their support in making this CARDIS edition a great success.

December 2022

Ileana Buhan
Tobias Schneider

Organization

General Chair

David Oswald   University of Birmingham, UK

Program Committee Chairs

Ileana Buhan   Radboud University, The Netherlands
Tobias Schneider   NXP Semiconductors, Austria

Programm Committee

Melissa Azouaoui	NXP Semiconductors, Germany
Davide Bellizia	Telsy, Italy
Shivam Bhasin	Nanyang Technological University, Singapore
Jakub Breier	Silicon Austria Labs, Graz, Austria
Olivier Bronchain	UCLouvain, Belgium
Łukasz Chmielewski	Radboud University, The Netherlands & Masaryk University, Czech Republic
Siemen Dhooghe	KU Leuven, Belgium
Vincent Grosso	CNRS and UJM, France
Annelie Heuser	Univ Rennes, CNRS, Inria, IRISA, France
Elif Bilge Kavun	University of Passau, Germany
Juliane Krämer	University of Regensburg, Germany
Roel Maes	Intrinsic ID, The Netherlands
Ben Marshall	PQShield, UK
Debdeep Mukhopadhyay	Indian Institute of Technology Kharagpur, India
Colin O'Flynn	NewAE Technology Inc., Canada
Kostas Papagiannopoulos	University of Amsterdam, The Netherlands
Guilherme Perin	Radboud University, The Netherlands
Thomas Pöppelmann	Infineon Technologies, Germany
Romain Poussier	ANSSI, France
Thomas Roche	NinjaLab, France
Pascal Sasdrich	Ruhr-Universität Bochum, Germany
Patrick Schaumont	Worcester Polytechnic Institute, USA
Peter Schwabe	MPI-SP, Germany & Radboud University, The Netherlands
Johanna Sepúlveda	Airbus Defence and Space, Germany

Sujoy Sinha Roy	Graz University of Technology, Austria
Marc Stöttinger	RheinMain University of Applied Science, Germany
Srinivas Vivek	IIIT Bangalore, India
Yannick Teglia	Thales, France
Yuval Yarom	University of Adelaide and Data61, Australia
Nusa Zidaric	Leiden University, The Netherlands

Additional Reviewers

Thomas Aulbach

Guillaume Bouffard

Tim Fritzmann

Suvadeep Hajra

Daniel Heinz

Jan Jancar

Arpan Jati

Victor Lomne

Soundes Marzougui

Nimish Mishra

Guenael Renault

Debapriya Basu Roy

Patrick Struck

Contents

Evaluation Methodologies

Attacking NTRU

Next-Generation Cryptography

Physical Attacks

Time's a Thief of Memory

Breaking Multi-tenant Isolation in TrustZones Through Timing Based Bidirectional Covert Channels

Nimish Mishra[1]([✉]) [iD], Anirban Chakraborty[1] [iD], Urbi Chatterjee[2] [iD],
and Debdeep Mukhopadhyay[1] [iD]

[1] Indian Institute of Technology, Kharagpur, India
nimish.mishra@kgpian.iitkgp.ac.in, anirban.chakraborty@iitkgp.ac.in,
debdeep@cse.iitkgp.ac.in
[2] Indian Institute of Technology, Kanpur, India
urbic@cse.iitk.ac.in

Abstract. ARM TrustZone is a system-on-chip security solution that provides hardware guarantees to isolate the untrusted applications running in the normal world from sensitive computation and data by placing them in the secure world. In a multi-tenant scenario, such isolation is paramount to protect tenants from each other and is guaranteed by partitioning resources (memory, peripherals, etc.) between the tenants. Several third-party defence mechanisms add to this isolation through techniques like statically whitelisting communication channels. Consequently, two tenants cannot communicate with each other except using two legitimate channels: (1) shared memory and (2) legitimate API calls. However, we show that seemingly simple covert channels can be created to break this isolation and create a third illegitimate channel. We use simple thread counters and TrustZone configurations to demonstrate a break in Trustzone's isolation through a non-root, cross-core, cross-user, bidirectional (secure to normal world and vice-versa), cross-world covert channel on OP-TEE implementation on ARM TrustZone that, by design, places no limit on channel capacity. Our channel bypasses the established defence mechanisms (which mainly target cache occupancy measurements and performance monitoring units) in ARM TrustZone, achieves a maximum bandwidth of 130 KBps with a 2.5% error rate, and allows arbitrary code execution in the secure world leading to *denial-of-service attacks*.

Keywords: Trusted execution environments · ARM TrustZone · OP-TEE · Denial-of-service · Trusted computing platforms

1 Introduction

Internet-of-Things (IoTs) have opened up ubiquitous sensing, communicating and actuating network with information sharing across several hardware and software platforms, ideally infused in various dimensions of modern life. Normally, IoT nodes house a variety of applications that provide a multitude of

I. Buhan and T. Schneider (Eds.): CARDIS 2022, LNCS 13820, pp. 3–24, 2023.
https://doi.org/10.1007/978-3-031-25319-5_1

services to the end-users. In the early days of IoTs, these applications usually belonged to one vendor. But as the world moves into an increasingly inter-dependent IoT ecosystem, *multi-tenant* IoT nodes are becoming very common [31]. In a business setting, an IoT node can be home to multiple tenants providing a wide variety of services to end-users. A *tenant* is very much a service provider to the end-customer, and is considered to be inherently trustworthy.

However, a critical security question [4,25,30] arises in the case of multi-tenant scenarios: *how to protect tenants from each other?* More precisely, while each tenant is, in isolation, considered to be trustworthy, can we carry forward this trust when multiple tenants (possibly with competing interests) come together to deliver services to end users? A solution to this problem is hardware-backed *tenant-isolation*: one tenant is virtually isolated from another tenant in terms of memory used and peripherals interfaced with [3,10]. One such hardware-backed isolation mechanism is Trusted Execution Environment (TEE). As ARM dominates the mobile chipset and IoT market, the *ARM Trustzone* [2] provides the hardware trust anchor for different TEEs from multiple vendors. ARM Trustzone has been an integral part of ARM chipsets since ARMv6[1], including Cortex-A (for mobile devices) and Cortex-M (for IoT devices) families of processors. The Trustzone provides an execution context for security-critical applications, such as user authentication, mobile payment, etc. It essentially partitions the *system-on-chip* hardware and software into two virtual execution environments: *secure world* or **Trusted Execution Environment (TEE)** and *normal world* or **Rich Execution Environment (REE)**. Applications running in REE are called *client applications* (CAs), while the ones running in TEE are called *trusted applications* (TAs). The REE supports a complex software stack and thus can be prone to severe software bugs, leading to leakage sources. In this context, TEE provides the necessary isolation guarantee such that the integrity and confidentiality of sensitive data are not compromised. Shared hardware resources such as storage, peripherals, etc. are made private to each world through hardware-powered mechanisms. There are also third-party defence mechanisms like *SeCReT* [21] (or alternatively the modified version of *SeCReT* [22]) which force each tenant to declare/whitelist the CAs allowed to access their services.

1.1 Can Tenants Covertly Communicate in ARM TrustZone?

Although ARM Trustzone and TEEs have been in extensive use in a variety of IoT and mobile devices for years, in-depth security analysis of these systems, especially in a multi-tenant environment, is rarely available in literature. In this work, we motivate the following question: *given the tenant isolation guarantees in ARM TrustZone, can two tenants covertly communicate?* We note that our goal in this work, given the premise that all legitimate tenant communication is monitored, is to *establish a break in tenant isolation by creating an illegitimate communication channel.* While it is true that two tenants can simply encrypt

---

[1] The current ARM version used in a majority of processors is ARMv8.

communication over a legitimate communication channel, there still remains the fact that a communication is indeed *happening* between two tenants. Our methodology finds a way to *hide* this fact- that any sort of communication is happening at all between two tenants- from ARM TrustZone or any other third-party defence in place. Concretely, the entirety of system forensics depends on logging [23,24] - a logger that records events occurring in the system throughout the duration of the run. Legitimate communication channels will be recorded by good loggers since these channels emit function call traces which are easy to capture. Furthermore, there are mechanisms to prevent attackers from editing these logs [24], because of which, a cross-tenant communication channel (that we are targeting in this paper) cannot remain hidden. However, our covert channel defeats such logging mechanisms because, as we note in later sections (cf. Sect. 4 and Sect. 5), it does not require direct interaction between the channel actors at any point in time. Even for setting up the channel, the channel actors do not directly collaborate; they take help of a third *innocent* actor. As we note in later sections (cf. Sect. 4 and Sect. 5), this *innocent* actor is actually whitelisted by defence mechanisms like *SeCReT*, and thereby, no suspicion is raised even if its actions are logged. Note that even this *innocent* actor is oblivious of the fact that it is aiding in setting up a covert channel.

We consider a multi-tenant scenario with the following assumptions, enforced by ARM TrustZone to ensure security in multi-tenant execution model, borrowed from similar attack models discussed in literature [17,33].

– Each tenant is (in isolation) trustworthy, which means each TA and CA of a tenant is considered benign as far as its own execution is concerned.
– Defense mechanisms like *SeCReT* [21] are in place. More precisely, each tenant declares a whitelist of CAs that can be allowed to legitimately establish a communication channel with the tenant's TAs. Naturally, declaring any other tenant's CA as whitelisted, while not exactly being considered as malicious, is at least noted with interest by the defense mechanism and is a probable point of interest during forensics in light of a security incident.
– Two TAs belonging to different tenants may choose to communicate with each other, honoring the GlobalPlatform Internal Core API specification [15]. Since TAs are considered to be inherently trustworthy, if two TAs choose to communicate with each other, ARM TrustZone and additional defense mechanisms do not block this communication. However, they do note it with interest and it is a probable point of interest during forensics in light of a security incident.

In the stated attack model, no CA of one tenant should be able to interchange information with the TA of another tenant. Two TAs belonging to two tenants may establish communication, but only within the confines of the API specification. Thereby, *the **goal** of the entire setting is to establish a covert channel between the CA of one tenant and the TA of another tenant such that this channel remains invisible to ARM TrustZone and the defense mechanisms (like SeCReT [21]) in place, thereby breaking the guarantees of tenant isolation.*

1.2   Limitations of Related Prior Works

We now compare our approach with prior works that target ARM TrustZone in recent years and state the general conceptual improvements we make in our attack vector. We do not consider literature that: ① requires direct hardware access, ② does not have a *remote* adversary in its attack model, for e.g., EM leakage based microarchitectural attacks [5], ③ is not related to microarchitectural attacks in general [28], ④ demonstrates attacks like thermal covert channels [29] which have strong proposed defences [19].

We now review cache-based attacks on TrustZone. Zhang *et al.* [34] demonstrate a timing-based side-channel exploiting cache contentions through prior collected channel statistics. They further exploit last level cache flushes [36] and reloads to mount side-channel attacks. Other works like [35] rely on cache contentions too. Similarly, Lipp *et al.* [26] demonstrate practical non-rooted cache attacks and create covert channels through the same. In [11], a cross-world covert channel based on cache occupancy measurements has also been presented. However, there are tough defences in place against such attacks, for e.g, undocumented features such as *AutoLock* [16] and cache randomization mechanisms [32], that pose solid defences to the generality of cache-based attacks. *AutoLock* on ARM processors disallows any cross-core attempts to evict cache lines from last-level cache if the victim lines are present in lower level caches. The works summarised above do not take into account such countermeasures and are thus fairly pessimistic indicators of ARM's vulnerability to cache-based attacks.

We now discuss how the covert channels in [26] and [11] fail in our case. Lipp *et al.* [26] design a covert channel, established between two unprivileged processes, that transfer data depending on a cache hit or a cache miss of a shared library address. Notwithstanding the limit on the maximum information transfer through the covert channel (since the number of addresses in the library is limited) and the overhead of cache-based operations, such an approach is irrelevant in our case since REE and TEE share no library. Similarly, Cho *et al.* [11] not only bears the overhead involving cache operations but also involves a fairly restrictive synchronous attack model and requires access to ARM's *performance monitoring unit* (PMUs). Since a defender can monitor changes in *performance monitors event counter selection register* (PMSELR) and the *performance monitors selected event type register* (PMXEVTYPER), it increases the risk of getting detected. The approach proposed in [11] is easily detectable, hence is fairly limited in modern attack scenarios (especially on IoT edge nodes which are expected to have human incident response teams periodically looking at their security [6]).

We now present some more defence mechanisms that need to be accounted for while designing attacks on Trustzones - *SeCReT* [21] (or alternatively the modified version of *SeCReT* [22]) and function stub defences [8]. *SeCReT* [21] allows only authenticated/whitelisted REE applications to initiate a communication with TEE. In contrast, function stubs [8] encapsulate a sensitive function with stubs whose job is to ① pollute the cache to prevent cache attacks, and ② pad the run time of the sensitive function to prevent any data leakage through timing measurement. Some other defence mechanisms like disallowing cache flush instructions in instruction set architectures [27] also pose challenges since several cache based attacks rely on reliably flushing cache lines.

1.3   Attack Target and Our Contributions

We choose Open Portable Trusted Execution Environment (OP-TEE) [7] as our target. It is one of the most popular open-source implementations of ARM TrustZone with major industrial players involved in its design and development (otherwise known as the Trusted Firmware project) and moderate to large scale production systems like Apertis [1] and iWave systems [20] mentioning their use-cases with OP-TEE integrations. Before proceeding, we set up expectations for a basic security level from OP-TEE that has suffered several attacks in the past [13] prompting the developers to note that *the Raspberry Pi3 OP-TEE port does not have hardware setup enabling secure boot, memory, and peripherals*. Although true, this bears little significance to our work as we attack none of these. We rather show a fundamental problem: *break in the promised isolation* through a covert channel that is set up through legitimate REE and TEE operations. Briefly, we present the following contributions in this work.

1. We develop a non-root, cross-user, cross-core, cross-world covert channel that breaks OP-TEE's implementation of ARM Trustzone, while defeating the state-of-the-art defences in literature and allowing bidirectional transfer of an unbounded amount of information (from normal world to secure world and vice-versa, between two tenants).
2. We demonstrate that our covert channel allows breaking the promised isolation between tenants, in the presence of ARM TrustZone and third-party defences. We succeed in establishing an illegitimate bidirectional communication channel in addition to the legitimate communication channels like shared memory and API calls.
3. We show the execution of arbitrary shellcode (excluding unsupported POSIX syscalls by OP-TEE OS) sent from normal world of one tenant to secure world of another tenant, resulting in a novel *denial-of-service (DoS) attack*.

1.4   Paper Organization

The rest of the paper is organised as follows. In Sect. 2, we present the background on OP-TEE implementation on ARM TrustZone and how TrustZone isolates its tenants. In Sect. 3, we present our attack model as well as the design decisions in Sect. 3.2. In Sect. 4, we present the specifics of the secure world to normal world covert channel; while in Sect. 5, we present the specifics of the reverse directional channel. In Sect. 6, we present some experiments evaluating our attack and discuss potential countermeasure ideas in Sect. 7. Finally, we conclude our paper in Sect. 8.

2   Background

2.1   OP-TEE Implementation

OP-TEE is a TrustZone implementation for Cortex-A cores with design goals - isolation, small memory footprint, and portability. All architectural design decisions conform to GlobalPlatform's specifications for embedded applications on secure hardware. The REE is modelled by the GlobalPlatform Client API specifications [14], while the TEE is governed by the GlobalPlatform Internal Core API specification [15].

Intuition for the Isolation. ARMv8 adopts a similar convention of privilege rings as Linux. As referred in Fig. 1, the REE and TEE are divided into two layers for granular control: Exception layers (EL) 0 and 1, which provide granular control over the actions of the REE and TEE. EL0 usually houses the userspace applications while EL1 houses the kernel. On the REE side, Linux is used as the operating system; while on the TEE side, a OP-TEE OS is used as the operating system. We omit the shared cache region between TEE EL0 and REE EL0 (which all cache-based attacks on ARM TrustZone use) because it is not relevant to the paper. Exception Layer 2 houses the secure monitor call (SMC) handler allowing cross-world communication through software interrupts, triggering context switches. OP-TEE also has a collection of interrupts named *supervisor calls* (SVC) that allow the transfer of control from EL0 to EL1 in the same world. Briefly, apart from the shared memory and these interrupts, *there is no contact between the REE and the TEE*. This design implies that even if the REE has been compromised, TEE is still secure.

Intuition for the World Switching. OP-TEE ensures the SMC interrupt is non-maskable and controlled by EL2. On an SMC interrupt, EL2 changes the ownership of the processor unless the current world is done executing or EL2 receives a non-maskable interrupt from the idle world. Within EL1 in both worlds, normal scheduling policies apply. Additionally, the ARM Generic Interrupt Controller (GIC) can also signal either world. Without loss of generality, let us assume the GIC signals REE. If REE is in possession of the processor, it branches to the known interrupt service routine (ISR). If TEE is running, it first relinquishes control of the processor to the REE through a SMC interrupt, and then the REE jumps to the known ISR to handle GIC's signal.

2.2   How ARM TrustZone Isolates Tenants

In multi-tenant scenario, the tenants run as TAs in the TEE. REE runs a vanilla operating system with rich features and expanded functionality. TEE runs a restricted OS (a heavily stripped-down kernel) containing the core functionality needed to implement a secure environment, manage TAs, and interface with the normal world. We now elaborate on the ways in which ARM TrustZone protects TAs in general: both from the secure world and from other TAs [18].

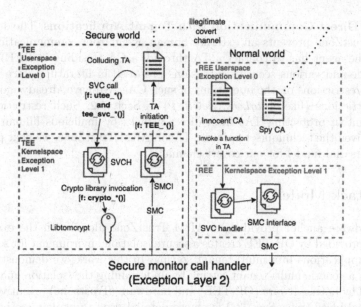

Fig. 1. OP-TEE Architecture and a one-way interaction between normal world and secure world actors. *[f:]* denotes the class of functions used. SVCH and SMCI denote Supervisor Call (SVC) Handler and Secure Monitor Call (SMC) Interface respectively.

Memory Partitioning. ARM TrustZone extends the ARM system-on-chip (SoC) architecture to introduce hardware-based decision making on the *owner* of memory under question. Concretely, each tenant runs as a TA and *owns* memory inaccessible to everyone with *lesser* privileges. This implies that no authorized tenant can access other tenants' data without their consent. Moreover, although the concept of shared memory exists in TrustZone, it's only allowed between the normal world and the secure world (with the secure world being the sole owner of that shared memory). *Two TAs can't map shared memory directly between themselves. This restriction prevents any communication between TAs except for normal message passing. Each TA can rest assured that no one else has access to it except for the legitimate message passing coming from other TAs.* This reinforces our assumption in Sect. 1.1 that two tenants cannot communicate with each other outside the boundaries of the API specifications.

Peripheral Partitioning. ARM TrustZone partitions the use of peripherals between the REE and the TEE. An Input/Output (I/O) device can be made to *belong* to a specific world. Any interrupt on that I/O device can be serviced by its respective world only (e.g. an interrupt on an I/O device assigned to the secure world can not be serviced by the normal world), thereby ensuring the normal world can not access interrupt data belonging to the secure world. Moreover, the TAs can still not steal critical data from each other since any TA interfaces with an I/O device with non-critical data only. Therefore, no TA can seriously compromise the secure world through interrupt hijacking.

Lack of Direct Communication with Tenant Applications. The design of ARM TrustZone prevents an external human agent from communicating with a TA. The agent has to contact a TA through a CA (residing in REE). ARM TrustZone and various security extensions proposed in literature place various kinds of restrictions on the operations of such CAs. We have already motivated such restrictions (like *AutoLock*, *SeCReT*) in Sect. 1.2. Such restrictions are paramount in protecting TAs from direct attacks by malicious human adversaries. Given the techniques to protect TAs in ARM TrustZone, we next propose the adversarial model for our covert channel.

3   Attack Model

The hardware anchor provided by ARM TrustZone along with the execution context provided by OP-TEE creates a secure isolation environment for security-critical applications in a multi-tenant setting. In this work, we demonstrate an attack in a generic multi-tenant setting by undermining the isolation guarantees provided by both software (OP-TEE) and hardware (Trustzone) primitives. The conceptual ideas (refer Sect. 3.2 for more details) used to exploit are based on the design decisions made in isolating TEE and REE. We leverage functionalities like `pthreads` belonging to Linux user-space (in the normal world) and the ability to observe REE timing, invoke other TAs, and maintaining TA persistence belonging to the GlobalPlatform API specification (in the secure world) to create such a channel. Since both Linux and GlobalPlatform API specification are independent of OP-TEE implementation, we are able to fundamentally compromise ARM TrustZone by establishing a covert channel without needing any OP-TEE vulnerability. We also succeed in launching *DoS attacks*.

3.1   Covert Channel Assumptions

We consider a multi-tenant scenario along the assumptions presented in Sect. 1.1. The goal of a tenant here is to create a channel that covertly leaks data to another tenant across the two worlds of ARM TrustZone by forcing *transparency in cross-world timing measurements* to capture intentional differential timing operations done on either world. Precisely, we want an illegitimate covert channel between a TA (of one tenant) as the sender and a *spy CA* (of another tenant) as the recipient (reverse directional channel has *spy CA* as sender and TA as the recipient). Our colluding TA is a persistent application that services RSA decryptions. We assume lowest level privileges for the *spy CA* in a *cross-user* and *cross-core* setting. We also assume presence of defensive mechanisms that monitor and add noise to the cache (like *AutoLock* [16]), and try to detect accesses to ARM's performance monitoring units (through `PMSELR` and `PMXEVTYPER` registers), and allow cross-world communication to authenticated CAs only (like *SeCReT* [21]).

We mainly exploit two properties of TAs in TEE for the proposed covert channel attack: a) longevity and b) persistent storage. First, a TA's usual life is equal to a CA's session duration. However, the TA can use an OP-TEE flag

TA_FLAG_INSTANCE_KEEP_ALIVE to remain alive even after the CA's initial session is over, until the TEE is rebooted. Second, a TA has access to a non-POSIX *trusted storage* through TEE APIs that allow it to store data in transient or persistent objects (as the use-case may require). It is worth mentioning that all trusted storage units are encrypted.

3.2 Attack Methodology Design Motivations

To motivate intended properties of our covert channels, we raise an important question: *Is it possible to create a* ① *non-root,* ② *cross-user,* ③ *cross-core,* ④ *cross-world* ⑤ *covert channel leaking information to other tenants in the system and* ⑥ *resisting established defence mechanisms; while posing, by construction,* ⑦ *unlimited channel capacity?* The answer is an astounding *yes* and we conceptually explain how we achieve each of the mentioned features.

We create a ⑤ *covert channel* from a *spy CA* of one tenant to a *colluding TA* of another tenant (as well as in the reverse direction). As the communication between the CA (in the normal world) and TA (in the secure world) takes place by a timing channel, usual detection mechanisms (suited to cache-based and performance monitoring unit based attacks) of ARM TrustZone and other third-parties are not able to detect the communication. Hence, we are able to achieve ④ *cross-world* property of the covert channel.

Our attack does not access performance monitoring unit (PMU). Hence we do not load kernel modules, thereby classifying the attack as ① *non-root*. The ② *cross-user* property is inherent to the design of Trustzone as for a single TEE, there could be multiple REEs. Since our threat model has one colluding TA in TEE, we are able to successfully establish a *cross-user* communication by allowing two users/tenants to communicate illegitimately. The ③ *cross-core* property is also achieved since thread counters are private to each program and, unlike PMUs, do not get affected by other programs running on the same or separate cores. Moreover, use of thread counters circumvents the defence mechanisms against both caches and PMUs discussed in Sect. 1.2. Finally, since we do not rely on channels with *limited capacity* (like using a shared library as in [26]), our channel's capacity is ⑦ virtually limitless and is only constrained by the will of the channel actors.

We now elaborate on ⑥ *established defense mechanisms* discussed in Sect. 1.2. We navigate against defences involving cache randomisation [8] by not relying on caches. Our covert channel is also not affected against [8]'s tendency to pad the running time of sensitive functions. For every estimated run-time t by the defence in [8], we can force the sensitive function to run for $t + \Delta t$ time (since our TA is already a colluding party to the covert channel). This cat-and-mouse game continues until the processor stops the defence due to computational demand. On the other hand, *SeCReT* [21] allows only whitelisted REE applications to initiate a communication with TEE. The design of our covert channel is such that the REE to TEE communication is initiated not by channel actors but by *innocent CA*, which is actually present in the CA whitelist generated by *SeCReT*. This *innocent* CA has no idea about the covert channel and, is consequently, allowed by *SeCReT* [21] to initiate communication with TEE.

Portability of the Attacks to Other TEEs. Here, we ask the question: *how portable is our covert channel design?* We create covert channels through simple Linux based `pthreads` on the normal world side and through legitimate GlobalPlatform TEE Internal Core API specification (v1.1.2) [15] functions. Therefore, our attack is not restrictive to OP-TEE, but portable to any system based on GlobalPlatform's API specification, for example Huawei's TEE [9].

4   TA-to-CA Covert Communication

In this section, we present a novel method to set up a covert channel from the secure world to the normal world. Briefly, the colluding TA performs RSA decryptions as requested by *innocent CA*, and also leaks sensitive information (like *innocent CA*'s plaintexts or a secure RAM dump) to the *spy CA* which in turn further leaks this information to an external adversary. The working principle of the covert channel can be divided into two phases: *data collection phase* and *attack phase* as discussed next in the following subsections.

4.1   Data Collection/Target Profiling Phase

Here we first summarize the strategy of profiling our target to generate a template to be used later during attack phase. As summarized in Fig. 2, the *spy CA* monitors the time *innocent CA* takes to complete execution, and creates a template from obtained timing measurements in order to predict leaked bytes during the attack phase. Unaware of this profiling, the *innocent CA* gets its ciphertext decrypted by the colluding TA which in turn covertly performs differential timing operations: extra RSA decryptions after legitimately decrypting CA's ciphertext. This differential timing operation allows the TA to covertly transfer one byte of data to the *spy CA*. Repeating these steps allows the adversary to build a template or *regression model*. We achieve this in two steps: *setup* phase that initialises the system and *measurements* phase which does the actual timing measurements.

Setup. The victim first executes a bash command from normal world UART terminal that runs the *innocent CA* binary (Fig. 2, step 1). By default, all OP-TEE CA binaries are in user path including the binaries of the *spy CA* and the *innocent CA*, and thereby executing a bash command is a permissible operation. Thereafter, through a chain of function calls, *innocent CA* transfers control to the colluding TA through the secure monitor call handler (refer Sect. 2.1 and Fig. 1). To build the template, the *spy CA* needs timing measurements of 8-bit numbers in sequence, considering byte-level leakage. For this, the TA maintains a counter $count_{TA}$ ranging from 0 to 255. For the first invocation of the TA, $count_{TA}$ is initialized to 0 and for subsequent invocations, $count_{TA}$ is incremented by 1.

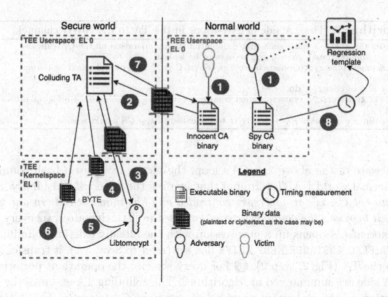

Fig. 2. TA to CA covert channel operation. $BYTE$ denotes the byte to be leaked.

Algorithm 1: The CA side of executions in the TA to CA covert channel.

Input: timer: 64 bit thread counter, incremented parallely by another thread. $count_{spyCA}$:
       Byte to be received from TA

while *true* **do**
    /*Wait for *innocent CA* to be invoked*/
    set *initial_timer* = timer
    if *innocent CA is in* ps -A *output* **then**
        | break
    end
end
while *innocent CA is in* ps -A *output* **do**
    /* Wait for *innocent CA* to finish execution */
    set *final_timer* = timer
end
set *diff* = *final_timer* − *initial_timer* = timer
/* Log *diff* in an array or into a file as time taken to transfer byte $count_{spyCA}$*/
set $count_{spyCA}$ = $count_{spyCA}$ + 1 /*Byte to be received in *spy CA*'s next session */

Measurements. Concurrently, as depicted by Fig. 2 step 1, the *spy CA* opens a new thread p, initialises a 64 bit integer *timer* as 0, and increments it as a thread counter in an infinite loop. It also initialises another variable $count_{spyCA}$ as 0. To achieve asynchronous granular timing measurements of the *innocent CA*'s entry and exit in the process table, the *spy CA* takes the following steps (summarized in Algorithm 1). It first ❶ records the value of *timer* as *initial_timer*. It then ❷ checks the presence of *innocent CA* in the list of active processes in the system, through ps -A. If absent, it repeats the measurement and continues checking the status of ps -A. Finally, ❸ if *innocent CA* binary is found in the output of ps -A, the *spy CA* enters into a loop as long as it continues finding *innocent CA* binary in the output of ps -A.

Algorithm 2: The TA side of executions in the TA to CA covert channel.

Input: threshold: a value such that differential timing operation on the CA side are
     differentiated. $count_{TA}$: The byte to be leaked
TEE_AsymmetricDecrypt(ciphertext from *innocent* CA) /*decrypt ciphertext as normal
  operation */
for $i = 0$ to $count_{TA}$ **do**
 | TEE_AsymmetricDecrypt(ciphertext from *innocent* CA) /*Differential timing operation */
end
set $count_{TA} := count_{TA} + 1$ /*Byte to be leaked in *spy* CA's next session */

Unaware of the above *spy CA*'s loop, the victim executes a bash command
from normal world UART terminal that invokes the *innocent CA* binary. *Since
the innocent CA is not adversary controlled and has innocent code in our threat
model, it bypasses the countermeasure proposed in* [21]. Hereafter, ❹ every time
the *innocent CA* opens up a new session with the already existing colluding TA
with TA_FLAG_INSTANCE_KEEP_ALIVE flag set (refer to Sect. 3.1), it transfers control to the TA (Fig. 2, step 2). ❺ For every session, the operations performs on
the TA side are summarized in Algorithm 2. The colluding TA executes the RSA
decryption on the ciphertext provided by *innocent CA* honestly (Fig. 2, steps 3
and 4), and then performs RSA decryptions $count_{TA}$ number of times (Fig. 2,
steps 5 and 6). Once done, the colluding TA increments $count_{TA}$ and returns
control (Fig. 2, step 7) to *innocent CA*. The *innocent CA* exits and subsequently
quits the process table. Now, ❻ the *Spy CA*'s loop on ps -A output notices the
removal of *innocent CA*'s entry and breaks the loop. ❼ *spy CA* records the value
of *timer* as *final_timer* and gets its first measurement (Fig. 2, step 8) as the
difference between *final_timer* and *initial_timer*. The *spy CA* logs this measurement, increments $count_{spyCA}$, and repeats the measurement process unless
all 8-bit numbers have corresponding measurements. ❽ Once done, the *spy CA*
sends a *SIGKILL* signal to its counter thread p to teardown and exit.

4.2   Attack Phase

Using the measurements collected in previous phase, the adversary trains a
regression model and deploys it to *spy CA*. On the secure world side, we assume
the TA maintains a buffer (let's call it *data* without loss of generality) which
is assumed to hold the data to be covertly transmitted and a variable *index* to
denote the index of the byte in *data* that is transferred in one iteration of the
attack. To transfer an arbitrary byte indexed by *index*, the TA performs the
decryptions *data*[*index*] number of times and increments the *index*. The *spy CA*
uses the trained regression model to predict *data*[*index*] and repeats the entire
process until entire *data* buffer has been transmitted. The exact implementation
and performance of regression model have been shown in Sect. 6.

4.3   Trade-Off Between Covert Channel Bandwidth and SNR

It is worth mentioning that the effect of noise from operations like context switching is of low timing variance and does not affect our signal drastically. Still, to

increase the signal-to-noise ratio, we can always sacrifice bandwidth and amplify the covert signal by doing RSA decryptions ($k \times count_{TA}$) times, for some scaling factor k which can be empirically selected (refer to Fig. 7a). This idea is useful in evading the function stub defences [8] that inserts function stubs around sensitive functions to pad the latter's run time (cf. Sect. 1). We also note that the TA does not know about covert channel teardown and continues its covert operation. The advantage is that no defence monitor can observe any change in the behaviour of the TA, thereby allowing it to remain *oblivious about the covert operation*. Additionally, OP-TEE provides access to process list through Busy-Box's[2] ps utility which allows monitoring processes by a separate user helping us to launch the proposed attack in a *cross-user* setting.

5   CA-to-TA Covert Communication

In the previous section, we successfully set up a cross-world covert channel from the secure world to the normal world using simple thread counters. We now develop another covert channel, *albeit in the reverse direction*, which allows transmission and execution of malicious programs sent by the spy CA to the TA side. The objective of the adversary is to execute arbitrary programs, such as executing *shellcode*, to create *Denial-of-Service* situation inside the trusted world and thereby render the system useless[3]. It must be noted that, although the TA is colluding in this case, it cannot directly host potentially malicious program structures due to the TrustZone's authentication of loaded TAs in the system. Moreover, in multi-tenant scenario, one malicious tenant can jeopardise the service of other honest tenants by stalling the entire system.

Unlike the forward direction channel, the TA does not have the liberty to create regression template profiles for the reverse channel as it might look suspicious to TrustZone's OS. Therefore, we send *one bit instead of a byte* in one measurement. *The major challenge in this setting is that threading is not supported on the TA side.* Still, the TA can synchronize measurements through observing timing differences on the REE side (where the *spy CA* will force timing fluctuations) through TEE_GetREETime system call[4]. On the normal world side, to force differential timing measurements in order to transfer bit 1, the *spy CA* does a number of I/O calls (precisely POSIX printf) on the REE side; while for bit value 0, *spy CA* does nothing. The TA captures this timing difference and interprets the bit value. We assume the attacker uses a loop around ps -A to know when *innocent CA* starts so that *spy CA* can start the bit transfer process accordingly. Given this attack setting, we address the channel specifics into two phases: *setup phase* and *measurement phase*.

---

[2] BusyBox is a software suite providing Unix utilities in a single executable file.
[3] We demonstrate a practical denial of service, leading to a system stall, in Sect. 6.
[4] This is a legitimate syscall that a non-privileged TA can use to access REE clock.

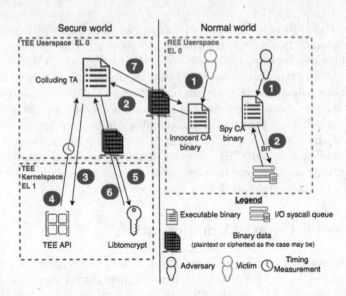

Fig. 3. CA to TA covert channel operation. *BIT* denotes the bit to be leaked. I/O syscall queue is filled up iff *BIT* is 1.

5.1 Setup

Similar to our previous attack, the victim executes a bash command from normal world UART terminal that executes the *innocent CA* binary, which transfers control to the corresponding colluding TA. Since this is the first time the colluding TA is invoked; it therefore initialises the counter $count_{TA}$ as 0. The TA also initializes a large buffer, say *info*, to null characters and an integer *val* to 0. The parameter *val* helps in recreating bytes from the transmitted bits that are leaked by the *spy CA*. And for each such recreated byte, the TA inserts it into the appropriate index in *info*[] buffer.

5.2 Measurements and Information Retrieval

The *innocent CA* executed by the victim (Fig. 3, step 1), assumed to be authenticated by [21], ❶ opens up a new session with the colluding TA, and transfers control to it (Fig. 3, step 2). Concurrently, ❷ the adversary executes *spy CA* (Fig. 3, step 1). If the bit b to be transmitted by *spy CA* is 1, it uses k' number of I/O system calls and exits (refer Sect. 6 for scaling factor discussion). If the bit b to be transmitted by *spy CA* is 0, it directly exits (Fig. 3, step 2). This scaling factor k' sets up an arbitrary high run-time for *spy CA* that defeats time pad based defences [8] (mentioned in Sect. 1). ❸ The TA side of executions are summarized in Algorithm 3. On the secure world side, the colluding TA invokes OP-TEE API call TEE_GetREETime (Fig. 3, steps 3 and 4) and stores the result in *initial_time*. It then does RSA decryptions on the ciphertext provided by the *innocent CA* honestly (Fig. 3, steps 5 and 6). Finally, the TA invokes

---

Algorithm 3: The TA side of executions in the CA to TA covert channel.

Input: *threshold*: a value such that differential timing operation on the CA side are differentiated. *val*: denotes the value of the byte to be generated. $count_{TA}$: denotes the number of bits received so far

Output: Generated covert channel data *info*

set *initial_time* := TEE_GetREETime() /* record time from REE clock */

TEE_AsymmetricDecrypt(ciphertext from *innocent* CA) /* decrypt normally */

set *final_time* := TEE_GetREETime() /* record time from REE clock */

set *diff* := TEE_TIME_SUB(*final_time*, *initial_time*) /* differential time difference */

if *diff* > *threshold* then

    | /* implies *spy* CA executed printf syscalls to transmit a 1*/

    | *val* := $2 \times val + 1$ /* Left-shift and insert a 1 to the right of *val* */

else

    | /* implies the *spy* CA did not execute printf syscalls, to transmit a 0*/

    | *val* := $2 \times val$ /* Left-shift and insert a 0 to the right of *val* */

end

if $count_{TA} \bmod 8 == 0$ then

    | /* 8 bits transmitted; ready to store it as byte in array *info* */

    | $info[\lfloor count_{TA}/8 \rfloor] = val$

    | *val* = 0

end

set $count_{TA}$:= $count_{TA} + 1$ /* Increment the total number of bits received */

TEE_GetREETime again (Fig. 3, steps 3 and 4) to store the result in *final_time* and gets the timing difference through TEE_TIME_SUB. Should the result be above a certain threshold (where threshold is empirically selected as the time taken to complete k' printf syscalls), the TA records the transmitted bit as 1; else the transmitted bit is 0. The discussion on choosing value of k' is given in Sect. 6. To generate the value *val* of a byte from a sequence of bits, we use bitwise operations: for every new bit b received, first left-shift *val* by one bit through $2 \times val$ and add b to insert the new bit into *val*. Finally, the TA closes the session, and ④ returns control to *innocent CA* (Fig. 3, step 7). ⑤ The *innocent CA* exits and the entire process is repeated for more transfer. We note that the TEE API calls TEE_GetREETime and TEE_TIME_SUB are essential since OP-TEE does not allow threading. However, since these calls do not access the PMUs, we have not relaxed our assumptions in any way. Hence, using TEE_GetREETime to monitor REE timing differences, we are able to set up a cross-world covert channel from a CA in the normal world to a TA in the secure world.

6 Experimental Results

In this section, we discuss the implementational details of the bi-directional covert channel and illustrate the experimental results. For the experimental setup, we configure OP-TEE for Raspberry Pi 3 Model B. We use OP-TEE's default RSA implementation using libtomcrypt [12]. For the TA-to-CA communication, the results of data collection phase are plotted in Fig. 4, where the x-axis denotes the value of the byte being leaked by the TA, and the y-axis denotes the thread counter measurement done by *spy* CA. This would then be passed as input to the regression template to regenerate the transmitted byte on the *spy* CA side. As depicted in Fig. 4, more the number of bits to be leaked by the TA,

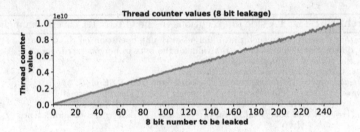

Fig. 4. Graphical visualisation of the template attack.

Table 1. Bandwidth ranges for the channel from Sect. 4. TC implies thread counter. Thread counter lower bound: 61690314 for decimal value 0.

Bits transferred	TC upper bound	Bandwidth	Error rate
1	96845983	130.2 Kbps	2.50%
2	226402659	43 Kbps	2.10 %
3	369514868	750 Bps	2.00 %
4	686169514	570 Bps	1.40 %
5	1334568373	437 Bps	1.30 %
6	2517649269	213 Bps	1.27 %
7	5010857993	184 Bps	1.23 %
8	10010838996	150 Bps	1.20 %

longer the thread counter needs to run to regenerate the leakage on *spy CA* side. Using Python's *sklearn* library, a train-test split of 30% with 10 times repetition of the experiments, our regression model has achieved a mean absolute error of 1.079 and a mean squared error of 2.185. This error rate signifies an acceptable performance of the trained model: for a transmitted byte n, the *spy CA* generates byte $n + 1$ or $n - 1$ in the worst case. Table 1 also reinforces this idea that more the number of leaked bits, larger the thread counter measurements, lower the bandwidth but better the prediction accuracy. For our proposed attack, we observe a wide range of bandwidth because of the differential timing. This in turn is based on the number of decryptions performed and the number of bits being leaked as well. Likewise, we observe a wide range of error rates because of larger thread counter measurements. This has also helped to increase the SNR of our covert channel significantly. Finally, to test the channel, our payload was a set of images (of varied sizes and resolutions) as depicted in Fig. 5. *We can see how the core structure of all images is successfully transmitted in all cases, denoting the transfer of information with high fidelity.*

Scaling Factor Selection. We first need to decide the value of the scaling factor as mentioned in Sect. 4 and 5. Section 4 involves a scaling factor k to control the number of decryptions to be performed on the TA side to leak a byte. Section 5, on the other hand, involves a similar scaling factor k' that controls the number

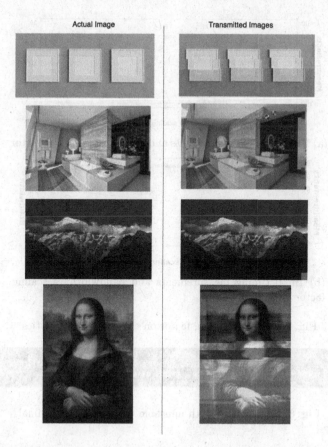

Fig. 5. Comparison of transmitted and retrieved images from the TA-to-CA channel. Refer scaling factor discussion in Sect. 6 to tweak the bandwidth of the channel in order to get clearer images, as the use case may be.

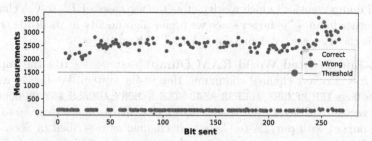

Fig. 6. Transfer of RAM dump (34 bytes). Measurements are in μs and bandwidth is nearly 42 Bps with error rate 0.36%.

of I/O `printf` syscalls made on the *spy CA* side to leak a bit. Figures 7a and 7b visually depict how varying the value of scaling factor impacts the bandwidth, error rates, timing measurements for bit value 1 (where scaling factor controls

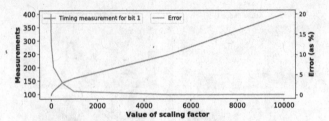

(a) Bit 1 timing measurements and error rate vs scaling factor

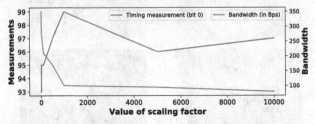

(b) Bit 0 timing measurements and bandwidth vs scaling factor

Fig. 7. Effect of scaling factor on channel characteristics

```
$ [  219.500489] rcu: INFO: rcu_preempt detected stalls on CPUs/tasks:
[  219.501376] rcu:    1-...!: (0 ticks this GP) idle=6fe/1/0x4000000000000000 softirq=569/569 fqs=0
[  219.501877]  (detected by 0, t=41273 jiffies, g=161, q=437)
```

Fig. 8. Stalled REE with unusable normal world terminal

amount of time wasted), and bit value 0 (which is independent of scaling factor). According to the requirements, a sweet spot can be chosen. We chose k as 2 and k' as 1000. k is a smaller value since we prefer high bandwidth in image transfer or RAM dump transfer (which we depict as the use cases of TA-to-CA channel). On the other hand, k' is larger since we prefer high fidelity in shellcode transfer (which is the use case we depict for CA-to-TA channel).

End-to-End Trusted World RAM Dump. Next we describe the impact of the *TA-to-CA* covert channel communication in the system. We use an exploit[5] that uses flags TEE_MEMORY_ACCESS_READ | TEE_MEMORY_ACCESS_ANY_OWNER to verify read access to arbitrary RAM addresses and C-like format string %p to dump them. Coupled with our TA-to-CA covert channel as described in Sect. 4, our *spy CA* is able to recreate RAM contents (as shown in Fig. 6). Figure 6 shows transfer of 34 bytes of secure world RAM dump from TA to *spy CA*. Out of 272 bits transmitted, only one bit is wrongly predicted. Hence we reiterate that *we achieve extremely low error rates through sacrificing bandwidth, asserting our channel's flexibility with respect to the bandwidth and error rates.*

---

[5] https://github.com/teesec-research/optee_examples.

End-to-End DoS Attack. This is our most significant result of the *CA-to-TA* covert channel attack. The TrustZone policies do not allow the functioning of the TAs to affect each other. We, however, subvert this protection by stalling the REE itself and thus making all TAs in the system useless. OP-TEE OS does not allow POSIX related functionality of our interest. Still, we can execute shellcode that does not require POSIX syscalls. Our *spy CA* sends shellcode bytes {0x00, 0x00, 0x00, 0x14} that translate to an infinite loop in ARMv8 assembly. OP-TEE REE has Busybox's `nice -n0` (POSIX command to alter scheduling priority) which is dangerous because TEE relies on REE scheduler to schedule the TAs. The adversary can create a new binary *spy CA* (which requests RSA decryptions from TEE) and execute it through a `nice -n0` command. This transfers control to the colluding TA which now executes the shellcode through a function pointer and stalls the entire REE (refer to the UART terminal output in Fig. 8) denying services of all TAs in the system to any willing CA.

7   Countermeasures

In this section, we motivate countermeasures to our attack vector - the covert channel. We first note that obvious defences such as monitoring too much use of `ps -A` and Linux userspace threading on the userspace or modifying API specification to remove TEE functionality are not good solutions due to their restrictive nature. For example, `ps -A` and Linux threading are basic Linux userspace related functionality and we cannot make Linux more restrictive by taking away these from the userspace. Likewise, removing TA related functionality from the GlobalPlatform API specification is also very restrictive since these functionalities are used by innocent TAs in various settings (like exchanging messages, monitoring REE clocks, maintaining persistence etc.).

For the covert channels, since the core of the attack is *transparency of timing measurements* in a cross-world setting, countermeasures like function stubs, REE authentication, and cache-based defences do not detect or stop our covert operation. Hence designing a countermeasure for this attack is equally challenging and can be considered a direction for future work. Some potential ideas include *static methods* like compilation checks for differential execution paths in OP-TEE binaries and *dynamic methods* like fuzzers that fuzz the TA binaries with varied inputs and try to observe the differential timing behaviours in them. Such countermeasures can flag TA or CA binaries as *potentially colluding*.

8   Conclusion

The use of normal world, secure world isolation in IoT devices has enabled practicality of multi-tenant executions since TrustZones are assumed to ensure tenant isolation. Even though prior research has focused on security of a variety of trusted execution environments (like ARM TrustZone), not much focus has been given to analysing security in multi-tenant scenarios. In this paper, we demonstrate a break in the promised tenant-isolation on ARM TrustZone by covertly communicating between two tenants. We establish a bidirectional covert channel between two tenants, allowing cross-tenant communication in a setting where such communication is not whitelisted. Moreover, our bidirectional channel has disastrous consequences like secure world RAM dump leakage and arbitrary non-POSIX shellcode execution leading to DoS attack. The channels are able to achieve low error rates with reasonable bandwidth, making them practical for any embedded device. Moreover, the attack is portable to any trusted execution environment conforming to GlobalPlatform API specifications. Overall, in a multi-tenant scenario, malicious tenants can break the *isolation* promised by ARM TrustZone's core principles.

Acknowledgements. The work is partially supported by the project entitled 'Development of Secured Hardware And Automotive Systems' from the iHub-NTIHAC Foundation, IIT Kanpur. The authors would also like to thank MeitY, India for the grant for 'Centre on Hardware Security - Hardware Security Entrepreneurship Research and Development (HERD)'.

References

1. Apertis: Integration of OP-TEE in Apertis. https://www.apertis.org/concepts/op-tee/
2. ARM: ARM TrustZone. https://www.arm.com/technologies/trustzone-for-cortex-m
3. Arnautov, S., et al.: Scone: secure Linux containers with Intel SGX. In: 12th USENIX Symposium on Operating Systems Design and Implementation (OSDI 2016), pp. 689–703 (2016)
4. Banerjee, S., et al.: Sesame: software defined enclaves to secure inference accelerators with multi-tenant execution. arXiv preprint arXiv:2007.06751 (2020)
5. Benhani, E.M., Bossuet, L.: DVFS as a security failure of TrustZone-enabled heterogeneous SoC. In: 2018 25th IEEE International Conference on Electronics, Circuits and Systems (ICECS), pp. 489–492. IEEE (2018)
6. Bernal, A.E., et al.: Methodology for computer security incident response teams into IoT strategy. KSII Trans. Internet Inf. Syst. (TIIS) **15**(5), 1909–1928 (2021)
7. OP-TEE Blog: OP-TEE Blog (2021). https://www.trustedfirmware.org/blog/
8. Braun, B.A., Jana, S., Boneh, D.: Robust and efficient elimination of cache and timing side channels. arXiv preprint arXiv:1506.00189 (2015)
9. Busch, M., Westphal, J., Müller, T.: Unearthing the TrustedCore: a critical review on Huawei's trusted execution environment. In: 14th USENIX Workshop on Offensive Technologies (WOOT 2020) (2020)
10. Chen, L., et al.: EnclaveCache: a secure and scalable key-value cache in multi-tenant clouds using intel SGX. In: Proceedings of the 20th International Middleware Conference, pp. 14–27 (2019)
11. Cho, H., et al.: Prime+ count: novel cross-world covert channels on arm TrustZone. In: Proceedings of the 34th Annual Computer Security Applications Conference, pp. 441–452 (2018)
12. OTC Documentation: Cryptographic implementation. https://optee.readthedocs.io/en/latest/architecture/crypto.html

13. F-secure: OP-TEE TrustZone bypass on multiple NXP i.MX models (2021). https://labs.f-secure.com/advisories/op-tee-trustzone-bypass-on-multiple-nxp-i-mx-models/

14. GlobalPlatform: TEE Client API specification (2010). https://globalplatform.org/specs-library/tee-client-api-specification/

15. GlobalPlatform: TEE Internal Core API specification (2018). https://globalplatform.org/wp-content/uploads/2016/11/GPD_TEE_Internal_Core_API_Specification_v1.2_PublicRelease.pdf

16. Green, M., et al.: AutoLock: why cache attacks on arm are harder than you think. In: 26th USENIX Security Symposium (USENIX Security 2017), pp. 1075–1091 (2017)

17. Gruss, D., et al.: Strong and efficient cache side-channel protection using hardware transactional memory. In: 26th USENIX Security Symposium (USENIX Security 2017), pp. 217–233 (2017)

18. Hua, Z., et al.: VTZ: virtualizing arm TrustZone. In: 26th USENIX Security Symposium, pp. 541–556 (2017)

19. Huang, H., et al.: Detection of and countermeasure against thermal covert channel in many-core systems. IEEE Trans. Comput.-Aided Des. Integr. Circuits Syst. **41**, 252–265 (2021)

20. iWave: Securing Edge IoT devices with OP-TE. https://www.iwavesystems.com/news/securing-edge-iot-devices-with-op-tee/

21. Jang, J.S., et al.: Secret: secure channel between rich execution environment and trusted execution environment. In: NDSS, pp. 1–15 (2015)

22. Jang, J., Kang, B.B.: Securing a communication channel for the trusted execution environment. Comput. Secur. **83**, 79–92 (2019)

23. Lee, S., et al.: Fine-grained access control-enabled logging method on arm Trust-Zone. IEEE Access **8**, 81348–81364 (2020)

24. Lee, S., Choi, W., Jo, H.J., Lee, D.H.: How to securely record logs based on arm TrustZone. In: Proceedings of the 2019 ACM Asia Conference on Computer and Communications Security, pp. 664–666 (2019)

25. Liang, Q., Shenoy, P., Irwin, D.: AI on the edge: characterizing AI-based IoT applications using specialized edge architectures. In: 2020 IEEE International Symposium on Workload Characterization (IISWC), pp. 145–156. IEEE (2020)

26. Lipp, M., et al.: ARMageddon: cache attacks on mobile devices. In: 25th USENIX Security Symposium (USENIX Security 2016), pp. 549–564 (2016)

27. Liu, N., Yu, M., Zang, W., Sandhu, R.S.: Cost and effectiveness of TrustZone defense and side-channel attack on arm platform. J. Wirel. Mob. Netw. Ubiquit. Comput. Dependable Appl. **11**(4), 1–15 (2020)

28. Machiry, A., et al.: BOOMERANG: exploiting the semantic gap in trusted execution environments. In: NDSS (2017)

29. Masti, R.J., et al.: Thermal covert channels on multi-core platforms. In: 24th USENIX Security Symposium (USENIX Security 2015), pp. 865–880 (2015)

30. Novković, B., Božić, A., Golub, M., Groš, S.: Confidential computing as an attempt to secure service provider's confidential client data in a multi-tenant cloud environment. In: 2021 44th International Convention on Information, Communication and Electronic Technology (MIPRO), pp. 1213–1218. IEEE (2021)

31. Stoyanova, M., et al.: A survey on the internet of things (IoT) forensics: challenges, approaches, and open issues. IEEE Commun. Surv. Tutor. **22**(2), 1191–1221 (2020)

32. Wang, H., et al.: Mitigating cache-based side-channel attacks through randomization: a comprehensive system and architecture level analysis. In: 2020 Design,

Automation & Test in Europe Conference & Exhibition (DATE), pp. 1414–1419 (2020)

33. Zeitouni, S., Dessouky, G., Sadeghi, A.R.: SoK: on the security challenges and risks of multi-tenant FPGAs in the cloud. arXiv preprint arXiv:2009.13914 (2020)

34. Zhang, N., et al.: TruSpy: cache side-channel information leakage from the secure world on arm devices. Cryptology ePrint Archive (2016)

35. Zhang, N., et al.: TruSense: information leakage from TrustZone. In: IEEE Conference on Computer Communications, IEEE INFOCOM 2018, pp. 1097–1105 (2018)

36. Zhang, X., Xiao, Y., Zhang, Y.: Return-oriented flush-reload side channels on arm and their implications for Android devices. In: Proceedings of the ACM SIGSAC Conference on Computer and Communications Security, pp. 858–870 (2016)

Combined Fault Injection and Real-Time Side-Channel Analysis for Android Secure-Boot Bypassing

Clément Fanjas[1,2]([✉]), Clément Gaine[1,2], Driss Aboulkassimi[1,2], Simon Pontié[1,2], and Olivier Potin[3]

[1] CEA-Leti, Centre CMP, Equipe Commune CEA Leti- Mines Saint-Etienne, 13541 Gardanne, France
{clement.fanjas,clement.gaine,driss.aboulkassimi,simon.pontie}@cea.fr
[2] Univ. Grenoble Alpes, CEA, Leti, 38000 Grenoble, France
[3] Mines Saint-Etienne, CEA, Leti, Centre CMP, 13541 Gardanne, France
olivier.potin@emse.fr

Abstract. The Secure-Boot is a critical security feature in modern devices based on System-on-Chips (SoC). It ensures the authenticity and integrity of the code before its execution, avoiding the SoC to run malicious code. To the best of our knowledge, this paper presents the first bypass of an Android Secure-Boot by using an Electromagnetic Fault Injection (EMFI). Two hardware characterization methods are combined to conduct this experiment. A real-time Side-Channel Analysis (SCA) is used to synchronize an EMFI during the Linux Kernel authentication step of the Android Secure-Boot of a smartphone-grade SoC. This new synchronization method is called Synchronization by Frequency Detection (SFD). It is based on the detection of the activation of a characteristic frequency in the target electromagnetic emanations. In this work we present a proof-of-concept of this new triggering method. By triggering the attack upon the activation of this characteristic frequency, we successfully bypassed this security feature, effectively running Android OS with a compromised Linux Kernel with one success every 15 min.

Keywords: Secure-boot · Synchronization · Frequency detection · Fault injection

1 Introduction

Hardware attacks such as Side-Channel Analysis or Fault Injection represent an important threat for modern devices. An attacker can exploit hardware vulnerabilities to extract sensitive information or modify the target behavior. Among hardware attacks, there are two main kinds of attacks which can exploit a physical access to the target:

– Side-Channel Analysis (SCA) relies on the fact that data manipulated by the target can leak through a physical channel like power consumption or

I. Buhan and T. Schneider (Eds.): CARDIS 2022, LNCS 13820, pp. 25–44, 2023.
https://doi.org/10.1007/978-3-031-25319-5_2

Electromagnetic (EM) emanations. By performing a statistical analysis, an attacker may retrieve these data.
- Fault Injection attacks aims at disrupting the target during the execution of sensitive function. There are multiple Fault Injection methodologies, including optical injection, Electromagnetic Fault Injection (EMFI), voltage and clock glitching or body biasing injection.

For both Side-Channel Analysis and Fault Injection, the attacker needs the best possible synchronization in order to capture the data leakage or disrupt the target behavior. This synchronization issue is even more present on modern SoCs which are more complex than traditional micro-controllers. This paper introduces a new method called Synchronization by Frequency Detection (SFD). This method is based on frequency activity detection in a side-channel as a triggering event. The name of the designed tool to implement this method is the frequency detector. It should not be confused with a frequency synchronizer which is a component used in Software Defined Radio (SDR). A passive probe captures the EM emanations from the target. Then an SDR transposes 0 Hz a selected frequency band of 20 MHz located between 1 MHz and 6 GHz. The transposed band is then sampled and transmitted to an FPGA which applies a narrow band-filter to isolate a specific frequency. The system output is proportional to the energy of a frequency selected by the user in the target EM emanations. The whole process is performed in real-time. We validate this method by synchronizing an EMFI during the Android Secure-Boot of the same target as [GAP+20] which is a smartphone SoC on development board with 4 cores 1.2-GHz ARM Cortex A53. The technology node of this SoC is 28nm litography process. We identified a critical instruction in the Linux Kernel authentication process of the Secure-Boot. Then we found a characteristic frequency that only appears before this instruction. Synchronization of hardware attacks is an old issue which can be hard to overcome with no possibility to use the target I/Os. By using the occurrence of this frequency as triggering event, an EMFI was successfully synchronized with the vulnerable instruction, resulting in the bypass of the Linux Kernel authentication without using any I/Os.

This work provides several contributions:

- We present a novel synchronization methodology. This new method allows to trigger hardware attacks on complex target such as high speed System-on-Chip without using I/Os. The SFD concept could be used in black or grey box context.
- To our knowledge, we present the first successful bypass of a System-on-Chip Secure-Boot using an EMFI. This is also one of the first hardware attack on System-on-Chip which combine SCA and FI. This attack has a high repeatability rate, with one bypass every 15 min.
- Although our target is a developpment board, our usecase is still realistic since the targeted System-on-Chip is used in smartphones, and the targeted software is the Android Bootloader.

Section 2 provides the related works in term of synchronization methods and Secure-Boot attacks. Section 3 focuses on the frequency detector system. Sections 4 and 5 describe respectively the attack set-up and the final experimentation. Finally, the Sect. 6 and 7 are dedicated for discussion, conclusion and perspectives.

2 Related Works

2.1 Bootloader Attacks

Other works focus on Secure-Boot attacks which permit a privilege escalation. In [BKGS21] the authors explain how to successfully re-enable a hidden bootloader by using a voltage glitch. This bootloader was destined for testing purposes and grants high privileges to its user. A similar attack is performed in [CH17], an EMFI corrupt the SoC DRAM during the boot process, which causes the bootloader to enter in a debug state. Then the authors use a software exploit to gain privileges over the TEE. [TSW16] demonstrates how to load arbitrary value in the target PC. It describes a scenario using this vulnerability to bypass the security provided by the Secure-Boot and execute a malicious code on the target. In [VTM+18], the authors provide a methodology for optical fault injection on a smartphone SoC, targeting the bootloader.

2.2 Attack Synchronization Issue

The issue related to the synchronization of hardware attacks has already been identified in several works such as [MBTO13, GAP+20]. The delay between the triggering event and the occurrence of the vulnerability needs to be as stable as possible to maximize the attack success rate. The variation of this delay is called jitter, it is highly correlated with the target complexity. Modern architectures make use of optimizations such as speculative execution and complex strategies of cache memory management. Although providing high performance, these mechanisms bring highly unpredictable timing, which may cause an important jitter with the accumulation of operations. One of the most commonly used techniques to overcome the synchronization issue is based on the I/Os target exploitation to generate the trigger signal [BKGS21, TSW16, MDH+13, RNR+15, DDRT12, SMC21, GAP+20]. This method minimizes the amount of operations performed by the target during the interval between the triggering event and the vulnerability. Therefore it reduces the jitter associated with these operations. However other triggering methods are required in scenarios where self-triggering is not possible or in which the synchronization quality is not enough.

2.3 Existing Synchronization Methods

One alternative to self-triggering is to use essential signals to the target, such as reset or communication bus [BKGS21, SMC21, VTM+18]. However these signals are not always available or suitable for triggering purpose.

Side-Channel attacks require aligned traces to perform an efficient statistical analysis to retrieve the secret information manipulated by the target. Traces alignment can be done during the attack with an online synchronization, but also with signal post-processing [DSN+11]. In [CPM+18] the authors presented an offline synchronization method to pre-cut traces before alignment[1]. They made only one capture using an SDR during the execution of several AES encryption. They identified a frequency in the orignal capture, which only appears before the AES encryptions. By cutting the original capture accordingly with this frequency occurrence, they were able to extract each· AES and roughly align the traces in post-processing. This method is close to the SFD method presented in this paper. The main difference is that [CPM+18] presents an offline method which is dedicated for side-channel analysis, whereas the SFD method is an online method, which is also efficient for fault injection.

[HHM+12] presents an attack based on the injection of continuous sinusoidal waves via a power cable to disrupt an AES implemented on an FPGA. This attack does not need synchronization since it is a continuous injection that affects the target during all the encryption. However in [HHM+14] the same authors present a method to improve this attack by injecting the sinusoidal waves only during the last round of the AES. The injection is triggered after the occurrence of an activity in the target EM emanations.

In [VWWM11] the authors use a method called "pattern based triggering" to synchronize an optical fault injection on a secure microcontroller. It consists in a real-time comparison of the target power signal with a known pattern which appears before the vulnerability. The FPGA board sampling frequency is 100 MHz. However there can be some higher frequency patterns between two samples. To detect these patterns, the system uses a frequency conversion filter which outputs an envelop of the target power signal. Although being at a lower frequency, this envelop signal represents the high frequency pattern occurring between two samples. The concept of "pattern based triggering" or "pattern matching" is explored in details in [BBGV16], the authors also propose a method based on an envelop signal to represent high frequency pattern in lower frequency. The pattern based triggering method feasibility is strongly correlated with the sampling rate of the monitored signal. Therefore this method is not suited for triggering attacks on target such as SoC which runs at high frequency (i.e. GHz level) unless there is a frequency conversion method between the input signal and the pattern comparison. An alternative would be to use ADC with high sampling rate, however it means that the processing behind the ADC needs to be able to handle more data.

3   Frequency Detector

This section proposes a new device for synchronizing hardware security characterization benches. The goal of this tool is to perform real-time analysis of the

---

[1] This method is partially inspired by https://github.com/bolek42/rsa-sdr which is an offline synchronization method to align SCA traces.

EM activity allowing the detection of events that happen inside the SoC. Targeted specifications for the frequency detector are listed bellow as requirements:

R.1 Be able to detect an event that happens as close as possible to the execution of an instruction vulnerable to fault injection. The number of instructions between the detected event and the vulnerable instruction should be limited to minimize the temporal uncertainty of the vulnerable instruction execution.

R.2 The event must be detected before the fault injection vulnerability. Due to causality issue, the fault injection setup must be triggered before the EM shot. Moreover, the use-case studied in this paper is more restrictive because our fault injection setup must be triggered at least 150 ns before. For SCA use-cases, the causality constraint is relaxed because an oscilloscope can record the past.

R.3 Provide a real-time detection. The delay between an event and its detection must be the most constant as possible. This delay is the latency of the EM activity analysis. In practice, this delay will not be constant. The variation of this delay is the temporal uncertainty inserted by the detection operation. To minimize this delay, implementation of this operation must fit a real-time constraint. For example, using a classical computer to perform the detection would insert too temporal uncertainty to fit this constraint. For SCA use-cases, an offline post-treatment can improve synchronization of side-channel traces if the Signal to Noise Ratio (SNR) is acceptable. This post-process relaxes the temporal uncertainty constraints but can not be applied in fault injection use-cases.

R.4 The kind of event EM signature is specific to the use-case. The requirement for this study is to be able to detect the computation of cryptographic operations executed before, during, or just after a RSA signature check. For example, long-integer arithmetic or hash computation can be targeted. Requirements for the SoC Secure-Boot use-case are:

R.4.1 To be able to detect repetitive events.

R.4.2 To be able to detect events that require few microseconds of computation as long-integer arithmetic or hash. The duration of events detected in this work is 20 μs.

R.4.3 To be able to detect event from the EM activity of an high speed SoC. Knowing that the CPU of the studied SoC runs at clock rates between 800 MHz and 1.2 GHz.

In the next, methodological and technological choices are detailed and the performances of the frequency detector are evaluated.

3.1 Frequency Detection Methodology

We consider that using an FPGA to perform analysis can meet the real-time constraint R.3. However, this analysis must be able to detect events in a large bandwidth signal (R.4.3). High speed Analog to Digital Conversion (ADC) and the data analysis from high sampling rate signal represent a challenge. To improve the feasibility of our solution, we associated the FPGA with an SDR. As illustrated by the Fig. 1, our tool is based on a passive probe and an amplifier to measure the EM emanations from the target. The SDR shifts one frequency range of this signal to the baseband. The SDR outputs are transmitted to the FPGA. These signals are an image of 20-MHz band selected by the user in the RF signal spectrum. This solution is simpler with an SDR because the FPGA can perform analysis at low sampling rate signals ($f_s = 20$ Msamples/s). It is compatible with requirement R.4.3 because these signals are an image of an high frequency band.

The choice of SDR output sampling rate is a trade-off. Designing an implementation that meets the real-time constraint R.3 is easier with a smaller sampling rate. Nevertheless, low temporal resolution increases the temporal uncertainty.

Regarding the requirement R.4.1, limiting the detection to repetitive events is acceptable. Execution of loops emits EM activities. If the loop step is regular and associated to a timing period T_{loop} then the EM emanations should be significant for frequency $F_{loop} = \frac{1}{T_{loop}}$ and its harmonics. Observing EM activity of a SoC around a specific frequency should allow to detect a repetitive event.

In this work, we propose to monitor in real-time the activity in a narrow band around a characteristic frequency in order to trigger the fault injection. The frequency is selected by the user, it should be characteristic from a loop in the code executed by the target a short time before the vulnerability. The methodology to identify a characteristic frequency is described in Sect. 4.4. A digital signal processing is performed by the FPGA on SDR output signals in order to focus on the activity in a narrow band. To select this band, a pass-band filter is implemented in the FPGA. This digital filter has a fixed band. Combining the SDR and this filter is equivalent to a high frequency band-pass filter. An user can control the central frequency of the band by configuring the SDR. Digital processing offers the possibility to design more selective filters than analog filters. To respect R.1 and R.2 requirements, the delay introduced by the signal processing in the detection is limited to $3.2\,\mu s$ by using 64th-order filters. The goal is to only detect the targeted event to avoid false-positive. The output of the frequency detector is an image of the power in a narrow band around the frequency selected by the user in the EM emanations. This image is sent to a Digital to Analog Converter (DAC).

3.2   Frequency Detector Design

In this section, we describe the design of the frequency detector. The two main components are an SDR and an FPGA. The Fig. 1 illustrates the system.

The SDR is a HackRF One from Great Scott Gadgets[2]. It has a half-duplex capability but its reception mode is only used in this work. The input is a RF signal and the output is pair of 8-bit samples to be sent to a computer through

---

[2] HackRF One: https://greatscottgadgets.com/hackrf/one/.

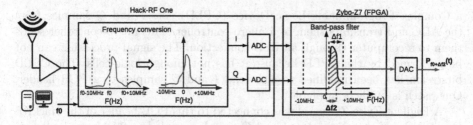

Fig. 1. Block diagram of the frequency detector

an USB connection. In our work, the input is an amplified image of the electro-magnetic activity of the SoC target. The output is composed by two sampled signals. These signals are result an IQ demodulation. User defines a f_0 frequency between 1 MHz and 6 GHz. An analog f_0 sinusoidal signal from an oscillator is mixed to the RF signal with a multiplier to generate I (superheterodyne receiver). I is the *In phase* signal. It is sampled by an ADC after a Low-Pass Filtering (LPF). The mixer shifts the RF signal from one frequency range to another. For example, mixer shifts activity at frequency $|f_{RF}|$ to activities at $|f_{RF} + f_0|$ and $|f_{RF} - f_0|$ (1).

$$\sin\left(2\pi f_{RF}t\right).\sin\left(2\pi f_0 t\right) = \frac{1}{2}\cos\left(2\pi(f_{RF} - f_0)t\right) - \frac{1}{2}\cos\left(2\pi(f_{RF} + f_0)t\right) \ (1)$$

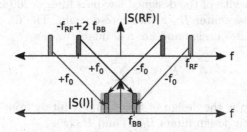

Fig. 2. Issue of image frequency

When only signal I is used, there is a an issue of image frequency. Figure 2 illustrates this problem. If $|f_{RF}|$ is shifted to $|f_{BB}|$ in the baseband with $f_0 = f_{RF} - f_{BB}$, then the image frequency $|-f_{RF} + 2f_{BB}|$ is also shifted to $|-f_{RF} + 2f_{BB} + f_0| = |f_{BB}|$ and produces interference. To solve this issue the signal $I + j.Q$ must be considered instead of only I. A sinusoidal signal with a 90° phase shift is mixed to the RF signal with a multiplier to generate Q. Q is the *Quadrature* signal. Sampled signal I and Q are sent to a computer. In classical application of an SDR, a software signal processing is performed on $I + j.Q$ to demodulate this shifted RF signal. As explained in Sect. 3.1, I and Q samples must be sent to an FPGA. Therefore, we modified the SDR. In an HackRF One, a Complex Programmable Logic Device (CPLD) gets I and Q samples from the ADC and transmits them to a micro-controller. This micro-controller sends them to a computer through an USB connection. The signal processing can not be performed by the CPLD because it has not enough logic-cells. The CPLD bitstream has been modified to also send I and Q samples to a PCB header. Our patch is publicly available[3].

A Digilent Zybo-Z7 board was connected to this PCB header of the HackRF One to receive these samples. I and Q signals sampled at 20 Msamples/s rate are received by the programmable logic of the Xilinx Zynq 7010 FPGA. This complex signal $I + j.Q$ is an image of a 20-MHz bandwidth around the user-defined frequency f_0.

The requirement is to select only activities of a small bandwidth. To fit this constraint, a narrow band-pass filter was designed. This digital filter will be used by the FPGA to filter I and Q samples. DC offsets in radio system is a known issue [Abi95]. Therefore, the band-pass filter has been centered around the 8-MHz frequency to avoid the system output to be impacted by a ghost DC offset. The band-pass filter was designed as an one-side filter. The filter should select frequencies around 8 MHz and attenuate frequencies around -8 MHz. This is possible because the FPGA can discriminate negative and positive frequencies in $I + j.Q$.

The digital filter was designed in two steps. The first step consist in designing of a low-pass filter. Targeted characteristics are: a 5-kHz pass band, a cut after 750 kHz, and a 64th-order. The MathWorks Matlab tool was used to design a linear-phase Finite Impulsion Response (FIR) filter targeting the characteristics. The effective bandwidth of the designed low-pass filter is 209 kHz. Equation (2) describes the low-pass filter $H_a(z)$ in the Z-domain. The 65 coefficients a_i are the output of the filter design and are real double values.

$$H_a(z) = \sum_{i=0}^{64} a_i z^{-i} \tag{2}$$

The second step is the design of two complex filters from $H_a(z)$. Shifts (3) and (4) were used to design filters $H_0(z)$ and $H_{\frac{\pi}{2}}(z)$.

$$H_0(z) = \sum_{i=0}^{64} b_i z^{-i} = \sum_{i=0}^{64} a_i . e^{j2\pi(i+1)\frac{8\,MHz}{f_{sampling}}} . z^{-i} \tag{3}$$

$$H_{\frac{\pi}{2}}(z) = \sum_{i=0}^{64} c_i z^{-i} = \sum_{i=0}^{64} a_i . e^{j2\pi(i+1)\frac{8\,MHz}{f_{sampling}}+j\frac{\pi}{2}} . z^{-i} \tag{4}$$

Equation (5) and (6) show how to filter input $I_{in} + j.Q_{in}$ to compute $I_{out} + j.Q_{out}$ with b_i and c_i the coefficients from (3) and (4) respectively.

$$I_{out}(n) = \sum_{i=0}^{64} Re(b_i).I_{in}(n-i) - \sum_{i=0}^{64} Im(b_i).Q_{in}(n-i) \tag{5}$$

---

[3] HackRF One, CPLD patch: https://github.com/simonpontie/hackrf_cpld_patch/.

$$Q_{out}(n) = \sum_{i=0}^{64} Im(c_i).Q_{in}(n-i) - \sum_{i=0}^{64} Re(c_i).I_{in}(n-i) \tag{6}$$

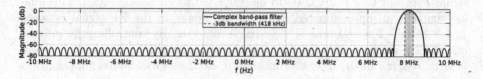

Fig. 3. Complex band-pass digital filter

Figure 3 shows performances of the filter. It is an one-side digital filter around 8 MHz (Δf_2 in Fig. 1) with a 418-kHz (Δf_1 in Fig. 1) bandwidth. This signal processing is composed by four FIR filters with real coefficients: $Re(b_i)$, $Im(b_i)$, $Re(c_i)$, and $Im(c_i)$. FIR filters were quantified and implemented with the "FIR compile" tool from the Xilinx Vivado tool suite. A hardware implementation was designed to use these filters and to compute I_{out} and Q_{out}. An image of the $I_{out} + j.Q_{out}$ power is approximated by implementing the Eq. (7).

$$\widetilde{P}(n) = I_{out}(n)^2 + Q_{out}(n)^2 \tag{7}$$

$I_{out} + j.Q_{out}$ is an image of a frequency band between $f_0 + 8$ MHz $- 209$ kHz and $f_0 + 8$ MHz $- 209$ kHz of the RF signal. Because it is a narrow band, the RF signal might be regarded as a sinusoidal signal RF_a (8). The power of RF_a is $\frac{A_a^2}{2R}$ (9). Equations (10), (11), and (12) show $\widetilde{P}$ as an image of the RF signal power with this sinusoidal assumption.

$$RF_a(t) = A_a sin\Big(2\pi(f_0 + 8\,\text{MHz} + f_a)t + \phi_a\Big), \quad f_a \in [-209\,\text{kHz}, 209\,\text{kHz}] \tag{8}$$

$$P_{RF_a} = \frac{\Big\langle A_a^2 sin^2\big(2\pi(f_0 + 8\,\text{MHz} + f_a)t + \phi_a\big)\Big\rangle}{R} = \frac{A_a^2}{2R} \tag{9}$$

$$I_{out}(t) = RF_a(t).\sin\Big(2\pi f_0 t\Big) \simeq \frac{A_a}{2}\cos\Big(2\pi(8\,\text{MHz} + f_a)t + \phi_a\Big) \tag{10}$$

$$Q_{out}(t) = RF_a(t).\sin\Big(2\pi f_0 t + \frac{\pi}{2}\Big) \simeq \frac{A_a}{2}\sin\Big(2\pi(8\,\text{MHz} + f_a)t + \phi_a\Big) \tag{11}$$

$$\widetilde{P}(t) \simeq \frac{A_a^2}{4}\Big(\cos^2\big(2\pi(8\,\text{MHz} + f_a)t + \phi_a\big) + \sin^2\big(2\pi(8\,\text{MHz} + f_a)t + \phi_a\big)\Big) = \frac{A_a^2}{4} \tag{12}$$

An uncontrolled ϕ_a can delay the time between activation of the frequency and when the output is maximal. IQ demodulation is important because output of our system will be the same regardless of the ϕ_a value (12).

The approximated image of the power in the narrow band (7) can be efficiently computed because it only requires two squares and one addition. This signal $\widetilde{P}$ is sent to a R-2R DAC to be converted as an analog signal. This analog signal is an approximated image of the power in the RF signal between $f_0 + 8$ MHz $- 209$ kHz and $f_0 + 8$ MHz $+ 209$ kHz. By controlling f_0, an user can observe an approximated image of the power in a 418-kHz band chosen in [8 MHz, 6 GHz]. In the next we continue to use the sinusoidal assumption, thus we refer to this narrow band as a frequency.

3.3   Frequency Detector Performances

This system red is able to detect the activation of a specific frequency between 8 MHz and 6 GHz. The system output is updated by the frequency shift and the signal processing. These operations are stream processes but require a delay to propagate information from the input to the output. To characterize this latency we used a Low Frequency Generator (LFG) to generate an Amplitude-Shift Keying (AFK) modulated signal with a carrier frequency F_i. It is emitted by a probe situated near the probe of the frequency detector. The frequency detector has been set to trigger upon the F_i frequency activation. An oscilloscope is used to measure the delay Δt between the frequency activation (modulation signal) and the output of our system. The standard deviation $\sigma_{\Delta t}$ of this delay corresponds to the jitter induced by our system. The oscilloscope triggers upon the rise of the frequency activation signal. The delay is measured between the oscilloscope trigger time and when the frequency detector output exceed 50% of its maximal value. Several measures (within ten thousand) have been performed. The mean time is equal to an average $\langle \Delta t \rangle$ of 2.56 µs and its standard deviation $\sigma_{\Delta t}$ is equal to 60.9 ns. The standard deviation $\sigma_{\Delta t}$ fits the requirement R.3 because 60.9 ns is a temporal uncertainty close to the temporal resolution. This resolution is 50 ns and it is corresponding to the software radio sampling period of the I/Q signals. 95% of the value belongs to an interval of $\Delta t_{mean} \pm 2\sigma$ which corresponds to the interval [2.44 µs; 2.68 µs]. To fit the causality requirement R.2, user must explore only characteristic frequency of events that happen 2.83 µs ($\simeq 2.68$ µs $+ 150$ ns) before the targeted vulnerability. In addition, the low-pass filter introduced by the DAC limits the minimum period of detectable activity. The frequency activity needs to stay active long enough to let the frequency detector output rise to the desired level. The measured rise-time value for the frequency detector output (between 5% and 95%) is 922 ns. We measure Δt as the delay between the rise of the frequency activation signal and the rising edge of the frequency detector output at 50% of the maximum value. Thus the frequency needs to stay active at least 461 ns (i.e. 50% of the rise-time) in order to be detected. This fits the requirement R.4.2.

4   Attack Environment Setup

The target used in this work is already described in Sect. 1. [GAP+20] presents a methodology to bypass one of the security mechanisms of Linux OS by targeting a specific core with a clock frequency fixed at 1.2 GHz. The fault model proposed by the authors is based on instruction skipping. This paper reproduces a similar experience with three main differences:

- The software targeted is the Android Secure-Boot.
- The core targeted is different.
- The frequency during the boot phase is set to 800 MHz.

This section describes preliminary experiments to tune fault injection and SFD setups. This exploration includes the search of an EMFI vulnerability in the Android Secure-Boot. For the EM analysis we used a probe (RF-B 0.3-3) and a preamplifier (PA303/306) from Langer. The injection probe is based on the same design as describe in [GAP+20]. To move the probe at the SoC surface, we used an XYZ motorized axis from Owis.

4.1   Electromagnetic Fault Injection

A pulse generator delivers a pulse up to 400 V into an EM injection probe. The target communicates with a host PC by UART. The PC configures the pulse generator voltage and controls an XYZ motorized stage to move the probe at the chip surface. The purpose of this experiment consists in characterizing the EMFI regardless of the triggering method. Therefore, a target with the Secure-Boot disabled was used to validate fault injection experiments on a fully controlled software code. The code used to observe the fault injection effect is composed by a sequence of SUB instructions, which are surrounded by GPIO toggles. This program has deterministic inputs and outputs in a scenario without injecting faults. The GPIO triggers the EMFI during the SUB sequence. The results are sent by the target through the UART bus. These results are compared with the expected value to determine whether a fault has been injected. This experiment is repeated 50 times for each position of the probe with a step of 500 μm between two positions. We scanned all the chip which corresponds to an area of 13.5 mm by 11 mm. The results of the global scan is superimposed on the chip IR imaging as shown in Fig. 4(A). We observe the presence of faults in a small area. Consequently, a more accurate scan of this faulty area with a small step of 50 μm was performed to identify the best position. This scan result is shown in Fig. 4(B). The best fault rate was achieved with a 400 V pulse voltage.

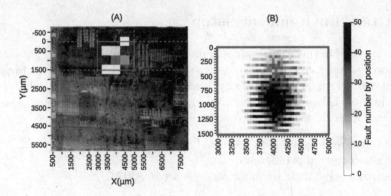

Fig. 4. Fault injection sensitivity scan for 400 V pulse.

4.2 Electromagnetic Leakage Measurement

A scan of the chip was performed during the execution of a code leaking at a predetermined frequency. The power around frequency was measured for each position above the chip. This experiment was also applied on the other side of the PCB, above the decoupling capacitors. Eventually, the best passive probe position appeared to be above one of the decoupling capacitors.

4.3 Secure-Boot Vulnerability

The Secure-Boot is a crucial security feature in a mobile device. It ensure that the running OS can be trusted. It is a chain of programs loaded successively in memory. There is an authentication of each program before executing it to ensure that it is legitimate. In our experimentations, we used a development board with a partially enabled Secure-Boot to start Android. Figure 5 describes the Secure-Boot architecture implemented on our target. The First Stage Bootloader (FSBL) is stored in Read Only Memory (ROM). The FSBL loads the Secondary Stage Bootloader (SSBL) from external memory. The SSBL starts and loads the Trusted OS executed in secure-mode. Then the SSBL loads and runs Little Kernel, which is the Android Bootloader of the target. Little Kernel

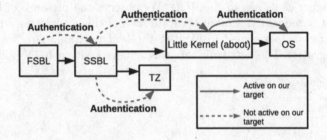

Fig. 5. Secure-boot architecture

loads the Linux Kernel of Android. Since the target is a development board, the Secure-Boot is partially enabled. The authentication of the SSBL by the FSBL is not active, there is also no authentication of Little Kernel by the SSBL. To the best of our knowledge, there is no publicly procedure for activating these authentications. However, the Linux Kernel authentication by Little Kernel can be easily activated by recompiling the Little Kernel code with the right compilation settings. This paper only focuses on the Linux Kernel authentication. During the Linux Kernel compilation, the SHA256 digest of the image is computed and signed with a private key using the RSA algorithm. The signed hash value is stored in the Kernel image. The authentication process is detailed in Fig. 6. Little Kernel has the public key which allows to decrypt the signature. During the authentication, Little Kernel computes the SHA256 digest of the current image (i.e. HASH_1). Little Kernel decrypts the signed digest available in the image (i.e. HASH_2) with the public key, then it compares the two hash values. A comparison result not equal to 0 means that the image is corrupted or the signature is invalid.

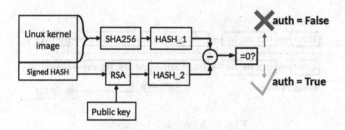

Fig. 6. Authentication process.

The comparison result is used to set the value of the *auth* variable. To load an image, the *auth* variable needs to be set to 1, it is set to 0 by default. By exploring the Little Kernel code, it appears that the comparison of the two hash values is performed by the function *memcmp(HASH_1,HASH_2)*. This function computes the difference between each byte of *HASH_1* and the corresponding bytes of *HASH_2*. The return value of this function is used by a conditional *if* to determine the image validity. This means that faulting the comparison result or the conditional *if* would be interesting to modify the program control flow to avoid setting the *auth* value to 0. We compiled Little Kernel to search a vulnerability in the assembly code. The conditional *if* which verifies the result of *memcmp(HASH_1, HASH_2)* is identified in the ASM code in Algorithm 1. The register *r6* is allocated by the compiler to represent the image authenticity (i.e. *auth*). The result of memcmp is stored in the register *r0*.

Algorithm 1. ASM and C pseudocodes

C pseudocode	ASM pseudocode
$ret \leftarrow memcmp(HASH\_1, HASH\_2)$	$r0 \leftarrow$ **bl** *memcmp*
if $ret == 0$ **then**	**CLZ** $r6, r0$
$\quad auth = 1$	**LSR** $r6, r6, \#5$
end if	

Figure 7 represents the paths the assembly code can follow after the comparison. The CLZ instruction[4] returns in the output register the number of bits equal to 0 before the first bit equal to 1 in the input value. The LSR instruction (see Footnote 4) translates each bits of the register value on the right by a specified number of bits given as input. By analyzing the behavior of this code, it seems that skipping the LSR instruction would keep the value of CLZ $r6,r0$ in $r6$. In such case the value in $r6$ is in the range $[0, 31]$ if the two digests are different.

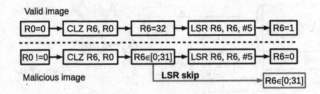

Fig. 7. Algorithm behavior

A modified Linux Kernel with only one different byte from the original was used to validate the potential exploitation of this vulnerability. Since the signed hash did not change, it should be different from the computed hash of the modified image. When the authentication is activated, Little Kernel rejects the corrupted image. However, the image is accepted if the LSR instruction is replaced by a NOP instruction. This confirm that skipping the LSR instruction could be exploited by an attacker to load successfully a corrupted image.

4.4   Characteristic Frequency Research

Little Kernel is modified to toggle a GPIO state a short time after the vulnerable instruction. An oscilloscope and a passive EM probe are used to measure the target EM emanations. This experience aims at finding a characteristic frequency suitable for triggering purpose. Figure 8 provides the spectrogram generated thanks to the EM measurements from the target. It is possible to identify several characteristic frequencies in the SHA256 computation and in the RSA decryption. The purpose of this methodology is to find a characteristic frequency which

---

[4] See "ARM Architecture Reference Manual ARMV7-A and ARMv7-R edition".

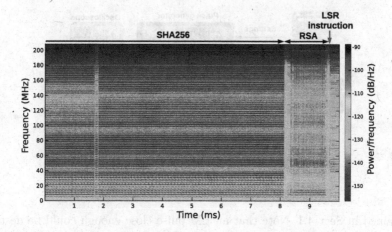

Fig. 8. Target EM emanations spectrogram around the LSR instruction.

happens a short time before the LSR instruction. The frequency at 124.5 MHz was sufficiently detectable by our frequency detector. This frequency appears during the execution of a loop in the function *BN_from_montgomery_word* from openssl which is used by the RSA decryption function. We set up the frequency detector to generate a trigger signal upon this frequency activation. We measured the delay between the vulnerability and the trigger signal by rising a GPIO a short time after the vulnerable instruction. The mean value of the delay between the rise of the GPIO and the rise of the frequency detector output is 80.57 µs, the standard deviation measured is 476 ns which corresponds to 381 clock cycles at 800 MHz. This result is used to set the delay between the frequency detection and the EM pulse. This value is an approximation since the mean delay of 80.57 µs is measured between the frequency detection and the GPIO, not between the frequency detection and the targeted instruction.

5   Linux Kernel Authentication Bypassing on Android Secure-Boot

The previous section shows that it is possible to modify the target control flow by skipping instructions using EMFI. Moreover, a characteristic frequency is identified before a vulnerability in the Android Secure-Boot. Section 5 presents an experiment using all these settings to bypass the Linux Kernel authentication.

5.1   Experimental Setup

An oscilloscope generate the trigger signal upon the rise of the frequency detector output. The power supply which reboot the board after each experiment is controlled by the PC as described in Fig. 9. The UART bus allows to monitor the results. The injection probe is placed above the SoC at the best location

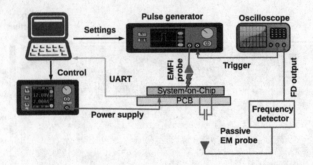

Fig. 9. Experimental setup

determined in Sect. 4.1. Note that an EM pulse close enough could be destructive for the frequency detector components. If the two probes are too close then a RF switch should protect the frequency detector acquisition path during the pulse. An alternative is to place the passive probe below the PCB near the decoupling capacitors. The PCB and the chip act as a shield between the two probes as described in Fig. 9. Unfortunately, the Linux Kernel authentication is not activated in the Little Kernel binary of the target. Therefore, an unmodified Little Kernel has been compiled to activate the authentication. Using a modified Linux Kernel image with only one modified byte confirms that the authentication works properly. When loading this image, the authentication fails and the board reboot in recovery mode. The goal of this work is to boot Android with the modified Linux Kernel, which should be impossible. The frequency detector is configured to trigger the injection upon the 124.5 MHz frequency activation. The pulse generator voltage and the delay between the frequency detection and the EMFI are also configured according to the parameters of the Sect. 4. The experiment follows two steps:

– Step 1: The board boots.
– Step 2: The PC gets a message from the UART logs which attests if the authentication succeed or failed.

The authentication happens between these two steps. During the authentication, the 124.5 MHz frequency is activated. It is detected by the frequency detector which triggers the pulse after the 80 μs fixed delay. The step 2 allows discriminating the following scenarios:

1. The "timeout" scenario: the board stops to print log on the UART, the PC never receives the message of step 2. This probably means that the board has been crashed.
2. The "recovery" scenario: the PC gets a message which indicates that the authentication has failed and the board will reboot in recovery mode. This is the expected behavior of the board when no fault has been injected.
3. The "false positive" scenario: the PC gets a message which indicates that the authentication succeeded, but for unknown reasons the board stops printing log just after sending this message.

4. The "success" scenario: The PC gets a first message confirming that the authentication succeeded and the board continues to print logs after this message. It means that the authentication has been successfully bypassed.

5.2 Experimental Results

15000 injections has been performed. The total campaign duration is 18 h. For the "success" case, the boot proceeds during 15 s before rebooting the board. It is unlikely that an error would propagate during 15 s and cause the board crash before the end of the boot process. To confirm this hypothesis we performed a new campaign, stopping after the first "success" case and letting the boot proceed. It confirmed that Android has been correctly started with the modified Linux Kernel. 83 "success" over 15000 injections corresponds to a 0.55% success rate. This experiment can succeed in less than 15 min if all the settings are properly fixed (Table 1).

Table 1. Campaign results

Scenario	Timeout	Recovery	False positive	**Success**
Number of attempts	6754 (45.03%)	7912 (52.75%)	251 (1.67%)	**83 (0.55%)**

6 Discussion

In this section, we propose a discussion about our results and methodology. Firstly, we want to highlight that we used a smartphone SoC implemented on a development board. Therefore, we have a quite high control over the target, which may not be possible with a true smartphone. For example, we have a physical access to the GPIO through the board connectors, for the setup validation. We also have the board schematics and the target code. Moreover, this attack depends from the target. The vulnerable instruction and the measured delay may change with an other compiler. Also the code leaking at 124.5 MHz may change on an other target. Therefore the attack settings need to be adapted for each target.

Secondly, we note that a similar approach based on a commercial solution exists such as the icWaves associated to the Transceiver from Riscure company. The icWaves solution applies a pattern matching method on its input signal to trigger the attack. The Transceiver is used before the icWaves to capture informations in high frequency. This system is based on an SDR coupled with an FPGA as same as our frequency detector but both solutions differ on signal processing. The first difference lays in the signal analyzed by the FPGA: the complex signal $(I + j.Q)$ for the frequency detector versus the module of this signal for the icWave ($|I + j.Q|$). Thus, the frequency detector has the ability to differentiate frequencies above and below the local oscillator frequency (f_0). In the signal processing chain of the Riscure solution, a rectifier is used as envelope detection. Furthermore, the envelope detection requires some constraints as the presence of the carrier in the RF signal and a modulation index that is less or equal than 100% to avoid losing information. The signal processing of the SFD method uses a high Q factor filter to increase the selectivity. This method is useful to build a system to only detect activities from a sub part of the SDR output bandwidth. It allows the user to increase the sampling rate of the SDR in order to reduce the temporal uncertainty of the detection by maintaining a narrow sensitive band ensuring the best selectivity. We also note that the SDR used in the frequency detector designed in this work is limited to a 20 MHz sampling rate versus 200 MHz for the Riscure Transceiver. The jitter of our system is correlated with the sampling rate, an higher sampling rate would reduce this jitter. Therefore we could implement our method with better hardware such as the Ettus USRP X310 used in the Riscure Transceiver.

Currently the Riscure solution does not implement our SFD method. The Riscure Transceiver bandwith is selectable between 390 kHz and 160 MHz. At 2 MHz[5] their tool has a 17.3 μs delay against a 2.56 μs delay for our tool. There is no information about the jitter of the Riscure Transceiver. However it is important to note that the Riscure setup can be optimized and a fair comparison should be based on experiments.

Other triggering methods may exists, such as PCB signals, communication bus or other events in side-channel. For comparison purpose, we measured the delay between other basics events and the vulnerability. The mean delay between the rise of the target power supply and the rise of the GPIO is around 1.248 s ($\langle \Delta t \rangle$) and its jitter is approximately 5 ms ($\sigma_{\Delta t}$), which is 10000 greater that the frequency detector temporal uncertainty. This signal is unusable for our attack. We also used an oscilloscope to generate a trigger signal upon the detection of a known message on the UART. We choose the closest UART message to the vulnerability. This message appears in the Little Kernel logs sent over the UART. The mean delay between this trigger signal and the rise of a GPIO after the targeted instruction is around 113 ms ($\langle \Delta t \rangle$) and has a jitter of 2 μs ($\sigma_{\Delta t}$). It is 4 times greater than the temporal uncertainty of our frequency detector with the 124.5 MHz frequency. This trigger signal is usable. The setup is limited to an oscilloscope and it is a simpler setup than our SFD method. The experiment can succeed in a short amount of time with this jitter (estimated to 1 h for a success). However, the UART logs can be easily disabled during the compilation of Little Kernel. If avoiding printing logs on the UART is quite easy, it is much more difficult to hide completely the EM emanations from the target. The advantages of the SFD method is that it detects internal activity of the SoC and is not limited to I/O.

---

[5] https://riscureprodstorage.blob.core.windows.net/production/2017/07/transceiver_datasheet.pdf.

7   Conclusion

In this paper, we present an hardware attack on a smartphone SoC. This is a combined attack using a real-time analysis of the target EM emanations to synchronize an EMFI. This attack allows to bypass the Linux Kernel authentication step of the Android Secure-Boot, therefore it is possible to load a malicious Linux Kernel despite the Secure-Boot being activated. The mean success rate of this experiment is around one bypass every 15 min. To our knowledge, this is the first System-on-Chip Secure-Boot bypass using EMFI. We also present a novel synchronization method for hardware security characterization. Our approach relies on the fact that reducing the delay between the triggering event and the targeted code vulnerability will decrease the jitter associated with this delay. Thus, it will increase the hardware attack success rate. The SFD method uses the activation of a characteristic frequency in the target EM emanations as triggering event. This method is based on an SDR and an FPGA to generate an output signal proportional to the power of the selected frequency. The selected frequency is included in the range between 8 MHz and 6 GHz and is identified thanks to EM emanations analysis. Our system introduces a mean delay of 2.56 µs between the input signal (i.e. the target EM emanations) and the output signal. This mean delay is associated to a jitter of 60.9 ns. Using this synchronization method, we were able to skip a critical instruction by triggering an injection upon the activation of a known frequency. This approach could be used in future work for synchronizing hardware attacks against other targets such as new SoC references or other Secure-Boots. A perspective consist in applying this methodology to bypass the SSBL authentication by the FSBL. Thus it would be possible to get privileges over the TEE.

Acknowledgment. The experiments were done on the Micro-PackS^{TM} platform in the context of EXFILES: H2020 project funded by European Commission (No. 88315).

References

[Abi95] Abidi, A.A.: Direct-conversion radio transceivers for digital communications. IEEE J. Solid-State Circuits **30**(12), 1399–1410 (1995)

[BBGV16] Beckers, A., Balasch, J., Gierlichs, B., Verbauwhede, I.: Design and implementation of a waveform-matching based triggering system. In: Standaert, F.-X., Oswald, E. (eds.) COSADE 2016. LNCS, vol. 9689, pp. 184–198. Springer, Cham (2016). https://doi.org/10.1007/978-3-319-43283-0_11

[BKGS21] Bittner, O., Krachenfels, T., Galauner, A., Seifert, J.-P.: The forgotten threat of voltage glitching: a case study on Nvidia Tegra X2 SoCs. In: 2021 Workshop on Fault Detection and Tolerance in Cryptography (FDTC), pp. 86–97. IEEE (2021)

[CH17] Cui, A., Housley, R.: BADFET: defeating modern secure boot using second-order pulsed electromagnetic fault injection. In: 11th USENIX Workshop on Offensive Technologies (WOOT 2017) (2017)

[CPM+18] Camurati, G., Poeplau, S., Muench, M., Hayes, T., Francillon, A.: Scream-
ing channels: when electromagnetic side channels meet radio transceivers.
In: Proceedings of the 2018 ACM SIGSAC Conference on Computer and
Communications Security, pp. 163–177 (2018)

[DDRT12] Dehbaoui, A., Dutertre, J.-M., Robisson, B., Tria, A.: Electromagnetic
transient faults injection on a hardware and a software implementations
of AES. In: 2012 Workshop on Fault Diagnosis and Tolerance in Cryptography, pp. 7–15. IEEE (2012)

[DSN+11] Debande, N., Souissi, Y., Nassar, M., Guilley, S., Le, T.-H., Danger, J.-L.:
Re-synchronization by moments: an efficient solution to align side-channel
traces. In: 2011 IEEE International Workshop on Information Forensics
and Security, pp. 1–6. IEEE (2011)

[GAP+20] Gaine, C., Aboulkassimi, D., Pontié, S., Nikolovski, J.-P., Dutertre, J.-M.:
Electromagnetic fault injection as a new forensic approach for SoCs. In:
2020 IEEE International Workshop on Information Forensics and Security
(WIFS), pp. 1–6. IEEE (2020)

[HHM+12] Hayashi, Y., Homma, N., Mizuki, T., Aoki, T., Sone, H.: Transient IEMI
threats for cryptographic devices. IEEE Trans. Electromagn. Compat.
55(1), 140–148 (2012)

[HHM+14] Hayashi, Y., Homma, N., Mizuki, T., Aoki, T., Sone, H.: Precisely timed
IEMI fault injection synchronized with EM information leakage. In:
2014 IEEE International Symposium on Electromagnetic Compatibility
(EMC), pp. 738–742. IEEE (2014)

[MBTO13] Montminy, D.P., Baldwin, R.O., Temple, M.A., Oxley, M.E.: Differential
electromagnetic attacks on a 32-bit microprocessor using software defined
radios. IEEE Trans. Inf. Forensics Secur. **8**(12), 2101–2114 (2013)

[MDH+13] Moro, N., Dehbaoui, A., Heydemann, K., Robisson, B., Encrenaz, E.:
Electromagnetic fault injection: towards a fault model on a 32-bit microcontroller. In: 2013 Workshop on Fault Diagnosis and Tolerance in Cryptography, pp. 77–88. IEEE (2013)

[RNR+15] Riviere, L., Najm, Z., Rauzy, P., Danger, J.-L., Bringer, J., Sauvage,
L.: High precision fault injections on the instruction cache of ARMv7-
M architectures. In: 2015 IEEE International Symposium on Hardware
Oriented Security and Trust (HOST), pp. 62–67. IEEE (2015)

[SMC21] Spruyt, A., Milburn, A., Chmielewski, Ł.: Fault injection as an oscilloscope: fault correlation analysis. IACR Trans. Cryptographic Hardware
Embed. Syst. 192–216 (2021)

[TSW16] Timmers, N., Spruyt, A., Witteman, M.: Controlling PC on ARM using
fault injection. In: 2016 Workshop on Fault Diagnosis and Tolerance in
Cryptography (FDTC), pp. 25–35. IEEE (2016)

[VTM+18] Vasselle, A., Thiebeauld, H., Maouhoub, Q., Morisset, A., Ermeneux, S.:
Laser-induced fault injection on smartphone bypassing the secure boot-
extended version. IEEE Trans. Comput. **69**(10), 1449–1459 (2018)

[VWWM11] Van Woudenberg, J.G.J., Witteman, M.F., Menarini, F.: Practical optical
fault injection on secure microcontrollers. In: 2011 Workshop on Fault
Diagnosis and Tolerance in Cryptography, pp. 91–99. IEEE (2011)

A Practical Introduction to Side-Channel Extraction of Deep Neural Network Parameters

Raphaël Joud[1,2(✉)], Pierre-Alain Moëllic[1,2], Simon Pontié[1,2], and Jean-Baptiste Rigaud[3]

[1] CEA Tech, Centre CMP, Equipe Commune CEA Tech - Mines Saint-Etienne, 13541 Gardanne, France
`{raphael.joud,pierre-alain.moellic,simon.pontie}@cea.fr`
[2] Univ. Grenoble Alpes, CEA, Leti, 38000 Grenoble, France
[3] Mines Saint-Etienne, CEA, Leti, Centre CMP, 13541 Gardanne, France
`rigaud@emse.fr`

Abstract. Model extraction is a major threat for embedded deep neural network models that leverages an extended attack surface. Indeed, by physically accessing a device, an adversary may exploit side-channel leakages to extract critical information of a model (i.e., its architecture or internal parameters). Different adversarial objectives are possible including a fidelity-based scenario where the architecture and parameters are precisely extracted (*model cloning*). We focus this work on software implementation of deep neural networks embedded in a high-end 32-bit microcontroller (Cortex-M7) and expose several challenges related to fidelity-based parameters extraction through side-channel analysis, from the basic multiplication operation to the feed-forward connection through the layers. To precisely extract the value of parameters represented in the single-precision floating point IEEE-754 standard, we propose an iterative process that is evaluated with both simulations and traces from a Cortex-M7 target. To our knowledge, this work is the first to target such an high-end 32-bit platform. Importantly, we raise and discuss the remaining challenges for the complete extraction of a deep neural network model, more particularly the critical case of biases.

Keywords: Side-channel analysis · Confidentiality · Machine Learning · Neural network

1 Introduction

Deep Neural Network (DNN) models are widely used in many domains with outstanding performances in several complex tasks. Therefore, an important trend in modern Machine Learning (ML) is a large-scale deployment of models in a wide variety of hardware platforms from FPGA to 32-bit microcontroller. However, major concerns related to their security are regularly highlighted with milestones works focused on availability, integrity, confidentiality and privacy

I. Buhan and T. Schneider (Eds.): CARDIS 2022, LNCS 13820, pp. 45–65, 2023.
https://doi.org/10.1007/978-3-031-25319-5_3

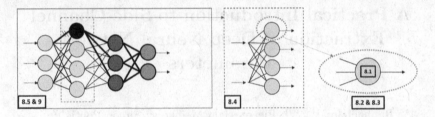

Fig. 1. The basic elements of a MLP model studied in the paper with the related Sections (black rectangles). (Left) The snowball effect (Sect. 7.1): an extraction error on a weight (first bold connection) misleads the neuron (black) output, propagates through the model and drastically impacts the recovery of all the other weights of the next neurons (dark gray).

threats. Even if adversarial examples are the flagship of ML security, confidentiality and privacy threats are becoming leading topics with mainly *training data leakage* and *model extraction*, the latest being the core subject of this work.

Model Extraction. The valuable aspects of a DNN model gather its architecture and internal parameters finely tuned to the task it is dedicated to. These carefully crafted parameters represent an asset for model owners and generally must remain secret. Jagielski *et al.* introduce an essential distinction between the objectives of an attacker that aims at extracting the parameters of a target model [9], by defining a clear difference between **fidelity** and **accuracy**:

- *Fidelity* measures how well extracted model predictions match those from the victim model. In that context, an adversary aims to precisely extract model's characteristics in order to obtain a *clone model*. In such a scenario, the extraction precision is important. Additionally to model theft, the adversary may aim to enhance his level of control over the system in order to shift from a black-box to a white-box context and design more powerful attacks against the integrity, confidentiality or availability of the model.
- *Accuracy* aims at performing well over the underlying learning task of the original model: the attacker's objective is to steal the *performance* of the model and, effortlessly, reach equal or even superior performance. In such a case, a high degree of precision is not compulsory.

Attack Surface. The large-scale deployment of DNN models raises many security issues. Most of the studied attacks target a model as an abstraction, exploiting theoretical flaws. However, implementing a model to a physically accessible device open doors toward a new attack surface taking advantage of physical threats [6,10], like side-channel (SCA) or fault injection analysis (FIA). This work is focused on fidelity-oriented attack targeting model confidentiality using SCA techniques.

Structure of the Paper. In Sect. 2, we first provide basic deep learning backgrounds that introduce most of our formalism. Related works are presented in Sect. 3, followed by an explanation of our positioning and contributions (Sect. 4).

Details on our experimental setups and comments on our implementations are presented in Sect. 5, before a description of the threat model setting, discussed in Sect. 6. As an introduction to all our experiments, we expose the main challenges related to fidelity-based parameter extraction and describe our overall methodology in Sect. 7. Then, in Sect. 8, we detail our extraction method, our experiments and results with a progressive focus on: (1) one multiplication operation, (2) one neuron, (3) sign extraction, (4) several neurons and (5) successive layers. As future works (Sect. 9), we discuss the critical case of bias and the scaling up of our approach on state-of-the-art models. Figure 1 illustrate the structure of the experimental sections with respect to the basic structural elements of a model. Finally, we conclude with possible mitigations.

2 Background

2.1 Neural Networks

Formalism. This work is about supervised DNN models. Input-output pairs $(x, y) \in \mathcal{X} \times \mathcal{Y}$ depend on the underlying task. A neural network model $\mathcal{M}_W :$ $\mathcal{X} \to \mathcal{Y}$, with parameters W, predicts an input $x \in \mathcal{X}$ to an output $\mathcal{M}_W(x) \in \mathcal{Y}$ (e.g., a label for classification task). W are optimized during the training phase in order to minimize a *loss* function that evaluates the quality of a prediction compared to the ground-truth y. Note, that a model $\mathcal{M}$, seen as an *abstract algorithm*, is distinguished from its *physical implementations* M^*, for example embedded models in microcontroller platforms. From a pure functional point of view, the embedded models rely on the same abstraction but differ in terms of implementation along with potential optimization processes (e.g., quantization, pruning) to reach hardware requirements (e.g., memory constraints).

Perceptron is the basic functional element of a neural network. The perceptron (also called *neuron* in the paper) first processes a weighted sum of the input with its trainable parameters w (also called *weights*) and b (called *bias*), then non-linearly maps the output thanks to an *activation function* (noted σ):

$$a(x) = \sigma\left(\sum_{j=0}^{n-1} w_j x_j + b\right) \tag{1}$$

where $x = (x_0, ..., x_j, x_{n-1}) \in \mathbb{R}^n$ is the input, w_j the weights, b the bias, σ is activation function and a the perceptron output. The historical perceptron used the sign function as σ but others are available, as detailed hereafter.

MultiLayer Perceptrons (MLP) are *deep* neural networks composed of many perceptrons stacked vertically, called a *layer*, and *multiple layers* stacked horizontally. A neuron from layer l gets information form all neurons belonging to the previous layer $l - 1$. Therefore, MLP are also called *feedforward fully-connected* neural networks (i.e., information goes straight from input layer to output one). For a MLP, Eq. 1 can be generalized as:

$$a_j^l(x) = \sigma\left(\sum_{i \in (l-1)} w_{i,j} a_i^{l-1} + b_j \right) \tag{2}$$

where $w_{i,j}$ is the weight that connects the neuron j of the layer l and the neuron i of the previous layer $(l-1)$, b_j is the bias of the neuron j of the layer l and a_i^{l-1} and a_j^l are the output of neuron i of layer $(l-1)$ and neuron j of layer l.

Activation Functions inject non-linearity through the layers. Typical functions maps the output of a neuron into a well-defined space like $[0, +\infty]$, $[-1, +1]$ or $[0, 1]$. The Rectified Linear Unit function (hereafter, ReLU) is the most popular function because of its simplicity and constant-gradient property. ReLU is piece-wise linear and defined as $ReLU(x) = max(0, x)$. We focus our work on ReLU but other activations are possible: *tanh*, *sigmoid* or *softmax* that is typically used at the end of classification models to normalize output to a probability distribution.

2.2 IEEE-754 Standard for Floating-Point Arithmetic

We study single-precision floating-point values on a 32-bit microcontroller. IEEE-754 standard details floating-point representation and arithmetic. Floating value are composed of three parts: Sign, Exponent and Mantissa as in Eq. 3 for a 32-bit single-precision floating-point value, a:

$$a = (-1)^{b_{31}} \times 2^{(b_{30}...b_{23})_2 - 127} \times \left(1.b_{22}...b_0 \right)_2 \tag{3}$$

$$= (-1)^{S_a} \times 2^{E_a - 127} \times \left(1 + 2^{-23} \times M_a \right)$$

This allows to represent values from almost 10^{-38} to 10^{+38} and considers specific case like infinity or *Not a Number* (NaN) values which are not considered here. We emphasize on the *usual case* when the exponent value belongs to $[1; 254]$. In this case, the final floating-point value a is as in Eq. 3 where S_a, E_a and M_a correspond respectively to the sign, exponent and mantissa values. With this representation, result of the multiplication operation $c = a \times b$ with b another single floating-point value, leads to the sign (S_c), exponent (E_c) and mantissa (M_c) detailed in Eq. 4. Note that these do not necessarily correspond the very final representation of c: depending on the value of M_c, some realignment can be performed affecting both M_c and E_c. However, it appears clearly that $M_a \times M_b$ have less impact on M_c value than $M_a + M_b$.

$$S_c = S_a \oplus S_b, E_c = E_a + E_b - 127, M_c = M_a + M_b + 2^{-23} \times M_a \times M_b \tag{4}$$

Table 1. Related state-of-the-art works. NS: Not Specified, μC: Microcontroller, AR: Architecture Recovery, PR: Parameters Recovery, TA: Timing Attack

Attack	Target	Technique	AR	PR	Specificity
Carlini *et al.* [3,9]	N.S	API		✓	Target ReLU in MLP
Oh *et al.* [13]	N.S	API	✓		Classifying DNN arch. from querries
Gongye *et al.* [7]	x86 proc.	TA		✓	Use IEEE754 subnormal values
Maji *et al.* [11]	μC	TA		✓	CNN recovery (1, 8 and 32 bits)
Xiang *et al.* [14]	μC	SPA + ML	✓		Classifying DNN arch. from traces
Batina *et al.* [1]	μC	TA + CPA	✓	✓	Arch. & low-fidelity param. extraction
Ours	*μC*	*CPA*		✓	*High-fidelity parameters extraction*
Hua *et al.* [8]	FPGA	SPA	✓	✓	Targeting memory-access pattern
Dubey *et al.* [5]	FPGA	CPA		✓	Advanced leak-model over BNN
Yu *et al.* [15]	FPGA	CPA + API	✓		Reconstruct BNN model
Breier *et al.* [2]	N.S	FIA		✓	Extract the last layer

3 Related Work

Table 1 presents works that are – to the best of our knowledge – references for the topic of model extraction. These works are distinguished through the adversary's objective (recover the architecture or recover the parameters) and the attack surface (API-based attacks or side-channel-based approaches). In this section, we detail works related to our scope. Interested readers may refer to surveys with a wider panorama such as [12] or [4][1].

3.1 API-Based Attacks

These approaches exploit input/output pairs and information about the target model. Carlini *et al.* consider the extraction of parameters as a *cryptanalytic* problem [3] and demonstrate significant improvements from [9]. The threat model sets an adversary that knows the architecture of the target model but not the internal parameters. The attack is only focused on ReLU-based multi-layer perceptron (MLP) models with one (scalar) output. The basic principle of this attack exploits the fact that the second derivative of ReLU is null everywhere except at a *critical point*, i.e. at the boundary of the negative and the positive input space of ReLU. By forcing exactly one neuron at this critical state thanks to chosen inputs and binary search, absolute values of weight matrix can be reconstructed progressively. Then, the sign is obtained thanks to small variations on the input and by checking activation output. Experimental results (state-of-the-art) show a complete extraction of a 100,000 parameters MLP (one hidden layer) with $2^{21.5}$ queries with a worst-case extraction error of 2^{-25}. Although the attack is an important step forward, limitations rely on its high complexity for deeper models and its strict dependence to ReLU.

---

[1] More particularly, cache-based attacks that are out of our scope.

3.2 Timing Analysis

In [7], Gongye *et al.* exploit, on a x86 processor, extra CPU cycles that significantly appear for IEEE-754 multiplication or addition with *subnormal* values. They precisely recover a 4-layer MLP models (weights *and* bias). However, a potential simple countermeasure against this attack is to enable *flush-to-zero* mode which turns subnormal values into zeros.

Maji *et al.* also demonstrate a timing-based parameter recovery that mainly rely on ReLU and the multiplication operation [11] with floating point, fixed point, and binary models deployed on three platforms without FPU (ATmega-328P, ARM Cortex-M0+, RISC-V RV32IM). Countermeasures encompass adapted floating-point representation and a constant-time ReLU implementation. However, they highlight the fact that even with constant-time implementations, correlation power analysis (CPA) may be efficient and demonstrate a CPA (referencing to [1]) on only one multiplication.

3.3 SCA-Based Extractions

[1] from Batina *et al.*, is a milestone work that covers the extraction of model's architecture, activation function and parameters with SCA. Two platforms are mentioned, Atmel ATmega328P (opened) and SAM3X8E ARM Cortex-M3 for which floating-point operations are performed without FPU.

Activation functions are characterized with a timing analysis that enables a clear distinction between *ReLU, sigmoid, tanh* and *softmax* and relies on the strong assumption that an adversary is capable of measuring precisely execution delay of each activation functions of the targeted model during inference.

The main contribution, for our work, is related to the parameter extraction method that is mainly demonstrated on the 8-bit ATmega328P. Bias extraction is not taken into account nor mentioned. The method is focused on a low-precision recovery of the IEEE-754 float32 weights. Correlation Electromagnetic Analysis is used to identify the Hamming Weight (HW) of multiplication result (STD instructions to the 8-bit registers). The weight values are set in a realistic range $[-N, +N]$ with a precision $p = 10^{-2}$ (therefore, $2N/p$ possible values). They extract the three bytes of the mantissa (three 8-bit registers) and the byte including the sign and the exponent[2]. There is no mention of an adaptation of this technique when dealing with the 32-bit Cortex-M3. Since desynchronization is strong (software multiplication and non-constant time activation function), the EM traces are resynchronized each time according to the target neuron. Note that, because the scope of [1] also encompass timing-based characterization and structure extraction, the scaling up from one weight to a complete deep model extraction and the related issues are not detailed.

Finally, model's topology is extracted during the weight extraction procedure: new correlation scores are used to detect layer boundaries, i.e. distinguish

---

[2] Due to IEEE-754 encoding, second byte of an encoded value contains the least significant bit of the exponent and the 7 most significant bits of mantissa.

Fig. 2. Experimental setup.

if currently targeted neuron belongs to the same layer as previously attacked neurons or to the next one.

Presented methods are confronted to a MLP trained on MNIST dataset and a 8-bit convolutional neural network (CNN) trained on CIFAR-10[3]. Original and recovered models have an accuracy difference of 0.01% and 0.36% respectively, with an average weight error of 0.0025 for the MLP. Implementations are not available and the compilation level is not mentioned.

4   Scope and Contributions

Our scope is a *fidelity-based* extraction of parameters of a MLP model embedded for inference purpose in an AI-suitable 32-bit microcontroller thanks to correlation-based SCA, such as CPA or CEMA. Our principal reference is the work from Batina *et al.* [1] (and to certain extend [3] as a state-of-the-art fidelity-based extraction approach) and we position our contributions as follow:

- Contrary to [1] (precision of 10^{-2}), we set in a fidelity scenario and aim at studying how SCA can precisely extract parameter values.
- Our claim is that the problem of parameter extraction raises several challenges, hardly mentioned in the literature, that we progressively describe. A wrong assumption may reduce this problem to a naive series of attacks targeting independent multiplications (that are actually not independent).
- From the basic operation (multiplication) to an overall model, we propose and discuss methods to extract the complete value of 32-bit floating point weights. Extraction error can reach IEEE-754 encoding error level.
- We do not claim to be able to fully recover all the parameters of a software embedded MLP model: we show that extraction of a secret weight absolute

---

[3] The specific features of CNN compared to MLP that should impact the leakage exploitation are not discussed in [1].

value from multiplication operation is necessary but not sufficient to generalize to the extraction of a complete MLP model. We discuss open issues preventing this generalization such as the extraction of bias values.

- We highlight the choice of our target, based on a ARM-Cortex M7, i.e. a high-end device particularly adapted to deep learning applications (STM32H7). To the best of our knowledge, such a target does not appear in the literature despite its DNN convenient attributes (e.g., FPU, memory capacity). Electromagnetic (EM) acquisitions have been made with an unopened chip which corresponds to a more restrictive attack context compared to literature.
- To foster further experiments and help the hardware security community to take on this topic, our traces and implementations are publicly available[4].

5   Experimental Setup

5.1   Target Device and Setup

Our experimental platform is a ARM Cortex-M7 based STM32H7 board. This high-end board provides large memories (2 MBytes of flash memory and 1 MByte of RAM) allowing to embed state-of-the-art models (e.g., 8-bit quantized MobileNet for image classification task). A 25 MHz quartz has been melted as part of the HSE oscillator to have more stable clock. CPU is running at 25 MHz as well, as its clock is directly derived from the melted quartz. EM emanations coming from the chip are measured with a probe from Langer (EMV-Technik LF-U 2,5 with a frequency range going from 100 kHz to 50 MHz) connected to a 200 MHz amplifier (Fento HVA-200M-40-F) with a 40 dB gain, as shown in Fig. 2. Acquisitions are collected and saved thanks to a Lecroy oscilloscope (4 GHz WaveRunner 640Zi).

To reduce noise and ease leakage exploitation, all traces acquired experimentally from Cortex M7 are averaged over 50 program executions.

5.2   Inference Program

Because of the scope, objective and methodology of this work, we need to perfectly master the programs under analysis to properly understand the leakage properties and their potential exploitation. Therefore, instead of attacking blackbox off-the-shelf inference libraries, we implement our own C programs for every experiments mentioned in this paper and compile them with O0 gcc optimization level to ensure each multiplication is followed by STR instruction saving result in SRAM. This point is discussed as further works in Sect. 10.

As in [11], some approaches exploit timing inconsistency to recover model information. In this work, we consider implementations protected from such kind of attacks as model inferences are performed in a timing constant manner. We claim that these choices represent more real-world applications, as for the selection of an high-end AI-suitable board:

---

[4] https://gitlab.emse.fr/securityml/SCANN-ex.git.

Listing 1.1. constant-time ReLU implementation

```
float layer_neuron_res; // input
int sign,mask,pre_v,post_v;
void *ppre_v,*ppost_v;

sign=(layer_neuron_res>0.0);
mask=0-sign;
ppre_v=(void*)(&(layer_neuron_res));
pre_v=*((int*)ppre_v);
post_v=pre_v & mask;
ppost_v=(void*)(&(post_v));
layer_neuron_res=*((float *)ppost_v); // output
```

- We use the floating-point unit (FPU) module that performs floating-point calculations in a single cycle rather than passing through C compiler library. When available, usage of such *hardware* module is preferred to its *software* counterpart as it speeds up program execution and relieve CPU.
- ReLU function has been implemented in a timing constant way as in Listing 1.1. It has been confirmed by checking on thousands of execution that its delay standard deviation is lower than one clock cycle.

6   Threat Model

Adversary's Goal. Considered adversary aims at reverse-engineering an embedded MLP model as closely as possible by cloning the targeted model with a fidelity-oriented approach. This objective implies that parameters values resulting from target model training phase have to be estimated.

Adversary's Knowledge. The attack context corresponds to a gray-box setting since the adversary knows several information about the target system: (1) the model architecture, (2) the used activation function is ReLU, (3) model parameters are stored as single-precision float following the IEEE-754 standard. With an appropriate expertise in Deep Learning, the attacker may also carry out upstream analyses more particularly on the typical distribution of the weights (ranges, normalization...) he aims at extracting.

Adversary's Capacity. The adversary is able to acquire side-channel information (in our case, EM emanations with an appropriate probe), leaking from the system embedding the targeted DNN model. The collected traces only results from the usual inference and the attacker does not alter the program execution. We assume a classical linear leakage model: the leakage captured is linearly dependent of the *HW* of the *processed* secret value (e.g., a floating-point multiplication between the secret w and an input coming from the previous layer). Typically, a gaussian noise encompasses the intrinsic and acquisition noise.

The adversary can feed the model with crafted inputs, without any limitation (nor normalization), allowing to control the distribution of the inputs according to the chosen leakage model. However, these chosen inputs belong to the *usual*

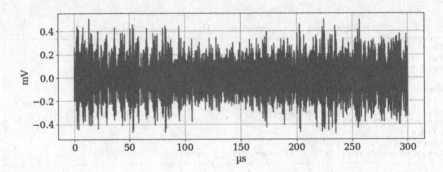

Fig. 3. An averaged trace of a 5 layers-deep model composed of 64 positive weights and no bias.

values according to the IEEE-754 standard. Contrary to API-based attacks, the attacker does not need to access the outputs of the model.

To simplify the scope of this introductory work, we set in a *worst* case scenario according to defender. Attacker is considered able to access a clone device and have enough knowledge and expertise to take benefit of his own implementations to estimate the temporal windows in which he will perform his analysis. We discuss that point in Sect. 7 and 9.

7   Challenges and Overall Methodology

7.1   Critical Challenges Related to SCA-Based Parameter Extraction

Impact of ReLU. The assumption stating that inputs of targeted operation are controlled by the adversary is partially correct: if we focus on a single hidden layer, the inputs are the outputs of the previous one after passing through ReLU. Therefore, the input range is necessarily restricted to non-negative values.

Fully-Connected Model. MLP parameters are not shared[5]: the activated output of a neuron is connected to *all* neurons of the next layer. Then, an input is involved in as many multiplications as neurons in the considered layer. As such, when performing a CEMA at the layer-level, several hypothesis would stand out from the analysis and would likely be correct as they would correspond to the leakage of each neuron of the layer. However, these hypothesis would not stand out at the same time if neurons outputs are computed sequentially. Therefore, knowledge of the order of the neurons is compulsory to correctly associate CEMA results to neurons. That point is closely related to the threat model we defined in Sect. 6 and the profiling ability of the attacker is also discussed in Sect. 9.

Error Propagation Problem. Because of the feedforward functioning of a MLP, extraction techniques must be designed as well: the correct extraction of

---

[5] Contrary to Convolutional Neural Network models.

parameters related to a layer cannot be achieved without fully recovering the previous parameters. A strong estimation error in the recovery of a weight (and therefore in the estimation of the neuron output) will impact the extraction of remaining neuron weights. The impact of this error will spread to the weight extraction of all neurons belonging to forthcoming layers as illustrated in Fig. 1.

Temporal Profiling. In [1], side-channel patterns could be visually recognized on 8-bit microcontroller on which most of the results have been demonstrated. In our context, SPA is hardly feasible (e.g., see Fig. 3) and the localization of all the relevant parts of the traces is a challenging issue that we consider as out the scope of this work. As mentioned in our threat model (Sect. 6), we set in a *worst case* scenario where the attacker is able to perform a *temporal profiling* on a clone device to have an estimation of the parts of the traces to target since he has several secrets to recover spread all over the traces. This estimation can be more or less precise to enable attacks at neuron or layer-level.

Bias Values. Knowledge of the bias value is compulsory to compute the entire neuron output. This parameter is not involved in multiplications with the inputs but added to the accumulated sum between neuron inputs and its weights. Thus, leakage of bias and weight must be exploited differently. Bias extraction is treated in the API-based attack [3] and the timing attack from [11] but not mentioned in [1]. In this work, we clearly states that we keep the extraction of bias values as a future work but discuss this challenging point in Sect. 9.1.

7.2   Our Methodology

Our methodology starts with analyzing the most basic operation of a model – i.e. a multiplication – then, to widen our scope, to a neuron, one layer, then several layers as illustrated in Fig. 1. Corresponding steps are evaluated with both simulated and real traces. Dealing with an entire model means to recover parameters layer by layer, following the (feedforward) network flow: extractions of a layer l being used to infer the inputs of the layer $l + 1$.

Since our main objective is a practical fidelity-based extraction, we aims at crafting an efficient extraction method, faster than a *brute-force* CEMA on 2^{32} hypothesis, that enables a progressive precision. With this introduction work, in addition to expose challenges that suffer analysis in the literature, we assess the precision degree SCA can reach.

8   Extraction Method and Experiments

8.1   Targeting Multiplication Operation

We first focus on the multiplication $c = w \times x$, between two IEEE-754 single-precision floating-point operands: a secret weight (w) and a known input (x). We remind that we use hardware operations thanks to the FPU.

Our Approach is composed of multiple CEMA to extract the absolute value $|w|$. Importantly, the sign bit is not considered yet and is extracted later on (Sect. 8.3). With *fidelity-oriented* extraction as objective, our method has no fixed accuracy objective and avoids exhaustive analysis over 2^{32} hypothesis. It allows to see how accurate SCA-based extraction can perform. It relies on two successive steps. First one tends to recover as much information as possible in a single attack by targeting most significant bits from a variable encoded with IEEE-754 standard. The second step is made to correct possible approximations from the previous one and enhance extraction accuracy by refining the granularity of tested hypothesis. In this step, no focus is made on specific bits, entire variable with all 32-bit varying are considered. Figure 4 describes these two steps. The attack relies on different parameters:

- d_0: size of the initial interval that is centered on the value C. Thus, the tested hypothesis belong to $[C - d_0/2, C + d_0/2]$
- $\lambda_1 > \lambda_2$: two shrinking factors that narrow the interval of analysis (λ_1 for the first iteration of step 2, λ_2 for successive step 2 iterations)
- m: number of times the step 2 is repeated.
- N: number of kept hypothesis (after CEMA) at each extraction step.

Step 1 of the extraction process is as follows:

1. First, we generate an exhaustive set of hypothesis with all possible 2^{16} combinations of the 8 bits of exponent and the 8 most significant bits of mantissa (remaining bits are set to 0) and filter out unlikely values (in a DNN context) by keeping hypothesis in $[C-d_0/2, C+d_0/2]$. Kept hypothesis are not linearly distributed in this interval.
2. We compute the HW of the targeted intermediate values: here, the result of the products between inputs and weight hypothesis.
3. We perform a CEMA between EM traces and our HW hypothesis and keep the N best ones according to the absolute values of correlation scores.

Step 2 is processed in an iterative way and depends on the previous extraction that could be the output of Step 1 or from the previous Step 2 iteration:

1. For each best hypothesis $\hat{w}_i$ kept from the previous step ($i \in [\![0; N - 1]\!]$), a set of assumptions is linearly sampled around $\hat{w}_i$ with an narrower interval of size $d_1 = d_0/\lambda_1$ (if the previous step was step 1) or $d_{i+1} = d_i/\lambda_2$ (otherwise).

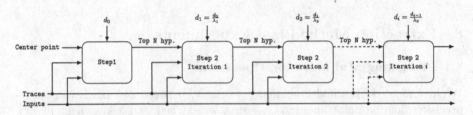

Fig. 4. Flowchart representing extraction process steps with their respective inputs, parameters and outputs.

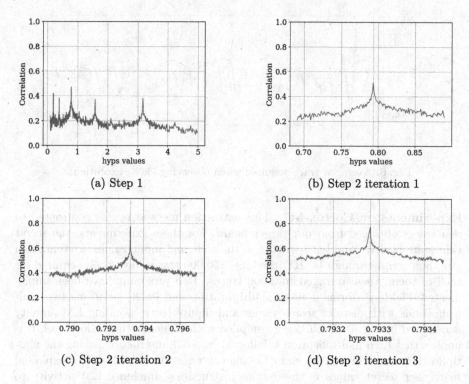

Fig. 5. Our extraction method applied to traces from PRGM$_2$. Step 1 (a) is focused on 8 bits of exponent and 8 bits of mantissa, then Step 2 (b–d) is repeated 3 times.

2. As in step 1, we compute the HW of the intermediate values and perform a CEMA to select the N best hypothesis among the N considered hypothesis sets (so that we always keep N hypothesis at each iteration) according to absolute value of the correlation scores.

Figure 5 shows the two steps of this extraction process ($w = 0.793281$, $d_0 = 5$ and $C = 2.5$) for 3,000 real traces on a STM32H7, obtained from the PRGM$_2$ experiment described hereafter. The second step is iterated three times and we progressively reach a high correlation score.

Validation on Simulated Traces. We first confirm our approach on simulated traces by computing the success rate of our extraction with respect to several extraction error (ϵ_{rr}) thresholds. We randomly generate 5,000 positive secret values w and for each of them, we craft 1,000 3-dimensional traces using random inputs x. At the middle sample, the trace value is the Hamming Weight of the multiplication: $HW(x \times w)$. A random uniform variable is used for the other samples. An additional gaussian noise ($\mu = 0$, σ) is applied on the entire trace. We set $N = 5$, $d_0 = 5$, $C = 2.5$, $m = 3$, $\lambda_1 = 100$ and $\lambda_2 = 50$. Results according to the noise level are presented in Table 2. We reach a significant success rate over 90% for the extraction process until a recovering error of 10^{-6}.

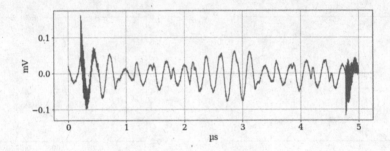

Fig. 6. Averaged trace acquired when observing PRGM$_2$ execution.

Experiments on Cortex-M7. This extraction method is also confronted to real traces obtained from our target board. For these experiments, the secret values are positive and hard-written in the code and input values are sent from a python script through UART interface, 150,000 traces have been acquired for each of them, then averaged to 3,000 traces. Two programs have been implemented: PRGM$_1$ performs a single multiplication and PRGM$_2$ performs two multiplications with distinct secret values and inputs (corresponding EM activity is depicted in Fig. 6). Both being compiled with OO gcc optimization level, this implies that each multiplication is followed by a STR instruction saving the multiplication result in SRAM as in Listing 1.2. One source of leakages exploited to recover secret values is these store instructions. Inference EM activity to be analyzed is framed by a trigger added by hand at assembler level. For this experiment, our extraction method allows to recover the secret values with high extraction level as presented in Table 3.

8.2   Extracting Parameters of a Perceptron

Neuron Computation Implementation. After extracting secret value from an isolated multiplication and studying success-rate of such attack, we scale up to a single neuron computation as described in Eq. 1. The most important difference is that the output of a neuron is the result of an accumulation of several multiplications. This accumulation is processed through two successive

Table 2. Attacking a single multiplication on simulated traces. Success-rate (SR) of the extraction ordered by estimation error (ϵ_{rr}) thresholds according to the noise level (σ).

$\epsilon_{rr} \leq$		10^{-1}	10^{-2}	10^{-3}	10^{-4}	10^{-5}	10^{-6}	10^{-7}	10^{-8}
SR	$\sigma^2 = 0.5$	100	100	98.4	98.3	96.4	94.2	81.8	77.1
	$\sigma^2 = 1$	100	99.9	98.8	98.6	96.9	94.8	81.2	75.4
	$\sigma^2 = 25$	99.9	99.2	97.0	96.8	94.9	91.6	78.0	73.2
	$\sigma^2 = 10^2$	99.9	98.2	94.3	93.0	89.8	86.5	69.6	62.4

Listing 1.2. Assembler code of multiplication execution using FPU

```
; load input
vldr     s14, [r7, #12]
ldr      r3, [pc, #76]
; load weight
vldr     s15, [r3, #4]
; perform product
vmul.f32 s15, s14, s15
; store result
vstr     s15, [r7, #24]
```

Table 3. Extraction results targeting multiplications.

Program	Correct value	ϵ_{rr}
PRGM$_1$	0.793281	4.09e−08
PRGM$_2$	0.793281	1.87e−08
	0.33147	1.27e−08

FPU instructions (`fmul` then `fadd`, as it is the case in our experiments with Listing 1.3) or a dedicated multiplication-addition instruction.

Listing 1.3. Assembler code of multiplication and accumulation using FPU

```
; layer1_neuron_res[i] += input[j] * weight_layer1[i][j];
[...]              ; r3 = accumulator address (SRAM)
vldr     s14, [r3] ; Load accumulator
[...]              ; r3 = input address (SRAM)
vldr     s13, [r3] ; Load input
[...]              ; r3 = weight address (FLASH)
vldr     s15, [r3] ; Load weight
vmul.f32 s15, s13, s15  ; Multiplication (FPU)
vadd.f32 s15, s14, s15  ; Addition (FPU)
[...]              ; r3 = accumulator address (SRAM)
vstr     s15, [r3] ; store result
```

That leads to two new challenges in the extraction of the secret values compared to our first experiments on isolated multiplications: (1) hypothesis have to be made on accumulated values, (2) the attacker needs to know the order in which multiplication are computed (i.e., how the accumulation evolves).

Extraction of Neuron Weights. We assume that the attacker knows the computation order. The first weight w_0 can be extracted as done before by exploiting the direct result of $w_0 \times x_0$. Then, the second weight w_1 can also be extracted with the same approach by targeting $w_0 \times x_0 + w_1 \times x_1$ because w_0 was recovered before. This process can be applied again for each weight value as long as all previous ones have been correctly extracted. Actually, that point is a critical one since the extraction quality of currently targeted weight strongly depends on the extraction accuracy of previously extracted weights.

Experiments on Cortex-M7. We apply that method on 2,000 averaged traces that capture the inference of a 4-input neuron. As presented in Table 4, we reach a very precise extraction of the four weights.

Table 4. Extraction error (ϵ_{rr}) from a 4-input neuron on a Cortex-M7 target (3,000 averaged traces)

Target weight	Correct Value	ϵ_{rr}
w_0	0.366193473339	5.96e−08
w_1	0.90820813179	3.58e−07
w_2	0.522847533226	5.96e−08
w_3	0.00123456	4.21e−08

8.3 Targeting the Sign

Problem Statement. As seen before, for a ReLU-MLP model, a neuron belonging to an hidden layer is fed with non-negative input values. An obvious but important observation is that, for $w \times x = c$, if the secret value w is multiplied with positive value $x \geq 0$ then $sign(c) = sign(w)$. Therefore, in such a context, CEMA is not able to distinguish sign by leveraging the input-weight product.

Extracting the Sign at the Neuron-Level. A way to overcome this issue is to set the sign extraction problem at the neuron-level, i.e. to build hypothesis on sign changes throughout the overall multiplication-accumulation process.

Let consider the accumulation of two successive multiplications: $acc = w_0 \times x_0 + w_1 \times x_1$ with inputs $x_0, x_1 \geq 0$. acc variations would change whether $sign(w_0) \neq sign(w_1)$ or $sign(w_0) = sign(w_1)$. Based on that, by focusing on variation of $|acc|$ value, it is possible to find out if w_0 and w_1 have an opposed sign or not. Thus, weight sign estimation can be done progressively, by checking if the sign associated to the weight currently extracted is similar or opposed to the sign of the previous weight. However, since the sign extraction is processed *relatively* to the sign of w_0, an additional verification has to be done to confirm which of the current extracted signs or the opposite is correct. This can be done thanks to ReLU output that matches with only one hypothesis.

Validation on Simulated Traces. As in Sect. 8.1, we craft simulated 50-dimensional traces for a m-input neuron with m randomly picked in $[\![2; 8]\!]$. We generate 5,000 neurons with m signed weights, no bias and fed by 3,000 positive inputs sets (i.e., 25M of traces). The generation process is similar to the previous experiment apart from the leakage placement which depends on m. Thus, leakages correspond to m product accumulations and one for the *ReLU* output. We uniformly place these $m + 1$ leakage samples in the traces with random uniform values for the other samples. To characterize the principle of the method, we set in a low-noise simulation ($\sigma^2 = 0.5$). We reach the following results:

- 78.8% of neurons have been extracted with all signs correctly assigned.
- For 91.6% of the weights, the sign is correctly assigned. Table 5 details the extraction success rates for these weights (consistent with Table 2).

Table 5. Attacking a neuron with simulated traces. Success-rate (SR) of the extraction of weights (with correctly recovered sign) ordered by estimation error (ϵ_{rr}) thresholds.

$\epsilon_{rr} \leq$	10^{-1}	10^{-2}	10^{-3}	10^{-4}	10^{-5}	10^{-6}	10^{-7}	10^{-8}
SR	99.9	99.1	96.2	92.8	86.5	79.3	65.4	61.0

Experiments on Cortex-M7. We use 5,000 averaged traces capturing the inference of one neuron with signed weights. With Table 6, we observe that the sign inversion and the relative value have been correctly affected. In addition to our previous experiences, our approach performs well at the neuron-level. We progressively scale-up in the next section at a layer-level.

Table 6. Attacking a neuron on Cortex-M7. Extraction error (ϵ_{rr}) for 4 weights.

Target weight	Correct Value	ϵ_{rr}	Sign match
w_0	−0.813444	1.38e−07	✓
w_1	0.0671324	3.88e−08	✓
w_2	0.107843	2.34e−07	✓
w_3	0.604393	6.50e−08	✓

8.4   Targeting One Layer

Previous structure has inputs involved in only one multiplication with weights. However, neural network interconnections between layers implies that an input value is passed to each neuron of the layer and thus is involved in several multiplications with different weights. If neurons are computed sequentially, this means that CEMA would likely bring out several hypothesis that would match weights of different neurons that leak at different moments.

In this context, to associate the extracted values to a specific neuron, we assume that neurons computation is made sequentially from top to bottom of the layer. To ensure an already extracted value is not associated again to another neuron, leaking time sample of tested hypothesis are filtered. Consider only leaking sample greater than the one from last extracted value prevents this.

Experiments on Cortex-M7. Two experiments have been made:

1. *2-neuron layer with 3 inputs each*: the 6 weights are positives and 3,000 traces have been captured by feeding the layer with random positive inputs (as for an hidden layer). The six weights have been recovered with an averaged estimation error $\epsilon_{rr} = 2.68e^{-6}$ (worst: $5.18e^{-6}$, best: $4.48e^{-8}$).
2. *5-neuron layer with 4 inputs each*: the 20 weights encompass positive and negative values and 5,000 traces have been acquired by feeding the layer with

random positive or negative values (as for an input layer). We reach a similar extraction error $\epsilon_{rr} = 1.03e^{-6}$ (worst: $3.10e^{-6}$, best: $1.55e^{-7}$). All weight signs have been correctly guessed.

8.5 Targeting Few Layers

DNN are characterized by layers stacked horizontally. Proposed method is able to extract weights from one layer and is supposed to be applied to each of them one after the other, by progressively reconstructing intermediate layer outputs.

Experiments on Cortex-M7. To verify this principle, we craft a MLP with 5 hidden layers with respectively 5, 4, 3, 2 and 3 neurons. The 64 corresponding weights are positive and the model is fed with 4-dimensional positive inputs. Every weights have been recovered with an estimation error lower than 10^{-6} ($SR = 95.31\%$ for $\epsilon_{rr} < 10^{-7}$, best $\epsilon_{rr} = 7.63e^{-10}$, worst $\epsilon_{rr} = 6.67e^{-6}$).

Note that sign is not considered in this experiments because the tested model has been crafted and is not functional (i.e., not the result of a training process). Such *none functional* models are likely to present too many *dead neurons* and even *dead layers* because of the accumulated ReLU effect. The scaling-up to a fully functional state-of-the-art model with signed weights *and* biases is planned for future works and discussed in the next section.

9   Future Works

9.1   What About Neuron Bias?

So far, biases have not been considered even though these values may significantly impact neuron outputs and also the way a neuron is implemented: the weighted sum between weights and inputs could be initialized to 0 or directly to the bias. In the latter case, our extraction method cannot be directly applied and needs an initial and challenging bias extraction based on IEEE-754 addition.

To better explain this challenge, lets consider that the accumulation is well initialised to 0 (i.e., bias is added *after* the weighted sum). Using simulated traces, we perform our extraction method to recover the weights *and* the bias by focusing on the final accumulation $\sum_j (w_{i,j} \times x_j) + b$.

5,000 neurons with m secret weights $w_{0..m}$ (m is randomly picked in $[\![2; 8]\!]$) and one secret bias b have been generated. 5,000 simulated traces have been crafted for each neuron with random positive inputs. Success rates (SR) are presented in Table 7. These SR only concern weight and bias for which sign has correctly been recovered. This corresponds to 93.25% over 24956 attacked weights and 92.14% over 5000 bias. While SR related to weight extraction remains consistent with previous simulations, SR corresponding to bias extraction significantly drops (e.g., $SR = 35.8$ for 10^{-3}).

A possible explanation relies on the IEEE-754 addition that requires a strict exponent realignment contrary to multiplication. If $a \gg b$ then $a + b = a$ because

b value *disappears* in front of a. In our context, as inputs x (controlled by the attacker that aims at covering as much as possible the float32 range) are defined randomly, then multiplied by weights, it is likely that $\sum_{j=0}^{n} w_j x_j \gg b$. Thus, secret information related to bias could be hardly recovered by exploiting our EM traces. Therefore, we need to develop a different strategy (including a coherent selection of the inputs) to exploit potential IEEE-754 addition leakages.

Table 7. Neurons extraction (signed weight & bias) success-rate

$\epsilon_{rr} \leq$	10^{-1}	10^{-2}	10^{-3}	10^{-4}	10^{-5}	10^{-6}	10^{-7}	10^{-8}
SR weight	99.9	99.5	96.8	93.7	87.2	79.8	64.9	61.3
SR bias	81.7	56.3	35.8	19.7	7.44	2.6	0.7	0.2

9.2   Targeting State-of-the-Art Functional Models

Further experiments will encompass compressed embedded models thanks to deployment libraries (e.g., TFLite, NNoM) with a focus on Convolutional Neural Network (CNN) models. Indeed, for memory constrains, deep embedded models in 32-bit microcontrollers are usually stored with parameters quantized in 8-bit integers with training-aware or post-training quantization methods. For the most straightforward quantization and embedding approaches, this quantization should simplify the extraction process with only 2^8 hypothesis for each weight and bias as well as an additional extraction of a scaling factor that enables the mapping from 8-bit to 32-bit values. Regarding CNN, these models also rely on multiplication-accumulation operations (and the same activation principle), but the fact that parameters are shared across the inputs should interestingly impact the way leakages could be exploited for a practical extraction.

10   Conclusion

Side-channel analysis is a well-known, powerful, mean to extract information from an embedded system. However, with this work, we clearly question the practicability of a complete parameters extraction with SCA when facing state-of-the-art models *and* real-world platforms. By demonstrating promising results on a high-end 32-bit microcontroller on a high fidelity-based extraction scenario, we do not claim this challenge as *impracticable* but we aim at inciting further (open) works focused on the exposed challenges as well as bridging different approaches with combined API and SCA-based methods.

   An additional outcome from our experiments concerns defenses. Classical *hiding* countermeasures, already demonstrated in other context (e.g., cryptographic modules), should be relevant (as also mentioned in [1]). More precisely, randomizing multiplication and/or accumulation order (including the bias) should significantly impact an adversary. An efficient complementary defense could be to randomly add fake or neutral operations at a neuron or layer-level. We keep as future works, the proper evaluation of such state-of-the-art protections in a model extraction context.

Acknowledgements. This work is supported by (CEA-Leti) the European project InSecTT (ECSEL JU 876038) and by the French ANR in the framework of the *Investissements d'avenir* program (ANR-10-AIRT-05, irtnanoelec); and (Mines Saint-Etienne) by the French program ANR PICTURE (AAPG2020). This work benefited from the French Jean Zay supercomputer with the AI dynamic access program.

References

1. Batina, L., Bhasin, S., Jap, D., Picek, S.: CSI NN: reverse engineering of neural network architectures through electromagnetic side channel. In: 28th USENIX Security Symposium (USENIX Security 2019), pp. 515–532 (2019)
2. Breier, J., Jap, D., Hou, X., Bhasin, S., Liu, Y.: SNIFF: reverse engineering of neural networks with fault attacks. IEEE Trans. Reliab. (2021)
3. Carlini, N., Jagielski, M., Mironov, I.: Cryptanalytic extraction of neural network models. In: Micciancio, D., Ristenpart, T. (eds.) CRYPTO 2020. LNCS, vol. 12172, pp. 189–218. Springer, Cham (2020). https://doi.org/10.1007/978-3-030-56877-1_7
4. Chabanne, H., Danger, J.L., Guiga, L., Kühne, U.: Side channel attacks for architecture extraction of neural networks. CAAI Trans. Intell. Technol. **6**(1), 3–16 (2021)
5. Dubey, A., Cammarota, R., Aysu, A.: MaskedNet: the first hardware inference engine aiming power side-channel protection. In: 2020 IEEE International Symposium on Hardware Oriented Security and Trust (HOST), pp. 197–208 (2020)
6. Dumont, M., Moëllic, P.A., Viera, R., Dutertre, J.M., Bernhard, R.: An overview of laser injection against embedded neural network models. In: 2021 IEEE 7th World Forum on Internet of Things (WF-IoT), pp. 616–621. IEEE (2021)
7. Gongye, C., Fei, Y., Wahl, T.: Reverse-engineering deep neural networks using floating-point timing side-channels. In: 2020 57th ACM/IEEE Design Automation Conference (DAC), pp. 1–6 (2020). ISSN 0738-100X
8. Hua, W., Zhang, Z., Suh, G.E.: Reverse engineering convolutional neural networks through side-channel information leaks. In: 2018 55th ACM/ESDA/IEEE Design Automation Conference (DAC), pp. 1–6 (2018)
9. Jagielski, M., Carlini, N., Berthelot, D., Kurakin, A., Papernot, N.: High accuracy and high fidelity extraction of neural networks. In: 29th USENIX Security Symposium (USENIX Security 2020), pp. 1345–1362 (2020)
10. Joud, R., Moëllic, P.A., Bernhard, R., Rigaud, J.B.: A review of confidentiality threats against embedded neural network models. In: 2021 IEEE 7th World Forum on Internet of Things (WF-IoT). IEEE (2021)
11. Maji, S., Banerjee, U., Chandrakasan, A.P.: Leaky nets: recovering embedded neural network models and inputs through simple power and timing side-channels-attacks and defenses. IEEE Internet of Things J. (2021)
12. Méndez Real, M., Salvador, R.: Physical side-channel attacks on embedded neural networks: a survey. Appl. Sci. **11**(15), 6790 (2021)

13. Oh, S.J., Schiele, B., Fritz, M.: Towards reverse-engineering black-box neural networks. In: Samek, W., Montavon, G., Vedaldi, A., Hansen, L.K., Müller, K.-R. (eds.) Explainable AI: Interpreting, Explaining and Visualizing Deep Learning. LNCS (LNAI), vol. 11700, pp. 121–144. Springer, Cham (2019). https://doi.org/10.1007/978-3-030-28954-6_7
14. Xiang, Y., Chen, Z., et al.: Open DNN box by power side-channel attack. IEEE Trans. Circuits Syst. II Express Briefs **67**(11), 2717–2721 (2020)
15. Yu, H., Ma, H., Yang, K., Zhao, Y., Jin, Y.: DeepEM: deep neural networks model recovery through EM side-channel information leakage. In: 2020 IEEE International Symposium on Hardware Oriented Security and Trust (HOST), pp. 209–218. IEEE (2020)

Physical Countermeasures

A Nearly Tight Proof of Duc et al.'s Conjectured Security Bound for Masked Implementations

Loïc Masure[1]([⊠])[ID], Olivier Rioul[2][ID], and François-Xavier Standaert[1][ID]

[1] ICTEAM Institute, Université catholique de Louvain, Louvain-la-Neuve, Belgium
loic.masure@uclouvain.be
[2] LTCI, Télécom Paris, Institut Polytechnique de Paris, Palaiseau, France

Abstract. We prove a bound that approaches Duc et al.'s conjecture from EUROCRYPT 2015 for the side-channel security of masked implementations. Let Y be a sensitive intermediate variable of a cryptographic primitive taking its values in a set $\mathcal{Y}$. If Y is protected by masking (a.k.a. secret sharing) at order d (i.e., with $d + 1$ shares), then the complexity of any non-adaptive side-channel analysis—measured by the number of queries to the target implementation required to guess the secret key with sufficient confidence—is lower bounded by a quantity inversely proportional to the product of mutual informations between each share of Y and their respective leakage. Our new bound is nearly tight in the sense that each factor in the product has an exponent of -1 as conjectured, and its multiplicative constant is $\mathcal{O}\big(\log |\mathcal{Y}| \cdot |\mathcal{Y}|^{-1} \cdot C^{-d}\big)$, where $C \leq 2\log(2) \approx 1.38$. It drastically improves upon previous proven bounds, where the exponent was $-1/2$, and the multiplicative constant was $\mathcal{O}\big(|\mathcal{Y}|^{-d}\big)$. As a consequence for side-channel security evaluators, it is possible to provably and efficiently infer the security level of a masked implementation by simply analyzing each individual share, under the necessary condition that the leakage of these shares are independent.

Keywords: Mutual information · Masking · Convolution · Security bound

1 Introduction

Evaluating the side-channel security of a cryptographic implementation is a sensitive task, in part due to the challenge of defining the adversary's capabilities [ABB+20]. One approach to deal with this problem is to consider the worst-case security of the implementation, which is characterized by the mutual information between its sensitive intermediate variables and the leakage [SMY09]. Worst-case analysis can be viewed as a natural extension of Kerckhoffs' laws to side-channel security, where all the implementation details are given to the evaluator who can even profile (i.e., estimate the statistical distribution of the leakage) in an offline phase where he controls the implementation

© The Author(s), under exclusive license to Springer Nature Switzerland AG 2023
I. Buhan and T. Schneider (Eds.): CARDIS 2022, LNCS 13820, pp. 69–81, 2023.
https://doi.org/10.1007/978-3-031-25319-5_4

(including its keys and the random coins used in countermeasures). This approach has the advantage of leading to a simple definition of security matching the standard practice of modern cryptography. It has been known for a while that the link between the mutual information metric and the security of an unprotected implementation is nearly tight [MOS11]. It is also known that if an evaluator can estimate this metric for an implementation with countermeasures, the link with its security level is nearly tight as well [dCGRP19]. This state of the art essentially leaves evaluators with the problem of estimating the mutual information between sensitive intermediate variables and the leakage of an implementation protected with countermeasures, which can be much more challenging since countermeasures typically make leakage distributions more complex.

In this paper, we are concerned with the important case of the masking countermeasures [CJRR99]. Its high-level idea is to split any sensitive variable of an implementation into $d + 1$ shares and to compute on those shares only. As a result, evaluating the worst-case security of a masked implementation requires the characterization of high-order and multivariate distributions, which rapidly turns out to be expensive as the number of shares increases [SVO+10]. In order to mitigate this difficulty, a sequence of works has focused on the formal understanding of the masking countermeasure [PR13, DDF14], and its link with concrete evaluation practice [DFS15a]. In this last reference, a lower bound on the minimum number of queries $N_a^\star$ required to recover a target secret with a success rate at least β thanks to a side-channel attack was established as:

$$N_a^\star \geq \frac{\log(1 - \beta)}{\log\left(1 - \left(\frac{|\mathcal{Y}|}{\sqrt{2}}\right)^{d+1} \prod_{i=0}^{d} \mathsf{MI}(Y_i; \mathbf{L}_i)^{1/2}\right)} \approx \frac{\log(1 - \beta)}{\left(\frac{|\mathcal{Y}|}{\sqrt{2}}\right)^{d+1}} \prod_{i=0}^{d} \mathsf{MI}(Y_i; \mathbf{L}_i)^{-1/2},$$

where $|\mathcal{Y}|$ stands for the size of the group over which masking is applied. Such bounds are interesting since they reduce the assessment of the success rate of an attack to the evaluation of the $d + 1$ mutual information values between the shares and their corresponding leakage $\mathsf{MI}(Y_i; \mathbf{L}_i)$. Each of these mutual information values is substantially simpler to estimate than $\mathsf{MI}(Y; \mathbf{L})$ since the distribution of the leakage random variable $\mathbf{L}_i$ is a first-order one. Unfortunately, is was also shown in the same paper that this proven bound is no tight. More precisely, empirical attacks suggest that the $-1/2$ exponent for each $\mathsf{MI}(Y_i; \mathbf{L}_i)$ factor might be decreased to -1 and that the $|\mathcal{Y}|^d$ factor would actually be a proof artifact. Halving the exponent has a strong practical impact since it implies that the required number of shares needed to provably reach a given security level might be doubled compared to what is strictly necessary, and the implementation overheads caused by masking scale quadratically in the number of shares. As a result, Duc et al. conjectured that, provided that the shares' leakage is sufficiently small, the lower bound might be tightened as:

$$N_a^\star \geq f(\beta) \prod_{i=0}^{d} \mathsf{MI}(Y_i; \mathbf{L}_i)^{-1}, \tag{1}$$

where $f(\beta) = \mathsf{H}(Y) - (1 - \beta) \cdot \log_2(2^n - 1) - \mathsf{H}_2(\beta)$ is a function of the attack success rate β, given by Fano's inequality [CT06], as shown in [dCGRP19].[1]

In this note, we prove a lower bound on $N_a^\star$ that fulfills almost all the conditions of Duc et al.'s conjecture. More precisely, we establish a lower bound like the one in Eq. 1 with the function $f(\beta)$ divided by a factor $|\mathcal{Y}| \cdot C^d$, with $C = 2 \log(2) \approx 1.38$, regardless of the nature of the group $\mathcal{Y}$.

The proof is simple to establish. It mixes Chérisey et al.'s inequality and Dziembowski et al.'s *XOR lemma* [dCGRP19,DFS16], and holds for any group-based masking, such as Boolean or arithmetical masking. The former is expressed with the mutual information metric while the latter is expressed with the statistical distance. We bridge the gap between both by converting Dziembowski et al.'s XOR lemma into a variant that is based on the mutual information.

Related Works. Prest et al. used the Rényi divergence in order to improve the tightness of masking security proofs but do not get rid of the square root loss (i.e., the $-1/2$ exponent) on which we focus [PGMP19]. Nevertheless, their bound has a logarithmic dependency on the field size $|\mathcal{Y}|$. Liu et al. used the α-information in order to improve Chérisey's bound [LCGR21]. It could be used to improve our results (at the cost of a slightly less readable bound). In a paper published on the IACR ePrint at the same time as ours (now accepted at CCS 2022), Ito et al. independently obtained a result very similar to ours [IUH22]. Although it is only valid for binary fields (whereas ours is valid for any finite field), their bound is slightly tighter, as the $|\mathcal{Y}|$ is replaced by a $|\mathcal{Y}| - 1$ factor. Interestingly, they also conjecture through the derivation of another bound that is leakage-dependent, and through experimental verifications similar to ours, that the obtained MI-dependent bound is not completely tight in practical cases.

2 Background

2.1 Problem Statement

Let $\mathcal{Y}$ be a finite set. Let $Y \in \mathcal{Y}$ be a random variable denoting the sensitive intermediate variable targeted by a side-channel adversary. In the "standard DPA setting" we consider [MOS11], Y depends on both a uniformly distributed public plaintext and a secret chunk. We assume an implementation that is protected by a d-th order masking. This means that Y is encoded into $d + 1$ shares $Y_0, \ldots, Y_d$ such that $Y_1, \ldots, Y_d$ are drawn uniformly at random from the group $(\mathcal{Y}, \star)$ and $Y_0 = Y \star (Y_1 \star \ldots \star Y_d)^{-1}$, with $\star$ the operation over which the masking is applied (e.g., $\oplus$ for Boolean masking, modular addition $+$ for arithmetic masking, ...). As required by masking security proofs, we further assume that the shares' leakage vectors $\mathbf{L}_i$ are the output of a memoryless side-channel and depend only on the realization Y_i, so that the random vectors $\mathbf{L}_0, \ldots, \mathbf{L}_d$ are

---

[1] H_2 stands for the *binary* entropy function [dCGRP19].

mutually independent.[2] Intuitively, the goal of a worst-case side-channel security evaluation is to quantify the distance of the random variable Y to the uniform distribution over $\mathcal{Y}$ given the observation of $\mathbf{L} = (\mathbf{L}_0, \ldots, \mathbf{L}_d)$. To simplify our computations, we use (in the proof of Theorem 3) Dziembowski et al.'s reduction to random walks [DFS16, Proof of Lemma 3]. Namely, it is equivalent to consider Y_0 to be uniformly distributed over $\mathcal{Y}$ and to quantify the distance of the variable $Y = Y_0 \star \ldots \star Y_d$ to the uniform distribution given the observations of $\mathbf{L}$.

2.2 Quantifying the Distance to Uniform

To quantify the notion of distance to the uniform distribution over $\mathcal{Y}$, we will use two different metrics. The first one is widely known in information theory.

Definition 1 (Mutual Information). *Let* p, m *be two Probability Mass Functions (PMFs) over the finite set* $\mathcal{Y}$.[3] *We denote by* $\mathsf{D_{KL}}(\mathsf{p} \parallel \mathsf{m})$ *the Kullback-Leibler (KL) divergence between* p *and* m:

$$\mathsf{D_{KL}}(\mathsf{p} \parallel \mathsf{m}) = \sum_{y \in \mathcal{Y}} \mathsf{p}(y) \log_2 \left(\frac{\mathsf{p}(y)}{\mathsf{m}(y)} \right). \tag{2}$$

Then, we define the Mutual Information (MI) between a discrete random variable Y *and a continuous random vector* $\mathbf{L}$ *as follows:*

$$\mathsf{MI}(Y; \boldsymbol{L}) = \mathop{\mathbb{E}}_{\mathbf{L}} \left[\mathsf{D_{KL}}(\mathsf{p}_{Y \mid \mathbf{L}} \parallel \mathsf{p}_{Y}) \right], \tag{3}$$

where p_Y *and* $\mathsf{p}_{Y \mid \mathbf{L}}$ *respectively denote the PMF of* Y *and the PMF of* Y *given a realization* l *of the random vector* $\mathbf{L}$, *with the expectation taken over* $\mathbf{L}$.

The second metric is well-known in the cryptographic community.

Definition 2 (Statistical Distance). *Let* p, m *be two PMFs over the finite set* $\mathcal{Y}$. *We denote by* $\mathsf{TV}(\mathsf{p}; \mathsf{m})$ *the Total Variation (TV) between* p *and* m:

$$\mathsf{TV}(\mathsf{p}; \mathsf{m}) = \frac{1}{2} \sum_{y \in \mathcal{Y}} |\mathsf{p}(y) - \mathsf{m}(y)|. \tag{4}$$

Then, we define the Statistical Distance (SD) as follows:

$$\mathsf{SD}(Y; \boldsymbol{L}) = \mathop{\mathbb{E}}_{\mathbf{L}} \left[\mathsf{TV}\left(\mathsf{p}_{Y \mid \mathbf{L}}; \mathsf{p}_Y\right) \right]. \tag{5}$$

---

[2] This typically captures a software implementation manipulating the shares sequentially, but as discussed in [BDF+17], Lemma 1, it generalizes to parallel (e.g., hardware) implementations as long as the leakage due to the manipulation of shares in parallel can be written as a linear combination of the $\mathbf{L}_i$ vectors.

[3] We assume without loss of generality that both p and m have full support over $\mathcal{Y}$.

Note that both the MI and the SD are so-called *global* metrics and are similarly constructed as the expectation over the marginal leakage distribution of so-called *local* quantities, namely the KL divergence and the TV.

Remark 1. Equation 5 is not a distance between the distributions of **L** and Y *per se*, as both random variables are not even defined on the same support. Actually, the TV (local) metric is denoted as SD in the work of Dziembowski et al. [DFS16], whereas our definition of SD coincides with their definition of bias. Nevertheless, we keep this notation for the SD metric in order to keep consistency with the notations for the MI, as previously done by Prest et al. [PGMP19].

In previous works such as the ones of Prouff et al. [PR13], Duc et al. [DFS15a] or Prest et al. [PGMP19], all the inequalities used are stated in terms of global metrics. The idea of our proof is to rely on some similar inequalities between the local quantities, since they are arguably stronger. We introduce such inequalities hereafter. The first one is the well-known Pinsker's inequality.

Proposition 1 (Pinsker's inequality [CT06, Lemma 11.6.1]). *Let* p, m *be two distributions over the—not necessarily finite—set* $\mathcal{Y}$. *Then:*

$$\mathsf{TV}(p; m)^2 \cdot 2\log_2(e) \leq \mathsf{D_{KL}}(p \parallel m). \tag{6}$$

Pinsker's local inequality implies the global inequality

$$2\,\mathsf{SD}(Y; \mathbf{L})^2 \leq \mathsf{MI}(Y; \mathbf{L})$$

used by Duc et al. [DFS15a, Thm. 1], and Prest et al. [PGMP19, Prop. 2].

Remark 2. It is possible to find tighter distribution-dependent constants for Eq. 6 [OW05]. Nevertheless, the universal constant denoted in the inequality remains the tightest possible if m is the uniform distribution—which is our case of interest.

The second inequality we need is a reversed version of Pinsker's inequality.

Theorem 1 (Reversed Pinsker's Inequality [SV15, Thm. 1]). *Let* p *be a PMF over the finite set* $\mathcal{Y}$, *and let* u *denote the uniform PMF over* $\mathcal{Y}$. *Then:*

$$\mathsf{D_{KL}}(p \parallel u) \leq \log_2\!\left(1 + 2\,|\mathcal{Y}|\,\mathsf{TV}(p; u)^2\right) \leq 2\log_2(e)\,|\mathcal{Y}| \cdot \mathsf{TV}(p; u)^2. \tag{7}$$

Again, the reversed Pinsker's inequality is stronger than some previous results from Prest et al. as it implies the following global inequality established by the authors of [PGMP19]: $\mathsf{MI}(Y; \mathbf{L}) \leq 2\,\mathsf{RE}(Y; \mathbf{L}) \cdot \mathsf{SD}(Y; \mathbf{L})$, where $\mathsf{RE}(Y; \mathbf{L})$ stands for the Relative Error (RE) between Y and **L**, another global distance metric introduced in this reference. The reversed Pinsker's (local) inequality is even strictly stronger than Prest et al.'s global inequality, as the former one enables to show that the KL divergence and the squared TV are equivalent metrics (up to a multiplicative constant) whereas, to the best of our knowledge, it is not possible to state that the MI and the squared SD are equivalent global metrics, as the latter one is always bounded by 1, whereas the former one is only bounded by $\log_2 |\mathcal{Y}|$, which can be arbitrarily high. The third inequality we need is the so-called XOR lemma stated by Dziembowski et al. at TCC 2016 [DFS16, Thm. 2]. We next provide a slightly looser version of this result with a simpler proof.

Theorem 2 (XOR Lemma). *Let Y_1, Y_2 be independent random variables on a group $\mathcal{Y}$, and let u denote the uniform PMF over $\mathcal{Y}$. Then:*

$$\mathsf{TV}(\mathsf{p}_{Y_1 \star Y_2}; u) \leq 2 \cdot \mathsf{TV}(\mathsf{p}_{Y_1}; u) \cdot \mathsf{TV}(\mathsf{p}_{Y_2}; u). \tag{8}$$

Proof. Equation 8 is actually a corollary of Young's convolution inequality [Gra14, Thm. 1.2.10]. Denoting by $*$ the convolution product between two PMFs over $\mathcal{Y}$, we have:

$$\mathsf{TV}(\mathsf{p}_{Y_1 \star Y_2}; u) = \frac{1}{2} \left\| \mathsf{p}_{Y_1 \star Y_2} - u \right\|_1 = \frac{1}{2} \left\| \mathsf{p}_{Y_1} * \mathsf{p}_{Y_2} - u \right\|_1 \tag{9}$$

$$= \frac{1}{2} \left\| (\mathsf{p}_{Y_1} - u) * (\mathsf{p}_{Y_2} - u) \right\|_1 \tag{10}$$

$$\leq \frac{1}{2} \left\| (\mathsf{p}_{Y_1} - u) \right\|_1 \cdot \left\| (\mathsf{p}_{Y_2} - u) \right\|_1 \tag{11}$$

$$= 2 \, \mathsf{TV}(\mathsf{p}_{Y_1}; u) \cdot \mathsf{TV}(\mathsf{p}_{Y_2}; u), \tag{12}$$

where Eq. 10 comes from the fact that the uniform PMF is absorbing for the convolution, and Eq. 11 comes from Young's convolution inequality.   □

Theorem 2 is the core of the noise amplification result of Dziembowski et al. and will also be used to argue about it in our proofs. More precisely, we will use the following corollary that is straightforwardly implied by Theorem 2.

Corollary 1. *Let $Y_0, \ldots, Y_d$ be independent random variables on a group $\mathcal{Y}$. Denote $Y_0 \star \ldots \star Y_d$ by Y. Then, we have:*

$$\mathsf{TV}(\mathsf{p}_Y; u) \leq 2^d \cdot \prod_{i=0}^{d} \mathsf{TV}(\mathsf{p}_{Y_i}; u). \tag{13}$$

3   Nearly Tight Bounds

We now provide new provable bounds for the worst-case side-channel security of masked cryptographic implementations. We start with an upper bound on the mutual information and follow with a lower bound on the security level.

3.1   Upper Bounding the Mutual Information

We first establish noise amplification in terms of KL divergence.

Proposition 2. *Let $Y_0, \ldots, Y_d$ be independent but not necessarily identically distributed random variables over $\mathcal{Y}$, with PMFs respectively worth p_{Y_i}. Let $C = 2\log(2) \approx 1.3862$. Denote the PMF of $Y_0 \star \ldots \star Y_d$ as p_Y. Then:*

$$D_{KL}(\mathsf{p}_Y \parallel u) \leq \log_2\left(1 + |\mathcal{Y}| \cdot \prod_{i=0}^{d} \left(C \cdot D_{KL}(\mathsf{p}_{Y_i} \parallel u)\right)\right). \tag{14}$$

Proof. Using the inequalities introduced in Subsect. 2.2, we get:

$$D_{KL}(\mathsf{p}_Y \parallel u) \underset{(7)}{\leq} \log_2\left(1 + 2\,|\mathcal{Y}|\,\mathsf{TV}(\mathsf{p}_Y; u)^2\right) \tag{15}$$

$$\underset{(13)}{\leq} \log_2\left(1 + |\mathcal{Y}| \cdot \prod_{i=0}^{d} \left(2\,\mathsf{TV}(\mathsf{p}_{Y_i}; u)\right)^2\right) \tag{16}$$

$$\underset{(6)}{\leq} \log_2\left(1 + |\mathcal{Y}| \cdot \prod_{i=0}^{d} \left(\frac{2}{\log_2(e)} D_{KL}(\mathsf{p}_{Y_i} \parallel u)\right)\right). \tag{17}$$

$\square$

Having established an amplification result at a local scale, we can now extend it towards the (global) MI metric by taking the expectation over the marginal distribution of the leakage, as stated by the following theorem.

Theorem 3 (MI upper bound (main result)). *Let $Y_0, \ldots, Y_d$ be $d+1$ Independent and Identically Distributed (IID) shares uniformly distributed over $\mathcal{Y}$. Let $\mathbf{L}_0, \ldots, \mathbf{L}_d$ be the leakages occurred by each share. Denote $Y = Y_0 \star \ldots \star Y_d$, and $\mathbf{L} = (\mathbf{L}_0, \ldots, \mathbf{L}_d)$. Then:*

$$\mathsf{MI}(Y; \mathbf{L}) \leq \log_2\left(1 + |\mathcal{Y}| \cdot \prod_{i=0}^{d} C \cdot \mathsf{MI}(Y_i; \mathbf{L}_i)\right). \tag{18}$$

Proof. We apply Proposition 2 to the random variables

$$Y_0' = (Y_0 \mid \mathbf{L}_0), \ldots, Y_d' = (Y_d \mid \mathbf{L}_d).$$

Therefore, we introduce $Y' = Y_0' \star \ldots \star Y_d'$. As a consequence, each term $D_{KL}(\mathsf{p}_{Y'} \parallel u)$ becomes a random variable depending only on the realization of $\mathbf{L}_i$. Furthermore, as stated in Subsect. 2.1, thanks to Dziembowski et al.'s reduction to random walks, the random variables $\mathbf{L}_0, \ldots, \mathbf{L}_d$ are mutually independent. As a consequence:

$$MI(Y; \mathbf{L}) = \mathop{\mathbb{E}}_{\mathbf{L}} \left[D_{\mathsf{KL}}(p_{Y \mid \mathbf{L}} \parallel p_Y) \right] = \mathop{\mathbb{E}}_{\mathbf{L}} \left[D_{\mathsf{KL}}(p_{Y'} \parallel u) \right] \tag{19}$$

$$\underset{(14)}{\leq} \mathop{\mathbb{E}}_{\mathbf{L}} \left[\log_2 \left(1 + |\mathcal{Y}| \cdot \prod_{i=0}^{d} \left(C \cdot D_{\mathsf{KL}}(p_{Y_i'} \parallel u) \right) \right) \right] \tag{20}$$

$$\leq \log_2 \left(1 + |\mathcal{Y}| \cdot \mathop{\mathbb{E}}_{\mathbf{L}} \left[\prod_{i=0}^{d} \left(C \cdot D_{\mathsf{KL}}(p_{Y_i'} \parallel u) \right) \right] \right) \tag{21}$$

$$\leq \log_2 \left(1 + |\mathcal{Y}| \cdot \prod_{i=0}^{d} \left(C \cdot \mathop{\mathbb{E}}_{\mathbf{L}_i} \left[D_{\mathsf{KL}}(p_{Y_i'} \parallel u) \right] \right) \right) \tag{22}$$

$$\leq \log_2 \left(1 + |\mathcal{Y}| \cdot \prod_{i=0}^{d} C \cdot MI(Y_i; \mathbf{L}_i) \right), \tag{23}$$

where Eq. 21 comes from Jensen's inequality applied to the logarithm, as it is concave, Eq. 22 comes from the independence of the leakages, and Eq. 19, Eq. 23 come from the definition of MI in Eq. 3. □

Corollary 2. *Using the same notations as in Theorem 3, we have:*

$$MI(Y; \mathbf{L}) \leq 2 \cdot |\mathcal{Y}| \, C^d \prod_{i=0}^{d} MI(Y_i; \mathbf{L}_i). \tag{24}$$

Proof. Direct by applying the inequality $\log_2(1 + x) \leq x \log_2(e)$ to Eq. 18. □

Verification on Simulated Measurements. To verify the soundness of the previous bound, we consider a standard simulated setting where the leakage of each share corresponds to its Hamming weight with additive Gaussian noise with variance σ^2. We first estimate the exact $MI(Y; \mathbf{L})$ with Monte-Carlo simulations for one, two, four and eight shares [MDP20]. Then, we use the estimated $MI(Y_i; \mathbf{L}_i)$ of one share (assuming it is equal for all the shares) to derive an upper bound for two, four and eight shares. Figure 1 shows the resulting information theoretic curves in function of the variance of the additive Gaussian noise. It confirms that the bound is nearly tight for binary targets. By contrast, as the size of the masking field increases, the factor $|\mathcal{Y}|$ of Eq. 24 makes it less tight. Based on the results of [DFS15b], we expect it to be a proof artifact, of which the removal in our bound is an interesting open problem.

3.2   From a MI Upper Bound to a Security Lower Bound

Combining Chérisey's bound in [dCGRP19] with Eq. 18 leads to following corollary that bounds the number of measurement queries needed to guess a target secret with sufficient confidence thanks to side-channel leakage.

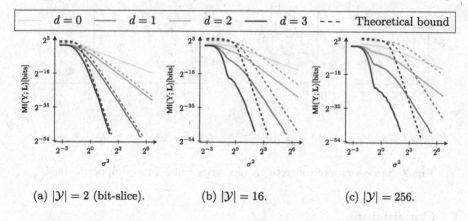

(a) $|\mathcal{Y}| = 2$ (bit-slice).   (b) $|\mathcal{Y}| = 16$.   (c) $|\mathcal{Y}| = 256$.

Fig. 1. MI (plain) and new MI upper bound (dashed) for different field sizes.

Corollary 3. *Let $\mathcal{A}$ be any random-plaintext side-channel adversary against an implementation masked at the order d, with each share leaking respectively an amount of information $\mathsf{MI}(Y_i; \mathbf{L}_i)$. Let $\frac{1}{|\mathcal{Y}|} \leq \beta \leq 1$. Then, for $\mathcal{A}$ to succeed in guessing the secret Y with probability higher than β, at least*

$$N_a^{\star} \geq \frac{f(\beta)}{\log_2\left(1 + |\mathcal{Y}| \cdot \prod_{i=0}^{d} C \cdot \mathsf{MI}(Y_i; \mathbf{L}_i)\right)} \geq \frac{f(\beta)}{2\,|\mathcal{Y}|\,C^d} \prod_{i=0}^{d} \mathsf{MI}(Y_i; \mathbf{L}_i)^{-1} \quad (25)$$

measurement queries to the target leaking implementation are needed.

Verification on Simulated Measurements. To verify the soundness of the proposed bound, we simulate a bit recovery using the same leakage model as in our previous simulations, for one, two and three shares, and different levels of noise—captured by the Gaussian noise variance σ^2.[4] The results are depicted in plain curves on Fig. 2. Based on the Monte-Carlo simulation of the MI for one share—still assuming that the shares verify the same leakage model—we also compute the right hand-side of Eq. 25, for $\beta \in [0.5, 1]$. This gives the dotted upper bounds of β depicted on Fig. 2. Figure 2a shows that the bound derived in Eq. 25 may not be tight when the MI is high. Nevertheless, we can see from Fig. 2b, Fig. 2c and Fig. 2d that the higher the noise variance (and the lower the MI) the tighter the expected upper bounds.

---

[4] The success rate is estimated with bootstrapping, which gives good estimations with a negligible bias provided that the number of simulated traces is far higher than the value of $N_a^{\star}$ such that $\beta = 1$. Due to memory constraints, not enough samples could be drawn to get a consistent simulation in the case where $\sigma^2 = 100, d = 2$.

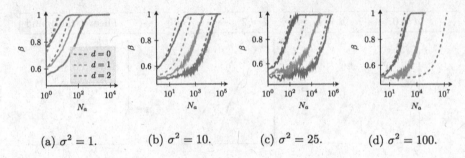

(a) $\sigma^2 = 1$.   (b) $\sigma^2 = 10$.   (c) $\sigma^2 = 25$.   (d) $\sigma^2 = 100$.

Fig. 2. Success rate of concrete bit recoveries and MI-based upper bounds.

4   Conclusions

We prove a new bound approaching by Duc et al.'s conjecture for the security of masked implementations. Our result is tight in $\mathbb{F}_2$, which makes it practically-relevant since bitslice masking is currently the most efficient way to implement masking for binary ciphers (especially lightweight ones) [GR17]. For larger field sizes, a factor corresponding to the field size $|\mathcal{Y}|$ makes it less tight. Getting rid of this last source of non-tightness therefore remains as an interesting direction for further improvements. We finally note that we can improve another bound from TCC 2016—which is stated in terms of statistical distance—as a side-effect of our investigations. We detail this last result in Sect. A.

Acknowledgments. François-Xavier Standaert is a Senior Research Associate of the Belgian Fund for Scientific Research (FNRS-F.R.S.). This work has been funded in part by the ERC project number 724725 (acronym SWORD).

A   Side-Effect: Improving TCC 2016's Bounds

In Subsect. 3.1, we have proven that the MI between Y and the whole leakage vector is bounded (up to a multiplicative constant) by the product of the shares' MIs. Since the squared TV and the KL divergence are consistent for finite sets like $\mathcal{Y}$, we may wonder whether an upper bound implying the SD as global metric can be derived from the XOR lemma as well. Dziembowski et al. actually provided such an upper bound as recalled hereafter.[5]

Proposition 3 ([DFS16, **Thm. 1(i)**], **restated**). *Let Y be a uniform random variable on a group $\mathcal{Y}$, encoded by the $(d + 1)$-sharing $Y_0, \ldots, Y_d$. Suppose that all the leakages are δ-noisy for $i = 0, \ldots, d$, i.e., for $0 \leq \delta < 1/2$,*

$$\mathsf{SD}(Y_i; \mathbf{L}_i) \leq \delta.$$

---

[5] The original version of the theorem is stated for non-uniform secrets. In order to avoid an unfair comparison with respect to Dziembowski et al.'s work, we present an intermediate result of their proof [DFS16, Sec. 3.1.5].

Define the noise *parameter* $\theta = 1/2 - \delta$. *Then, for all* $\epsilon > 0$, *in order to get* $\mathsf{SD}(Y;\mathbf{L}) \leq \epsilon$, *it is sufficient that the masking order verifies:*

$$d \geq 8\theta^{-2} \log\left(\frac{3}{2}\epsilon^{-1}\right). \tag{26}$$

Informally, Eq. 26 gives the sufficient masking order d in order to achieve a desired security level ϵ (expressed in terms of Statistical Distance (SD)), depending on the noise level θ that the developer may leverage—the higher θ, the noisier the leakage model. Unfortunately, the bound (26) is not tight, as the authors also derive the following necessary condition [DFS16, Eq. (12)]:

$$d \geq \frac{\log\left((2\epsilon)^{-1}\right)}{\log\left((1 - 2\theta)^{-1}\right)}, \tag{27}$$

which is asymptotically linear in θ^{-1} when $\theta \to 0$, whereas it is quadratic in Eq. 26. Actually, this is mostly due to the overhead term that occurs from the authors' so-called "reduction to unconditional random walks". We shall show that this reduction is not necessary, by leveraging the independence of the leakages to compute the expectation, as in our proof of Theorem 3. As a result, we end up with a tight upper bound, no longer involving the overhead term.

Proposition 4 (Improved bound). *With the same notations as in Proposition 3, it is sufficient that the masking order verifies Eq. 27:*

Proof. Starting from the definition of SD (Definition 2), we have:

$$\mathsf{SD}(Y;\mathbf{L}) = \underset{\mathbf{L}}{\mathbb{E}}\left[\mathsf{TV}\left(\mathsf{p}_Y; \mathsf{p}_{Y\,|\,\mathbf{L}}\right)\right] \tag{28}$$

$$\underset{(13)}{\leq} 2^d \underset{\mathbf{L}}{\mathbb{E}}\left[\prod_{i=0}^{d} \mathsf{TV}\left(\mathsf{p}_{Y_i}; \mathsf{p}_{Y_i\,|\,\mathbf{L}_i}\right)\right] \tag{29}$$

$$= 2^d \prod_{i=0}^{d} \underset{\mathbf{L}_i}{\mathbb{E}}\left[\mathsf{TV}\left(\mathsf{p}_{Y_i}; \mathsf{p}_{Y_i\,|\,\mathbf{L}_i}\right)\right] \tag{30}$$

$$= 2^d \prod_{i=0}^{d} \mathsf{SD}(Y_i; \mathbf{L}_i), \tag{31}$$

where Eq. 30 comes from the mutual independence of the leakages $\mathbf{L}_i$. Now, assuming that for all $i \in [\![0,d]\!]$ we have $\mathsf{SD}(Y_i;\mathbf{L}_i) \leq \delta < \frac{1}{2}$ and defining $\theta = \frac{1}{2} - \delta$, we have $\mathsf{SD}(Y;\mathbf{L}) \leq \epsilon$ if $(1 - 2\theta)^d \leq 2\epsilon$. Hence, the inequality $\mathsf{SD}(Y;\mathbf{L}) \leq \epsilon$ holds if Eq. 27 holds. $\square$

References

[ABB+20] Azouaoui, M., et al.: A systematic appraisal of side channel evaluation strategies. In: van der Merwe, T., Mitchell, C., Mehrnezhad, M. (eds.) SSR 2020. LNCS, vol. 12529, pp. 46–66. Springer, Cham (2020). https://doi.org/10.1007/978-3-030-64357-7_3

[BDF+17] Barthe, G., Dupressoir, F., Faust, S., Grégoire, B., Standaert, F.-X., Strub, P.-Y.: Parallel implementations of masking schemes and the bounded moment leakage model. In: Coron, J.-S., Nielsen, J.B. (eds.) EUROCRYPT 2017. LNCS, vol. 10210, pp. 535–566. Springer, Cham (2017). https://doi.org/10.1007/978-3-319-56620-7_19

[CJRR99] Chari, S., Jutla, C.S., Rao, J.R., Rohatgi, P.: Towards sound approaches to counteract power-analysis attacks. In: Wiener, M. (ed.) CRYPTO 1999. LNCS, vol. 1666, pp. 398–412. Springer, Heidelberg (1999). https://doi.org/10.1007/3-540-48405-1_26

[CT06] Cover, T.M., Thomas, J.A.: Elements of Information Theory, 2nd edn. Wiley, Hoboken (2006)

[dCGRP19] de Chérisey, E., Guilley, S., Rioul, O., Piantanida, P.: Best information is most successful mutual information and success rate in side-channel analysis. IACR Trans. Cryptogr. Hardw. Embed. Syst. **2019**(2), 49–79 (2019)

[DDF14] Duc, A., Dziembowski, S., Faust, S.: Unifying leakage models: from probing attacks to noisy leakage. IACR Cryptol. ePrint Arch., p. 79 (2014)

[DFS15a] Duc, A., Faust, S., Standaert, F.-X.: Making masking security proofs concrete. In: Oswald, E., Fischlin, M. (eds.) EUROCRYPT 2015. LNCS, vol. 9056, pp. 401–429. Springer, Heidelberg (2015). https://doi.org/10.1007/978-3-662-46800-5_16

[DFS15b] Dziembowski, S., Faust, S., Skorski, M.: Noisy leakage revisited. In: Oswald, E., Fischlin, M. (eds.) EUROCRYPT 2015. LNCS, vol. 9057, pp. 159–188. Springer, Heidelberg (2015). https://doi.org/10.1007/978-3-662-46803-6_6

[DFS16] Dziembowski, S., Faust, S., Skórski, M.: Optimal amplification of noisy leakages. In: Kushilevitz, E., Malkin, T. (eds.) TCC 2016. LNCS, vol. 9563, pp. 291–318. Springer, Heidelberg (2016). https://doi.org/10.1007/978-3-662-49099-0_11

[GR17] Goudarzi, D., Rivain, M.: How fast can higher-order masking be in software? In: Coron, J.-S., Nielsen, J.B. (eds.) EUROCRYPT 2017. LNCS, vol. 10210, pp. 567–597. Springer, Cham (2017). https://doi.org/10.1007/978-3-319-56620-7_20

[Gra14] Grafakos, L.: Classical Fourier analysis (2014)

[IUH22] Ito, A., Ueno, R., Homma, N.: On the success rate of side-channel attacks on masked implementations: information-theoretical bounds and their practical usage. In: Yin, H., Stavrou, A., Cremers, C., Shi, E. (eds.) Proceedings of the 2022 ACM SIGSAC Conference on Computer and Communications Security, CCS 2022, Los Angeles, CA, USA, 7–11 November 2022, pp. 1521–1535. ACM (2022)

[LCGR21] Liu, Y., Cheng, W., Guilley, S., Rioul, O.: On conditional alpha-information and its application to side-channel analysis. In: ITW, pp. 1–6. IEEE (2021)

[MDP20] Masure, L., Dumas, C., Prouff, E.: A comprehensive study of deep learning for side-channel analysis. IACR Trans. Cryptogr. Hardw. Embed. Syst. **2020**(1), 348–375 (2020)

[MOS11] Mangard, S., Oswald, E., Standaert, F.-X.: One for all - all for one: unifying standard differential power analysis attacks. IET Inf. Secur. **5**(2), 100–110 (2011)

[OW05] Ordentlich, E., Weinberger, M.J.: A distribution dependent refinement of Pinsker's inequality. IEEE Trans. Inf. Theory **51**(5), 1836–1840 (2005)

[PGMP19] Prest, T., Goudarzi, D., Martinelli, A., Passelègue, A.: Unifying leakage models on a rényi day. In: Boldyreva, A., Micciancio, D. (eds.) CRYPTO 2019. LNCS, vol. 11692, pp. 683–712. Springer, Cham (2019). https://doi.org/10.1007/978-3-030-26948-7_24

  [PR13] Prouff, E., Rivain, M.: Masking against side-channel attacks: a formal security proof. In: Johansson, T., Nguyen, P.Q. (eds.) EUROCRYPT 2013. LNCS, vol. 7881, pp. 142–159. Springer, Heidelberg (2013). https://doi.org/10.1007/978-3-642-38348-9_9

 [SMY09] Standaert, F.-X., Malkin, T.G., Yung, M.: A unified framework for the analysis of side-channel key recovery attacks. In: Joux, A. (ed.) EURO-CRYPT 2009. LNCS, vol. 5479, pp. 443–461. Springer, Heidelberg (2009). https://doi.org/10.1007/978-3-642-01001-9_26

  [SV15] Sason, I., Verdú, S.: Upper bounds on the relative entropy and rényi divergence as a function of total variation distance for finite alphabets. In: ITW Fall, pp. 214–218. IEEE (2015)

[SVO+10] Standaert, F.-X., et al.: The world is not enough: another look on second-order DPA. In: Abe, M. (ed.) ASIACRYPT 2010. LNCS, vol. 6477, pp. 112–129. Springer, Heidelberg (2010). https://doi.org/10.1007/978-3-642-17373-8_7

Short-Iteration Constant-Time GCD and Modular Inversion

Yaoan Jin[1]([⊠]) [iD] and Atsuko Miyaji[1,2] [iD]

[1] Graduate School of Engineering, Osaka University, Suita, Japan
jin@cy2sec.comm.eng.osaka-u.ac.jp
[2] Japan Advanced Institute of Science and Technology, Nomi, Japan

Abstract. Even theoretically secure cryptosystems, digital signatures, etc. may not be secure after being implemented on the Internet of Things (IoT) devices and PCs because of Side-Channel Attack (SCA). Since RSA key generation and ECDSA need GCD computations or modular inversions, which are often computed by Binary Euclidean Algorithm (BEA) or Binary Extended Euclidean Algorithm (BEEA), the SCA weakness of BEA and BEEA becomes serious. For countermeasures, the Constant-Time GCD (CT-GCD) and Constant-Time Modular Inversion (CTMI) algorithms are good choices. Modular inversion based on Fermat's Little Theorem (FLT) can work in constant time but it is not efficient for general inputs. Two CTMI algorithms, named BOS and BY in this paper, are proposed by Joppe W. Bos and Bernstein, Yang respectively, which are based on the idea of BEA. However, BOS has complicated computations during one iteration and BY uses more iterations. Small number of iterations and simple computations during one iteration are good characteristics of a constant-time algorithm. Based on this view, this paper proposes new short-iteration CT-GCD and CTMI algorithms over $\mathbb{F}_p$ borrowing a simple idea of BEA. Our algorithms are evaluated from the theoretical point of view. Compared with BOS, BY and the improved version of BY, our short-iteration algorithms are experimentally demonstrated to be faster than theirs.

Keywords: Constant-Time Modular Inversion (CTMI) ·
Constant-Time Greatest Common Divisor (CT-GCD) · Side Channel
Attack (SCA)

1 Introduction

Secure cryptosystems, digital signatures, etc. are threatened by Side Channel Attack (SCA) after being implemented on Internet of Things (IoT) devices and PCs. Secret information can be obtained by analyses of power consumption, implementation time, etc. during the executions of cryptographic algorithms. With the development of SCA, the endless emergence of SCA methods and the improvement of SCA methods make such attacks more and more dangerous [7, 15].

The Binary Euclidean Algorithm (BEA) computes Greatest Common Divisor (GCD). It can be used in RSA key generation to check whether the GCD of the

I. Buhan and T. Schneider (Eds.): CARDIS 2022, LNCS 13820, pp. 82–99, 2023.
https://doi.org/10.1007/978-3-031-25319-5_5

public key e and $\phi(n) = (p-1)(q-1)$, where p and q are randomly generated large primes, is equal to one. The Binary Extended Euclidean Algorithm (BEEA) can be used in RSA key generation, ECDSA, etc. to compute a modular inversion. Both of them are attractive because they consist of only shift and subtraction operations. However, BEA and BEEA are threatened by SCA. Specifically, Simple Power Analysis (SPA), Cache-Timing Attack (CTA), Machine Learning based Profiling Attack (MLPA) were conducted on BEA or BEEA to recover the secrets with high success rate [2–4,9,13].

For countermeasures, one can apply a constant-time algorithm. Modular inversion based on Fermat's Little Theorem (FLT) can be computed in constant time. However, it is not efficient for general inputs and only can compute modular inversions over prime numbers. Two Constant-Time Modular Inversion (CTMI) algorithms based on the basic idea of BEA are proposed by Bos [6] and Bernstein, Yang [5], respectively, which are denoted by BOS and BY in this paper. Their algorithms can also be constant-time GCD (CT-GCD). The number of iterations of CTMI by Bernstein and Yang is improved by Pieter Wuille in [14], denoted by hdBY in this paper. However, BY (even hdBY) still uses more iterations than BOS and the computations during one iteration of BOS are complicated.

In this paper, we propose Short-Iteration Constant-Time GCD (SICT-GCD) and CTMI (SICT-MI) on $\mathbb{F}_p$ with simple computations in each iteration and small number of iterations. The core computations (or the iteration formula) of SICT-GCD and SICT-MI on $\mathbb{F}_p$ are defined in Definition 1. The iteration formula based on the GCD computations in Lemma 2 consists of only shift and subtraction operations. Theorem 1 shows that the iteration formula converges to $a_n = \text{GCD}(a_0, b_0)$ and $b_n = 0$, where $a_0, b_0 \in \mathbb{Z}$, $a_0 \geq b_0 \geq 0$, and a_0 or b_0 is odd, with limited iterations. Theorem 2 shows that the necessary number of iterations is $\texttt{bitlen}(a_0) + \texttt{bitlen}(b_0)$. Based on Theorems 1 and 2, the number of iterations of SICT-GCD and SICT-MI algorithms on $\mathbb{F}_p$ is fixed to $2 \cdot \texttt{bitlen}(a_0)$. We show SI-GCD and SI-MI algorithms on $\mathbb{F}_p$ in Algorithms 6 and 7, whose computations in each branch are balanced, then branchless SICT-GCD and SICT-MI on $\mathbb{F}_p$ in Algorithm 8.

We theoretically analyze the efficiency of SICT-GCD and SICT-MI algorithms by comparing the number of iterations and $\mathbb{F}_p$-arithmetic. We also evaluate the efficiency of SICT-GCD and SICT-MI algorithms on $\mathbb{F}_p$, FLT-CTMI, BOS, CT-GCD and CTMI on $\mathbb{F}_p$ by BY and hdBY through experiments. The results show that our SICT-GCD on $\mathbb{F}_p$ saves 7.44%, 13.78%, 16.67%, and 28.3% clock cycles of hdBY on 224-, 256-, 384-, and 2048-bit GCD computation, respectively. Our SICT-MI on $\mathbb{F}_p$ saves

- 17.41%, 16.08%, 18.67%, and 17.76% clock cycles of hdBY on 224-, 256-, 384-, and 2048-bit modular inversion respectively.
- 10.86%, 16.44%, and 82.78% clock cycles of FLT-CTMI on 224-, 384-, and 2048-bit modular inversion respectively.
- 40.86%, 41.43%, 41.8%, and 37.66% clock cycles of BOS on 224-, 256-, 384-, and 2048-bit modular inversion respectively.

The remainder of this paper is organized as follows. The related work is introduced in Sect. 2. SICT-GCD and SICT-MI algorithms on $\mathbb{F}_p$ are described in Sect. 3. In Sect. 4, we evaluate our SICT-GCD and SICT-MI on $\mathbb{F}_p$. Finally, we conclude our paper in Sect. 5.

2   Related Work

2.1   BEA and BEEA

The methods to compute a modular inversion can be divided into two categories. One is based on Fermat's Little Theorem (FLT), whose basic idea is $a^{-1} = a^{p-2} \bmod p$ for a prime number p ($a \in \mathbb{Z}$ and $GCD(a, p) = 1$, which is the greatest common divisor of a and p). Another one is based on extended Euclidean algorithm, Algorithm 1, whose basic idea is $GCD(a, b) = GCD(b, a \bmod b)$ for $a, b \in \mathbb{Z}$. Note that $(b - r)/c$ means a quotient of $b - r$ divided by c in Algorithm 1. The most notable feature of FLT is that it can compute a modular inversion in constant time. In contrast, extended Euclidean algorithm cannot compute modular inversion in constant time generally. Although, it is more efficient and can be used to compute a Greatest Common Divisor (GCD), etc.

Algorithm 1: Extended Euclidean Algorithm

Input: a, p, where $GCD(a, p) = 1$.
Result: $y_1 = a^{-1} \bmod p$
Initialization:
$b = p$; $c = a$; $x_0 = 1$; $x_1 = 0$; $y_0 = 0$; $y_1 = 1$;
while $c \neq 1$ **do**
  | $r = b \bmod c$; $q = (b - r)/c$;
  | $b = c$; $c = r$; $t = x_1$; $x_1 = x_0 - q \cdot x_1$; $x_0 = t$; $t = y_1$; $y_1 = y_0 - q \cdot y_1$; $y_0 = t$;
end
$y_1 = y_1 \bmod p$;
return y_1;

One of the most frequently used variants of Euclidean algorithm (and extended Euclidean algorithm) is Binary Euclidean Algorithm (and Binary Extended Euclidean algorithm), which is called BEA (and BEEA) respectively. BEA and BEEA compute GCD and modular inversion, respectively, which consist of only shift and subtraction operations. They are attractive because shift and subtraction operations can be implemented more easily on both software and hardware. The basic idea of BEA and BEEA is shown as:

$$GCD(a, b) = \begin{cases} GCD(|a - b|, \min(a, b)), & a \text{ and } b \text{ are odd.} \\ GCD(a/2, b), & a \text{ is even}, b \text{ is odd.} \\ GCD(a, b/2), & a \text{ is odd}, b \text{ is even.} \\ 2 \cdot GCD(a/2, b/2), & a \text{ and } b \text{ are even.} \end{cases} \tag{1}$$

BEA computes $GCD(a, b)$ by using Eq. (1), which is shown in Algorithm 2. For modular inversion, the case that a and b are even does not exist, and the first "while" loop in Algorithm 2 can be removed because $GCD(a, p) = 1$ when $a^{-1} \bmod p$ exists. BEEA computes a modular inversion of $a^{-1} \bmod p$ ($GCD(a, p) = 1$), which is shown in Algorithm 3. Note that $>>$ indicates the right shift operation and $<<$ indicates the left shift operation. $A/2 \bmod p$ can be computed as $A >> 1$ when A is even and $(A + p) >> 1$ when A is odd.

Algorithm 2: BEA

Input: $a, b \in \mathbb{Z}$
Result: $r = GCD(a, b)$
Initialization:
$u = a$; $v = b$; $r = 1$;
while *u and v are even* **do**
  | $u = u >> 1$; $v = v >> 1$; $r = r << 1$;
end
while $u \neq 0$ **do**
  | **while** *u is even* **do**
  | | $u = u >> 1$;
  | **end**
  | **while** *v is even* **do**
  | | $v = v >> 1$;
  | **end**
  | **if** $u \geq v$ **then**
  | | $u = u - v$;
  | **else**
  | | $v = v - u$;
  | **end**
end
$r = r \cdot v$;
return r

2.2   SCA of BEA and BEEA

A method to recover the inputs of BEA (or BEEA) is first proposed in [1]. Specifically, the inputs can be recovered by inputting the complete operations flow of an execution of BEA (or BEEA) into the algorithm ([1] Fig. 7). Because u and v are updated to zero and $GCD(a, b)$ finally in Algorithms 2 and 3. With the help of the complete operations flow of an execution of BEA (or BEEA), we can inversely compute the initial values of u and v. The remaining question is that how to obtain the complete operations flow of an execution of BEA (or BEEA).

The complete operations flow of an execution of BEA (or BEEA) can be obtained by SCA. There are some "bad" characteristics in Algorithms 2 and 3, which can be used in SCA to predict the operations flow: 1) There is at most one shift loop of either "u" or "v" in each iteration. 2) There must be one subtraction operation in each iteration. The complete operations flow of an execution of BEEA is obtained by Simple Power Analysis (SPA) and used to attack ECDSA

Algorithm 3: BEEA

Input: a, p, where $\text{GCD}(a, p) = 1$
Result: $C = a^{-1} \bmod p$
Initialization:
$u = a; v = p; A = 1; C = 0;$
while $u \neq 0$ **do**
    while u *is even* **do**
        | $u = u \gg 1; A = A/2 \bmod p;$
    end
    while v *is even* **do**
        | $v = v \gg 1; C = C/2 \bmod p;$
    end
    if $u \geq v$ **then**
        | $u = u - v; A = A - C;$
    else
        | $v = v - u; C = C - A;$
    end
end
$C = C \bmod p;$
return C

---

in [2]. It is pointed out that the subtraction operations flow can be recovered from the shift operations flow in [2]. Thus, the complete shift operations flow of an execution of BEA (or BEEA) is enough to reveal the inputs.

In practical applications, one of the inputs of BEA (or BEEA) is always public. Some characteristics of the inputs of BEA (or BEEA) can be known in advance. All of these make SCA to BEA and BEEA easier. Take RSA key generation in OpenSSL as an example. $\text{GCD}(e, p - 1)$ and $\text{GCD}(e, q - 1)$, where e is a stable public key and p, q are random large prime numbers with the same bit length, are computed by BEA to check whether p and q are generated properly. Modular inversion of $d = e^{-1} \bmod (p - 1)(q - 1)$ is computed to get the secret key by BEEA. Based on the characteristics of RSA key generation: 1) e and $n = pq$ are public; 2) $(p-1)(q-1)$ has almost half the same Most Significant Bits (MSBs) as n, the secret $(p-1)(q-1)$ can be recovered by a part of the operations flow in the beginning of the modular inversion computation, $e^{-1} \bmod (p-1)(q-1)$. Referring to n, it is enough to recover some Least Significant Bits (LSBs) of $(p-1)(q-1)$ with the operations flow in the beginning and the public input e [3,4]. Benefiting from 1) the general setting of $e = 65537$ is much smaller than $(p-1)(q-1)$, $p-1$ and $q-1$; 2) $4|(p-1)(q-1)$, $2|(p-1)$ and $2|(q-1)$, the partial operations flow in the beginning of BEA (or BEEA) can be predicted in advance. Making use all of these characteristics, SPA, Cache-Timing Attack (CTA), Machine Learning based Profiling Attack (MLPA) threatened the security of RSA key generation [3,4,9].

For countermeasures, one can hide the secret inputs of BEA or BEEA by appropriate masking procedures, whose security is reported in [8]. Another choice is to use constant-time GCD (CT-GCD) or constant-time modular inversion (CTMI).

2.3 CT-GCD and CTMI Algorithms

Modular inversion by Fermat's little theorem, which is shown in Algorithm 4, can be computed in constant time because p is the same for each modular inversion, and "if statements" are executed in the same way for any input a. The number of iterations is fixed to $\texttt{bitlen}(p-2)-1$, where $\texttt{bitlen}(p-2)$ is the bit length of $p-2$. The computations in each iteration, which depend on each bit of $p-2$, are the same for any input a. The efficiency of a constant-time modular inversion based on FLT, denoted by FLT-CTMI, depends on the Hamming weight of $p-2$. The larger Hamming weight of $p-2$, the lower efficiency of FLT-CTMI. Thus FLT-CTMI is not efficient for general inputs. Moreover, FLT can compute neither GCD nor modular inversion on a composite number, which is required in RSA key generation such as $\mathsf{GCD}(e, p-1)$ and $e^{-1} \bmod (p-1)(q-1)$.

Algorithm 4: FLT-CTMI

Input: a, p, $a \in \mathbb{Z}$, $\mathsf{GCD}(a, p) = 1$ and p is a prime.
Result: $C = a^{-1} \bmod p$
Initialization:
$C = a$; $k = p - 2$ $(\sum_{i=0}^{n-1} k_i 2^i)$;
$\textbf{for } i = n - 2; \ i \geq 0; \ i = i - 1 \ \textbf{do}$
$\quad C = C^2 \bmod p$;
$\quad \textbf{if } k_i = 1 \textbf{ then}$
$\quad \quad | \quad C = C \times a \bmod p$;

end
return C

Joppe W. Bos proposed a CTMI algorithm by improving Kaliski's algorithm [10] to a constant-time version, called BOS in this paper [6]. The basic idea is that computing all the branches then selecting the correct values from them according to the defined signal variables. Because the sum of bit length of the inputs reduces by one in each iteration, BOS set the number of iterations as $2 \cdot \texttt{bitlen}(p)$. BOS can compute modular inversion, Montgomery inversion, and $\mathsf{GCD}(a, b)$, where a or b is odd.

A CTMI is proposed in [12] with inserted dummy computations. The number of iterations is the same as that of BOS. It is clear that inserting dummy computations is not a good solution of CTMI, which makes efficiency lower and has potential SCA security issues.

Bernstein and Yang proposed a CTMI algorithm on $\mathbb{F}_p$, Algorithm 5, called BY in this paper [5]. The iteration formula of their algorithm is shown in Eq. (2), where $\delta \in \mathbb{Z}$ is initialized to 1 and the input $b \in \mathbb{Z}$ should be odd.

$$F(\delta, b, a) = \begin{cases} (1 - \delta, a, \frac{a-b}{2}) \delta > 0 \text{ and } a \text{ is odd} \\ (1 + \delta, b, \frac{a + (a \bmod 2)b}{2}) \text{ otherwise} \end{cases} \quad (\delta, a, b \in \mathbb{Z}, b \text{ is odd}) \quad (2)$$

Algorithm 5: BY [5]

Input: $a, p, l = \lfloor(49 \cdot \text{bitlen}(p) + 57)/17\rfloor$, $pre\_com = 2^{-l} \bmod p$
Result: $q = a^{-1} \bmod p$
Initialization:
$u = a; v = p; q = 0; r = 1; \delta = 1;$
for $i = 0; i < l; i = i + 1$ **do**
$\quad z = u_{lsb}; s = \text{signbit}(-\delta); \delta = 1 + (1 - 2sz)\delta;$
$\quad u, v = (u + (1 - 2sz)zv) >> 1, v \oplus sz(v \oplus u);$
$\quad q, r = (q \oplus sz(q \oplus r)) << 1, (1 - 2sz)zq + r;$
end
$q = \text{sign}(v) \cdot q;$
$q = q \cdot pre\_com \bmod p;$
return q

The computations of their algorithm during one iteration are simpler than BOS. Bernstein and Yang analyzed the rate of shrinking of the transition matrix after each iteration and proved that the necessary number of iterations of BY is $\lfloor(49d + 57)/17\rfloor(d \geq 46)$, where d is the largest bit length of the inputs. Its number of iterations is much more than BOS. BY can compute modular inversion, Montgomery inversion and $\text{GCD}(a, b)$, where a or b is odd. In Algorithm 5, $\text{signbit}(a)$ returns zero when $a \geq 0$ and one when $a < 0$, $\text{sign}(a)$ returns -1 when $a < 0$ and 1 when $a > 0$. The first k iterations of BY are completely determined by the lowest k bits of b and a. This feature supports a divide and conquer strategy for BY. Bernstein and Yang also proposed a CTMI algorithm on $\mathbb{F}_{p^n}$, where $\mathbb{F}_{p^n} : \mathbb{F}_p[x]/(f(x))$. Then any element in $\mathbb{F}_{p^n}$ is represented by $g(x) \in \mathbb{F}_p[x]$. The iteration formula is shown in Eq. (3), where $f(0) \neq 0$.

$$F(\delta, f, g) = \begin{cases} (1 - \delta, g, (g(0)f - f(0)g)/x)\delta > 0 \text{ and } g(0) \neq 0 \\ (1 + \delta, f, (f(0)g - g(0)f)/x) \text{ otherwise} \end{cases} \quad (3)$$

Each branch has the computations of one addition (or subtraction) on $\mathbb{Z}$, one subtraction on $\mathbb{F}_{p^n}$, one shift on $\mathbb{F}_{p^n}$, and two multiplications on $\mathbb{F}_{p^n}$. The number of iterations is fixed to $2 \cdot \max(\deg(f), \deg(g))$.

Pieter Wuille kept track of convex hulls of possible (b, a) after each iteration to find the necessary number of iterations of BY and obtained similar results [14]. Pieter Wuille found out that the necessary number of iterations of BY for a 224-, 256-, 384-bit computation is 634, 724, 1086, respectively, which is 649, 741, 1110 by Bernstein and Yang. Moreover, Pieter Wuille proposed a variant of BY, denoted by hdBY, by initializing $\delta = 1/2$ or using the iteration formula in Eq. (4).

$$F(\delta, b, a) = \begin{cases} (2 - \delta, a, \frac{a-b}{2})\delta > 0 \text{ and } a \text{ is odd} \\ (2 + \delta, b, \frac{a+(a \bmod 2)b}{2}) \text{ otherwise} \end{cases} \quad (4)$$

The number of iterations of hdBY, defined by $\max(2\lfloor(2455\log_2(M) + 1402)/1736\rfloor, 2\lfloor(2455\log_2(M) + 1676)/1736\rfloor - 1)$, $M \geq 157$, $0 \leq a \leq b \leq M$, is smaller. The necessary number of iterations of hdBY for a 224-, 256-, and 384-bit computation is 517, 590, and 885, respectively.

3   Short-Iteration CT-GCD and CTMI

Both short iterations and simple computations during one iteration are good characteristics of a constant-time algorithm. We combine the basic idea of BEA and the Lemma 1 and propose new Short-Iteration GCD (SI-GCD) and Modular Inversion (SI-MI) algorithms, whose computations in each branch are balanced. Then we propose CT-GCD and CTMI algorithms, called Short-Iteration CT-GCD (SICT-GCD) and Short-Iteration CTMI (SICT-MI) with short iterations and simple computations.

3.1   Our Iteration Formula

Let us start from a simple lemma of GCD. Lemma 1 implies that GCD of any two of a, b, $a - b$, where $a, b \in \mathbb{Z}$ and $a \geq b \geq 0$, is the same. Lemma 1 is used to show Lemma 2.

Lemma 1. $GCD(a, b) = GCD(b, a - b) = GCD(a, a - b)$, where $a, b \in \mathbb{Z}$ and $a \geq b \geq 0$.

Proof. With the well-known fact, $GCD(a, b) = GCD(a - nb, b)$ for any n, it is clear that $GCD(a, b) = GCD(a - b, b)$ with $n = 1$ and $GCD(b, a - b) = GCD(a, a - b)$ with $n = -1$.

Based on the basic idea of BEA in Eq. (1) and Lemma 1, we show that GCD has equality targeting GCD between two numbers that are not both even and modular inversion as follows.

Lemma 2. GCD *has the following equality:*

$$GCD(a, b) = \begin{cases} GCD((a - b)/2, b), & a \text{ and } b \text{ are odd.} \\ GCD(a/2, a - b), & a \text{ is even and } b \text{ is odd.} \\ GCD(b/2, a - b), & a \text{ is odd and } b \text{ is even.} \end{cases}$$

$a, b \in \mathbb{Z}$, $a \geq b \geq 0$ *and* a *or* b *is odd.*

Proof. – Assume that a and b are odd. $GCD(a, b) = GCD(b, a - b)$ by Lemma 1 and $GCD(b, a - b) = GCD(b, (a - b)/2)$ by the idea of BEA.
– Assume that a is even and b is odd. $GCD(a, b) = GCD(a, a - b)$ by Lemma 1 and $GCD(a, a - b) = GCD(a/2, a - b)$ by the idea of BEA.
– Assume that a is odd and b is even. $GCD(a, b) = GCD(b, a - b)$ by Lemma 1 and $GCD(b, a - b) = GCD(b/2, a - b)$ by the idea of BEA.

According to Lemma 2, the iteration formula, which is the core computation in our SICT-GCD and SICT-MI, is defined as follows.

Definition 1. *The iteration formula,* $(a_{n+1}, b_{n+1}) = f(a_n, b_n)$*, for* $a_n, b_n \in \mathbb{Z}$*,* $a_n \geq b_n \geq 0$ *and* a_n *or* b_n *is odd is defined as follows:*

$$f(a_n, b_n) = \begin{cases} \text{Branch}_1 : \text{max.min}((a_n - b_n)/2, b_n), a_n \text{ and } b_n \text{ are odd} \\ \text{Branch}_2 : \text{max.min}(a_n/2, a_n - b_n), a_n \text{ is even, } b_n \text{ is odd} \\ \text{Branch}_3 : \text{max.min}(b_n/2, a_n - b_n), a_n \text{ is odd, } b_n \text{ is even.} \end{cases}$$

Definition 2. $(a, b) = \text{max.min}(a, b)$*, where* $a, b \in \mathbb{Z}$*, is defined as follows:*

$$\text{max.min}(a, b) = \begin{cases} (a, b) \text{ when } a \geq b, \\ (b, a) \text{ otherwise.} \end{cases}$$

Lemma 3. *Let* $a_n = d \in \mathbb{Z}^+$*,* $b_n = 0$*, where* d *is odd, be the inputs of* f *in Definition 1. Then, no matter how many* f *are iterated, the outputs are the same as the inputs, which are* $a_{n+i} = d$*,* $b_{n+i} = 0$*,* $i \geq 1$*. This is,* $f(f \cdots f(f(a_n, b_n))) = (a_n, b_n)$*.*

Proof. Assume that $a_n = d \in \mathbb{Z}^+$, $b_n = 0$, where d is odd. $a_{n+1} = d$, $b_{n+1} = 0$ are computed by Branch_3, which are the same as the inputs.

Lemma 4. *Sequence* $\{\mathbb{Z} \ni (a_i + b_i) > 0\}$ *is a monotonically decreasing sequence for every 2 step, where* a_0 *and* b_0 *are the initial inputs and* a_i *and* b_i *(* $\mathbb{Z} \ni i > 0$*) are updated by the iteration formula in Definition 1.*

Proof. Every two steps can be classified into nine patterns of $(\text{Branch}_i, \text{Branch}_j)$, where $i, j \in \{1, 2, 3\}$, as shown in Table 1. All the patterns are proved by the following (1)–(7).

(1) After Branch_1, $a_{i+1} + b_{i+1} < a_i + b_i$ holds in the Eq. (5).

$$\begin{aligned} & (a_{i+1} + b_{i+1}) - (a_i + b_i) \\ &= (\frac{a_i - b_i}{2} + b_i) - (a_i + b_i) \\ &= -\frac{a_i + b_i}{2} < 0 \end{aligned} \qquad (5)$$

(2) After Branch_2 when $4b_i > a_i \geq b_i$, $a_{i+1} + b_{i+1} < a_i + b_i$ holds in Eq. (6).

$$\begin{aligned} & (a_{i+1} + b_{i+1}) - (a_i + b_i) \\ &= (\frac{a_i}{2} + a_i - b_i) - (a_i + b_i) \\ &= \frac{a_i - 4b_i}{2} < 0 \end{aligned} \qquad (6)$$

(3) After $\mathbf{Branch_2}$ when $a_i \geq 4b_i$, $a_{i+2} + b_{i+2} < a_i + b_i$. $a_{i+1} = a_i - b_i$, $b_{i+1} = a_i/2$ are updated because of $a_i - b_i - a_i/2 = (a_i - 2b_i)/2 > 0$. a_{i+1} is odd because a_i is even and b_i is odd. Then a_{i+2}, b_{i+2} are computed by $\mathbf{Branch_1}$ and $a_{i+2} + b_{i+2} = (3a_i - 2b_i)/4$, when $a_i/2$ is odd. a_{i+2}, b_{i+2} are computed by $\mathbf{Branch_3}$ and $a_{i+2} + b_{i+2} = (3a_i - 4b_i)/4$, when $a_i/2$ is even. Finally, $a_{i+2} + b_{i+2} < a_i + b_i$ holds in Eq. (7).

$$
\begin{aligned}
&(a_{i+2} + b_{i+2}) - (a_i + b_i) \\
&= -\frac{a_i + 6b_i}{4} < 0 \text{ when } a_i/2 \text{ is odd.} \\
&(a_{i+2} + b_{i+2}) - (a_i + b_i) \\
&= -\frac{a_i + 8b_i}{4} < 0 \text{ when } a_i/2 \text{ is even}
\end{aligned}
\tag{7}
$$

(4) After $\mathbf{Branch_3}$, $a_{i+1} + b_{i+1} \leq a_i + b_i$ holds in Eq. (8).

$$
\begin{aligned}
&(a_{i+1} + b_{i+1}) - (a_i + b_i) \\
&= (\frac{b_i}{2} + a_i - b_i) - (a_i + b_i) \\
&= -\frac{3b_i}{2} \leq 0(= 0 \text{ when } b_i = 0)
\end{aligned}
\tag{8}
$$

(5) After $(\mathbf{Branch_1}, \mathbf{Branch_2})$, $a_{i+2} + b_{i+2} < a_i + b_i$ holds in Eq. (9).

$$
\begin{aligned}
&(a_{i+2} + b_{i+2}) - (a_i + b_i) \\
&= (\frac{a_i - b_i}{4} + \frac{a_i - 3b_i}{2}) - (a_i + b_i) \\
&= -\frac{a_i + 11b_i}{4} < 0
\end{aligned}
\tag{9}
$$

(6) After $(\mathbf{Branch_3}, \mathbf{Branch_2})$, $a_{i+2} + b_{i+2} < a_i + b_i$ holds in Eq. (10).

$$
\begin{aligned}
&(a_{i+2} + b_{i+2}) - (a_i + b_i) \\
&= (\frac{b_i}{4} + \frac{3b_i - 2a_i}{2}) - (a_i + b_i) \\
&= \frac{3b_i - 8a_i}{4} < 0
\end{aligned}
\tag{10}
$$

(7) After $(\mathbf{Branch_2}, \mathbf{Branch_2})$ when $2b_i \geq a_i \geq b_i$, $a_{i+2} + b_{i+2} < a_i + b_i$ holds in Eq. (11). Note that there is no pattern of $(\mathbf{Branch_2}, \mathbf{Branch_2})$ when $a_i > 2b_i$.

$$
\begin{aligned}
&(a_{i+2} + b_{i+2}) - (a_i + b_i) \\
&= (\frac{a_i}{4} + \frac{2b_i - a_i}{2}) - (a_i + b_i) \\
&= -\frac{5a_i}{4} < 0
\end{aligned}
\tag{11}
$$

In summary, every case is proved as shown in Table 1.

Table 1. Proved case

Case	Proved by
$(\text{Branch}_1, \text{Branch}_1)$	(1)
$(\text{Branch}_1, \text{Branch}_2)$	(5)
$(\text{Branch}_1, \text{Branch}_3)$	(1) and (4)
$(\text{Branch}_2, \text{Branch}_1)$	(1), (2) and (3)
$(\text{Branch}_2, \text{Branch}_2)$	(7)
$(\text{Branch}_2, \text{Branch}_3)$	(2), (3) and (4)
$(\text{Branch}_3, \text{Branch}_1)$	(1) and (4)
$(\text{Branch}_3, \text{Branch}_2)$	(6)
$(\text{Branch}_3, \text{Branch}_3)$	(4)

Theorem 1. *(Convergence) After enough iterations of the iteration formula defined in Definition 1, any valid inputs a_0, b_0 converge to $a_n = d$, $b_n = 0$, where $d = \text{GCD}(a_0, b_0)$ and d is odd because a_0 or b_0 is odd.*

Proof. By Definition 1 and Lemmas 2, 3, 4, Theorem 1 can be proved clearly.

As shown in Theorem 1, after enough iterations of the iteration formula, $a_n = \text{GCD}(a_0, b_0)$ and $b_n = 0$ are the outputs. Then, we construct our SICT-GCD and SICT-MI by finding out the necessary number of iterations.

3.2   The Number of Iterations

We have already shown our iteration formula in Definition 1, which has simple and the same computations (a shift and a subtraction) in each iteration. According to Theorem 2, the iteration formula converges after $\texttt{bitlen}(a_0) + \texttt{bitlen}(b_0)$ iterations. Thus, the number of iterations are 448, 512, and 768 for 224-, 256-, and 384-bit GCD computation and modular inversion. Our number of iterations is shorter than hdBY [14] and the same as BOS [6]. The necessary number of iterations is not intuitive in our iteration formula, because the decrease of total bit length of inputs a_i and b_i after Branch_2 can not be observed directly.

Lemma 5. *For any valid a_i and b_i, $\texttt{bitlen}(a_i) + \texttt{bitlen}(b_i)$ is reduced by at least one after Branch_1, Branch_2 $(2b_i \geq a_i \geq b_i)$ and Branch_3 of the iteration formula in Definition 1.*

Proof.

(1) It is clear that $\texttt{bitlen}(a_i) + \texttt{bitlen}(b_i)$ is reduced by at least one after Branch_1 or Branch_3 of the iteration formula in Definition 1.
(2) Assume that even a_i has x bits, odd b_i has y bits, and $2b_i \geq a_i \geq b_i$. After Branch_2 of the iteration formula, $a_{i+1} = a_i/2$ has $x - 1$ bits and $0 \leq b_{i+1} = a_i - b_i \leq b_i$ has y bits at most. Thus, $\texttt{bitlen}(a_i) + \texttt{bitlen}(b_i)$ is reduced by at least one after Branch_2 $(2b_i \geq a_i \geq b_i)$.

Lemma 6. *Assume that even a_i is of x bits, odd b_i is of y bits, and $a_i > 2b_i$, $x - y \geq 1$. $\mathtt{bitlen}(a_i) + \mathtt{bitlen}(b_i)$ is reduced by at least $x - y + 1$ bits after $x - y + 1$ iterations.*

Proof. a_{i+1} and b_{i+1} are outputted by $\mathtt{Branch_2}$ firstly. $a_{i+1} = a_i - b_i$, $b_{i+1} = a_i/2$ because of $a_i - b_i - a_i/2 = (a_i - 2b_i)/2 > 0$. When $x - y = 1$, take one more iteration into consideration. If $a_i/2$ is odd, then $a_{i+2} = a_i/2$ with $x - 1$ bits, $b_{i+2} = (a_i - 2b_i)/4$ with $x - 2$ bits at most by $\mathtt{Branch_1}$. If $a_i/2$ is even, then $a_{i+2} = a_i/4$ with $x - 2$ bits, $b_{i+2} = (a_i - 2b_i)/2$ with $x - 1$ bits at most by $\mathtt{Branch_3}$. Then $\mathtt{bitlen}(a_i) + \mathtt{bitlen}(b_i)$ is reduced by $x - y + 1 = 2$ bits at least.

When $x - y = 2$, take a_{i+3} and b_{i+3} into consideration. With the computations, $\mathtt{bitlen}(a_{i+3}) + \mathtt{bitlen}(b_{i+3})$ is at most $(x - 2) + (x - 3) = x + y - 3$. Thus, $\mathtt{bitlen}(a_i) + \mathtt{bitlen}(b_i)$ is reduced by at least $x - y + 1 = 3$ bits after $x - y + 1 = 3$ iterations.

For $x - y = 3 \cdots$, one can continue to calculate a_{i+4}, $b_{i+4} \cdots$ and find out that $\mathtt{bitlen}(a_{i+k}) + \mathtt{bitlen}(b_{i+k})$ is at most $(x - k) + (x - k + 1) = x + y - k$, where $x - y = k - 1$. Then, $\mathtt{bitlen}(a_i) + \mathtt{bitlen}(b_i)$ is reduced by at least $x - y + 1$ bits after $x - y + 1$ iterations.

Theorem 2. *After $\mathtt{bitlen}(a_0) + \mathtt{bitlen}(b_0)$ iterations of the iteration formula defined in Definition 1, the valid inputs (a_0, b_0) converge to $(a_n = d, b_n = 0)$, where $\mathrm{GCD}(a_0, b_0) = d$, d is odd.*

Proof. Theorem 1 shows that the valid inputs (a_0, b_0) converge to $(a_n = d, b_n = 0)$, where $\mathrm{GCD}(a_0, b_0) = d$, d is odd, with enough iterations. By Lemmas 5 and 6, $\mathtt{bitlen}(a_0) + \mathtt{bitlen}(b_0)$ is reduced by at least one after each iteration on average. Thus the iteration formula converges after at most $\mathtt{bitlen}(a_0) + \mathtt{bitlen}(b_0)$ iterations.

Algorithm 6: SI-GCD

Input: $a, b \in \mathbb{Z}$, $a \geq b \geq 0$, a or b is odd. $l = 2 \cdot \mathtt{bitlen}(a)$.
Result: $v = \mathrm{GCD}(a, b)$
Initialization:
$u = b$; $v = a$;
for $i = 0$; $i < l$; $i = i + 1$ **do**
 $t_1 = v - u$;
 if u *is even* **then**
  $u = u >> 1$; $(v, u) = \mathrm{max.min}(u, t_1)$;
 else
  if v *is odd* **then**
   $t_1 = t_1 >> 1$; $(v, u) = \mathrm{max.min}(t_1, u)$;
  else
   $v = v >> 1$; $(v, u) = \mathrm{max.min}(v, t_1)$;
  end
 end
end
return v

With the iteration formula in Definition 1 and the necessary number of iterations, our SI-GCD algorithm is shown in Algorithm 6. It is not difficult to find out transition matrices of a_i and b_i. According to transition matrices, our SI-MI algorithm is shown in Algorithm 7. The computations of each iteration consist of two shifts and two subtractions in Algorithm 7. Note that all transition matrices are multiplied by two to eliminate the multiplications by $1/2$. Thus we need to multiply $2^{-l} \mod p$, which can be precomputed, to the final result, where l is the number of iterations.

Montgomery inversion, $a^{-1} \times R \mod p$, can be computed by SI-MI, where R is $2^{\mathtt{bitlen}(p)}$ generally [11]. By setting $pre\_com = 2^{-\mathtt{bitlen}(p)} \mod p$, the output of Algorithm 7 is $a^{-1} \times 2^{\mathtt{bitlen}(p)} \mod p$. For different choices of R, one can change the pre_com and obtain $a^{-1} \times R \mod p$ by Algorithm 7.

Algorithm 7: SI-MI

Input: $a, p \in \mathbb{Z}$, $p > a > 0$, $\mathrm{GCD}(a,p) = 1$. $l = 2 \cdot \mathtt{bitlen}(p)$.
$pre\_com = 2^{-l} \mod p$.
Result: $q = a^{-1} \mod p$
Initialization:
$u = a$; $v = p$; $q = 0$; $r = 1$;
for $i = 0$; $i < l$; $i = i + 1$ **do**
    $t_1 = v - u$; $t_2 = q - r$;
    if u *is odd and* v *is odd* **then**
        $t_1 = t_1 >> 1$; $r = r << 1$;
        if $u > t_1$ **then**
          | $q = r$; $r = t_2$; $v = u$; $u = t_1$;
        else
          | $q = t_2$; $r = r$; $v = t_1$; $u = u$;
        end
    else if u *is odd and* v *is even* **then**
        $v = v >> 1$; $t_2 = t_2 << 1$;
        if $t_1 > v$ **then**
          | $r = q$; $q = t_2$; $u = v$, $v = t_1$;
        else
          | $q = q$; $r = t_2$; $v = v$; $u = t_1$;
        end
    else
        $u = u >> 1$; $t_2 = t_2 << 1$;
        if $t_1 > u$ **then**
          | $r = r$; $q = t_2$; $u = u$; $v = t_1$;
        else
          | $q = r$; $r = t_2$; $v = u$; $u = t_1$;
        end
    end
end
$q = q \times pre\_com \mod p$;
return q

Table 2. Data matrix of u and v.

Branch$_1$	t_1	u
Branch$_2$	v	t_1
Branch$_3$	u	t_1

Table 3. Data matrix of q and r.

Branch$_1$	t_2	r
Branch$_2$	q	t_2
Branch$_3$	r	t_2

For clarity, the computations flow in one iteration of Algorithm 7 can be described as:

(1) Update $t_1 = v - u$ and $t_2 = q - r$.
(2) According to the LSB of u and the LSB of v, decide the Branch$_i$.
(3) According to the decided Branch$_i$, right shift the data in the second column of Table 2 and left shift the data in the third column of Table 3.
(4) According to the decided Branch$_i$, compare the data in the second column of Table 2 and the data in the third column of Table 2. Then update u, v, q, r by the results of the comparison.

3.3   SICT-GCD and SICT-MI

Because the computations in each branch are balanced in Algorithms 6 and 7, the same computations in each branch, the conditional statements can be removed as shown in Algorithm 8. Note that it is not the only way to remove conditional statements. For instance, t_1 and t_2 can also be updated as $t_1 = szu \oplus (\bar{s} + \bar{z})(v - u)$ and $t_2 = [sz(v - u) \oplus s\bar{z}v \oplus \bar{s}zu] >> 1$, which uses three more XOR operations on F_p and one less addition. SICT-GCD can be obtained easily by removing the computations of q, r from Algorithm 8. Thus we omit to describe it. Function $\text{cmp}(a, b)$ returns one if $a \geq b$ and returns zero if $a < b$. There are two secrets-independent table look-ups of $sort_1[2]$ and $sort_2[2]$ according to the comparison result of t_1 and t_2. Without loss of generality, assume that the result of comparison is $t_2 < t_1$, which can happen in any branch. The operations flow mentioned in Sect. 2.2 can not be decided based on such information.

Table 4 shows the comparison between FLT-CTMI, BOS [6], BY [5], hdBY [14] and our algorithms, where S, M, xor, add, sub, shift represent a square, a multiplication, a xor, an addition, a subtraction and a shift on F_p, respectively. Table 5 shows the number of iterations for 224-, 256-, 384-, 2048-bit computations of FLT-CTMI, BOS, BY, hdBY and our algorithms. FLT-CTMI can only be used to compute modular inversion over prime numbers and is not efficient for general inputs. In contrast, BOS, BY, hdBY and SICT-GCD can compute GCD(a, b), where a or b is odd. BOS, BY, hdBY and SICT-MI can compute Montgomery inversion and modular inversion over any number if it exists. BOS has the same number of iterations as us. However its computations, not only the computations

---

Algorithm 8: SICT-MI

Input: $a, p \in \mathbb{Z}$, $p > a > 0$, $GCD(a, p) = 1$. $l = 2 \cdot \texttt{bitlen}(p)$.
$pre\_com = 2^{-l} \bmod p$.
Result: $q = a^{-1} \bmod p$
Initialization:
$u = a;\ v = p;\ q = 0;\ r = 1;$
$sort_1[2] = \{t_1, t_2\};\ sort_2[2] = \{t_3, t_4\};$
for $i = 0;\ i < l;\ i = i + 1$ **do**
> $s = u_{lsb};\ z = v_{lsb};$
> $t_1, t_2 = (s \oplus z)v + ((sz << 1) - 1)u, [sv + (2 - (s << 1) - z)u] >> 1;$
> $(sz << 1$ is the same as $2sz.)$
> $t_3, t_4 = [(s \oplus z)q + ((sz << 1) - 1)r] << 1, sq + (2 - (s << 1) - z)r;$
> $s = \texttt{cmp}(t_2, t_1);\ z =!s;$
> $v = sort_1[s];\ u = sort_1[z];$
> $q = sort_2[s];\ r = sort_2[z];$

end
$q = q \times pre\_com \bmod p;$
return q

---

on F_p but also the computations of control signals, in one iteration are more complicated than us clearly (Algorithm 2 in [6]). BY and hdBY have the similar computations in one iteration as us. However, the number of iterations of them, even hdBY, are larger than us. Actually, SICT-GCD and SICT-MI save 13.35%, 13.22%, 13.22%, and 13.18% the number of iterations for 224-, 256-, 384-, and 2048-bit computations of hdBY.

Table 4. Comparison of CTMI.

	#iterations	Computations in one iteration
FLT-CTMI	$\texttt{bitlen}(p - 2) - 1$	S (or $S + M$)
BOS [6]	$2 \cdot \texttt{bitlen}(p)$	add + 2sub + 6shift
BY [5]	$\lfloor (49\texttt{bitlen}(p) + 57)/17 \rfloor$	4xor + 2add + 2shift
hdBY [14]	$\max(2\lfloor (2455 \log_2(M) + 1402)/1736 \rfloor,$ $2\lfloor (2455 \log_2(M) + 1676)/1736 \rfloor - 1),$ $M \geq 157,\ 0 \leq g \leq f \leq M$	4xor + 2add + 2shift
SICT-MI	$2 \cdot \texttt{bitlen}(p)$	4add + 2shift

4 Experiments Analysis

Our experimental platform uses C programming language with GNU MP 6.1.2, which is a multiple precision arithmetic library, and Intel (R) Core (TM) i7-8650U CPU @ 1.90 GHz 2.11 GHz personal 64-bit computer with 16.0 GB RAM; the operating system is Ubuntu 20.04.3 LTS. We turn off Intel turbo boost to ensure that our computer works at 1.80 GHz.

Table 5. The number of iterations of CTMI for 224-, 256-, 384-, 2048-bit computations.

	224-bit	256-bit	384-bit	2048-bit
FLT-CTMI	223	255	383	2047
BOS [6]	448	512	768	4096
BY [5]	634	724	1086	5794
hdBY [14]	517	590	885	4718
SICT-MI	448	512	768	4096

We implement SICT-GCD, BOS, BY and hdBY to compare the efficiency of GCD computations. In our GCD computations experiments, we generate four sets of 10^5 random odd numbers. The numbers in each set are 224 bits, 256 bits, 384 bits, and 2048 bits, respectively. We compute the GCD of numbers in each set with P224 − 1, P256 − 1, P384 − 1, and P2048 − 1, where P224, P256, P384 are NIST primes, and P2048 is a random prime number of 2048 bits, respectively. Then the average clock cycles are measured by rdtsc. The results are shown in Table 6.

From Table 6, it is clear that our SICT-GCD is more efficient than BOS, BY and hdBY. SICT-GCD saves 7.44%, 13.78%, 16.67%, and 28.3% clock cycles of hdBY on 224-, 256-, 384-, and 2048-bit GCD computation, respectively.

Table 6. Comparison of average clock cycles of GCD computation.

	224 bits	256 bits	384 bits	2048 bits
BOS [6]	163085	190734	294382	2016499
BY [5]	107117	134523	211943	1591591
hdBY [14]	91997	116523	184972	1454694
SICT-GCD	85156	100464	154142	1042982

We implement SICT-MI, FLT-CTMI, BOS, BY, hdBY to compare the efficiency of modular inversion computations. Four sets of 10^5 random numbers with 224-, 256-, 384-, and 2048-bit are generated respectively. The modular inversions of them over P224, P256, P384, and P2048 are computed respectively. The average clock cycles are recorded in Table 7.

Table 7. Comparison of average clock cycles of modular inversion.

	224 bits	256 bits	384 bits	2048 bits
FLT-CTMI	206676	154477	389729	14650073
BOS [6]	311512	356102	559549	4046376
BY [5]	262478	295822	481934	3675057
hdBY [14]	223057	248520	400410	3067074
SICT-MI	184230	208553	325659	2522420

From Table 7, FLT-CTMI is the most efficient on 256-bit modular inversion because of small Hamming weight of P256 − 2, which is 128 exactly. SICT-MI is the most efficient on 224-, 384-, and 2048-bit modular inversion. SICT-MI saves 17.41%, 16.08%, 18.67%, and 17.76% clock cycles of hdBY on 224-, 256-, 384-, and 2048-bit modular inversion, respectively. SICT-MI saves 10.86%, 16.44%, and 82.78% clock cycles of FLT-CTMI on 224-, 384-, and 2048-bit modular inversion, respectively. SICT-MI saves 40.86%, 41.43%, 41.8%, and 37.66% clock cycles of BOS on 224-, 256-, 384-, and 2048-bit modular inversion, respectively.

5   Conclusion

To find out constant-time algorithms computing GCD and modular inversion with short iterations and simple computations during one iteration, we define the iteration formula in Definition 1. We prove that the iteration formula converges to $a_n = \text{GCD}(a_0, b_0)$, $b_n = 0$ with limited iterations in Theorem 1 and the necessary number of iterations needed by the iteration formula is $\text{bitlen}(a_0) + \text{bitlen}(b_0)$ in Theorem 2. Based on these, we propose SICT-GCD and SICT-MI algorithms and compare the number of iterations and the computations during one iteration of them with those of FLT-CTMI, BOS, BY and hdBY. Finally we evaluate the efficiency of SICT-GCD and SICT-MI algorithms compared with FLT-CTMI, BOS, BY and hdBY by experiments. The results show that SICT-GCD and SICT-MI algorithms are more efficient than hdBY. The number of iterations of hdBY is approximately 2.3 times the bit length of the larger input and the number of iterations of SICT-GCD and SICT-MI algorithms is 2 times the bit length of the larger input. Thus, the longer the bit length of the larger input, the more iterations and time our algorithms save. We plan to perform evaluations on FPGA in the future.

Acknowledgements. This work is partially supported by the JSPS KAKENHI Grant Number JP21H03443, Innovation Platform for Society 5.0 at MEXT, and JST Next Generation Researchers Challenging Research Program JPMJSP2138.

References

1. Acıiçmez, O., Gueron, S., Seifert, J.-P.: New branch prediction vulnerabilities in OpenSSL and necessary software countermeasures. In: Galbraith, S.D. (ed.) Cryptography and Coding 2007. LNCS, vol. 4887, pp. 185–203. Springer, Heidelberg (2007). https://doi.org/10.1007/978-3-540-77272-9_12
2. Aldaya, A.C., Sarmiento, A.J.C., Sánchez-Solano, S.: SPA vulnerabilities of the binary extended Euclidean algorithm. J. Cryptogr. Eng. **7**(4), 273–285 (2017)
3. Aldaya, A.C., Márquez, R.C., Sarmiento, A.J.C., Sánchez-Solano, S.: Side-channel analysis of the modular inversion step in the RSA key generation algorithm. Int. J. Circ. Theory Appl. **45**(2), 199–213 (2017)
4. Aldaya, A.C., García, C.P., Tapia, L.M.A., Brumley, B.B.: Cache-timing attacks on RSA key generation. Cryptology ePrint Archive (2018)

5. Bernstein, D.J., Yang, B.Y.: Fast constant-time GCD computation and modular inversion. IACR Trans. Cryptogr. Hardw. Embed. Syst. 340–398 (2019)
6. Bos, J.W.: Constant time modular inversion. J. Cryptogr. Eng. **4**(4), 275–281 (2014). https://doi.org/10.1007/s13389-014-0084-8
7. Chari, S., Rao, J.R., Rohatgi, P.: Template attacks. In: Kaliski, B.S., Koç, K., Paar, C. (eds.) CHES 2002. LNCS, vol. 2523, pp. 13–28. Springer, Heidelberg (2003). https://doi.org/10.1007/3-540-36400-5_3
8. Duc, A., Faust, S., Standaert, F.X.: Making masking security proofs concrete (or how to evaluate the security of any leaking device), extended version. J. Cryptol. **32**(4), 1263–1297 (2019)
9. de la Fe, S., Park, H.B., Sim, B.Y., Han, D.G., Ferrer, C.: Profiling attack against RSA key generation based on a Euclidean algorithm. Information **12**(11), 462 (2021)
10. Kaliski, B.S.: The Montgomery inverse and its applications. IEEE Trans. Comput. **44**(8), 1064–1065 (1995)
11. Montgomery, P.L.: Modular multiplication without trial division. Math. Comput. **44**(170), 519–521 (1985)
12. Sarna, S., Czerwinski, R.: RSA and ECC universal, constant time modular inversion. In: AIP Conference Proceedings, vol. 2343, p. 050004. AIP Publishing LLC (2021)
13. Sen, X., et al.: To construct high level secure communication system: CTMI is not enough. China Commun. **15**(11), 122–137 (2018)
14. Wuille, P., Maxwell, G., roconnor-blockstream: Safegcd-bounds. Github (2021). https://github.com/sipa/safegcd-bounds
15. Yarom, Y., Falkner, K.: FLUSH+ RELOAD: A high resolution, low noise, L3 cache side-channel attack. In: 23rd USENIX Security Symposium (USENIX Security 14), pp. 719–732 (2014)

Protecting AES

Guarding the First Order: The Rise of AES Maskings

Amund Askeland[2], Siemen Dhooghe[1], Svetla Nikova[1,2], Vincent Rijmen[1,2], and Zhenda Zhang[1(✉)]

[1] imec-COSIC, ESAT, KU Leuven, Heverlee, Belgium
{siemen.dhooghe,svetla.nikova,vincent.rijmen,
zhenda.zhang}@esat.kuleuven.be
[2] University of Bergen, Bergen, Norway
{amund.askeland,vincent.rijmen}@uib.no

Abstract. We provide three first-order hardware maskings of the AES, each allowing for a different trade-off between the number of shares and the number of register stages. All maskings use a generalization of the changing of the guards method enabling the re-use of randomness between masked S-boxes. As a result, the maskings do not require fresh randomness while still allowing for a minimal number of shares and providing provable security in the glitch-extended probing model. The low-area variant has five cycles of latency and a serialized area cost of 8.13 kGE. The low-latency variant reduces the latency to three cycles while increasing the serialized area by 67.89% compared to the low-area variant. The maskings of the AES encryption are implemented on FPGA and evaluated with Test Vector Leakage Assessment (TVLA).

Keywords: AES · Hardware · Probing security · Threshold implementations

1 Introduction

The Advanced Encryption Standard (AES) [8] is one of the most used cryptographic building blocks in practice. The cipher has secured many applications, including the world wide web. However, for some applications, like embedded devices, naive implementations of the AES are vulnerable to side-channel attacks such as Differential Power Analysis (DPA) due to Kocher *et al.* [18]. The current agreed-upon method to protect implementations against DPA is masking. In masking, each key-dependent variable is split into several random shares such that an adversary needs to view the power consumption of each share to gain information on the secret variable.

Several maskings of the AES appeared in the literature in the past twenty years. The efficiency and security of them have been significantly improved over time. Threshold implementations by Nikova *et al.* [21] allowed for maskings that protect against glitches in hardware. The uniformity aspect of threshold implementations allows for the reduction of randomness. The changing of the guards

I. Buhan and T. Schneider (Eds.): CARDIS 2022, LNCS 13820, pp. 103–122, 2023.
https://doi.org/10.1007/978-3-031-25319-5_6

technique by Daemen [7] showed how to make a masking uniform without significantly increasing costs. Canright [4] proposed an efficient tower field decomposition of the AES S-box in order to improve hardware costs. This decomposition was then used by De Cnudde *et al.* [9] to create efficient threshold implementation maskings of the AES. However, the authors noted that the randomness cost of their designs is high, which makes it infeasible to generate the randomness in a cryptographic secure way.

Contributions. In this paper, we generalize the changing of the guards technique to include maskings that use randomness in Sect. 3. The method allows the re-use of randomness between all masked S-boxes and retains first-order probing security. As a result, the generalization tackles the open question by De Cnudde *et al.* as we significantly reduce the randomness cost of their masking, at least for first-order security. We then provide three variants of the masking in Sect. 4, which show the trade-off between the number of register stages and the number of shares.

We apply these maskings of S-boxes to both the serialized and round-based hardware architecture of the AES-128 encryption. These constructions are proven secure in the first-order glitch-extended robust probing model. The low-area variation of the serialized AES, described in Sect. 4.2, has an area cost of 8.13 kGE and 5 cycles of latency. The latency is reduced to 4 cycles for the S-box in Sect. 4.3 and 3 cycles for the S-box in Sect. 4.4 with the area cost of the serialized AES increased by 56.75% and 67.89%, respectively. The serialized AES implementations are tested on our side-channel leakage assessment setup to show the first-order security of the designs. Our implementations amortize the cost of online randomness, at the same time provide provable and physically tested first-order probing security, and achieve one of the most efficient area versus latency trade-offs in the literature.

2   Preliminaries

In this section, we go over the used notation, introduce the AES, the probing side-channel security model, and threshold implementations.

2.1   Notation

We denote bits by subscript and shares by superscript. We denote the most significant bit by a bigger subscript. For example, given $(a_1, ..., a_8)$, a_8 denotes the most significant bit (MSB) and a_1 the least significant bit (LSB). Finally, the concatenation of bits is denoted by ⓪.

2.2 Description of AES

We quickly introduce the standardized AES cipher by Daemen and Rijmen [8]. There are three levels of security 128, 192, and 256. AES consists of a 128-bit state and 128, 192, or 256-bit key, respectively, divided into bytes. The cipher is composed of 10, 12, or 14 rounds, respectively, each applying an addition of a subkey, a bricklayer of S-Boxes, a ShiftRows operation, and a MixColumns operation. The AES S-Box consists of an inversion in the field $\mathbb{F}_{2^8}$ and the application of an affine layer. This is visually represented in Fig. 1.

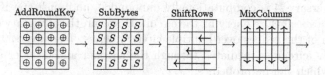

Fig. 1. Representation of the AES.

The key schedule for AES-128, which operates on 4 columns of 32 bits each, is depicted in Fig. 2. Each round of the AES state function has a parallel round of the key schedule. We provide a description for the AES-128 key schedule. Denote V_j, with $j \in \{1, ..., 4\}$, the jth word of the key state at round i and W_j the jth word of the key state at round $i + 1$. Then a round of the key schedule is defined as

$$W_1 = V_1 \oplus \texttt{RotWord}(\texttt{SubWord}(V_4)) + C_{i+1},$$
$$W_2 = V_2 \oplus W_1,$$
$$W_3 = V_3 \oplus W_2,$$
$$W_4 = V_4 \oplus W_3.$$

With RotWord, the left circular shift, SubWord, the application of four AES S-boxes, and C_{i+1}, the round constants for round $i + 1$.

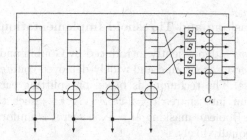

Fig. 2. The AES-128 key schedule. $C_i \in \mathbb{F}_2^{32}$ denotes the ith round constants, and S denotes the AES S-box.

2.3   The Threshold Glitch-Extended Probing Model

This section introduces the threshold probing model.

Threshold Probing. A dth-order threshold probing adversary $\mathcal{A}$, as first proposed by Ishai *et al.* [17], can view the values present on up to d gates or wires in a circuit implementing a cipher during a single execution (cipher evaluation). We note that by "probe" we do not mean a physical probe such as an EM probe. Instead, the word probe is used as an abstract concept through which an adversary can perfectly observe a part of the computation.

The adversary $\mathcal{A}$ is computationally unbounded, and must specify the location of the probes before querying the circuit. However, the adversary can change the location of the probes over multiple cipher queries. The adversary's interaction with the circuit is mediated through the encoder and decoder algorithms, neither of which can be probed.

The security model is a simulation model where the simulator needs to simulate the probed values from scratch, more specifically, the simulator is not given the input (including the key of a block cipher) of the circuit. The adversary needs to distinguish the probed values from the real circuit with the returned values from the simulator. A failure in doing so proves that the adversary can not learn anything from the circuit's input via the probes. In a security proof, this essentially comes down to proving that the probed values follow a distribution which is independent of the value of the circuit's input.

Glitches. The above model is extended to capture the effect of glitches on hardware. Whereas one of the adversary's probes normally results in the value of a single wire, a glitch-extended probe allows obtaining all the registered inputs leading to the gate/wire which is probed. This extension of the probing model has been discussed in the work of Reparaz *et al.* [22] and formalized by Faust *et al.* [12]. The formulation of the latter work is as follows: "For any ϵ-input circuit gadget G, combinatorial recombinations (aka glitches) can be modeled with specifically ϵ-extended probes so that probing any output of the function allows the adversary to observe all its ϵ inputs."

2.4   Boolean Masking and Threshold Implementations

Boolean masking was independently introduced by Goubin and Patarin [15] and Chari *et al.* [5]. It serves as a sound and widely-deployed countermeasure against side-channel attacks. The technique is based on splitting each secret variable $x \in \mathbb{F}_2$ in the circuit into shares $\bar{x} = (x^1, x^2, \dots, x^{s_x})$ such that $x = \sum_{i=1}^{s_x} x^i$ over $\mathbb{F}_2$. A random Boolean masking of a fixed secret is uniform if all maskings of that secret are equally likely.

There are several approaches to protect a circuit by masking. In this work, we make use of threshold implementations, proposed by Nikova *et al.* [21]. In particular, we focus on "first-order threshold implementations" as those which protect against first-order side-channel attacks. The interested reader is referred to the works by Bilgin *et al.* [2] and Beyne *et al.* [1] for more information on how to use threshold implementations to secure against higher-order attacks. In the following, the main properties of threshold implementations as introduced by Nikova *et al.* are reviewed.

A threshold implementation consists of several layers of Boolean functions, as shown in Fig. 3. As for any masked design, a black-box encoder function generates a uniform random masking of the input before it enters the masked circuit. At the end of each layer, synchronization is ensured by means of registers which stop the propagation of glitches.

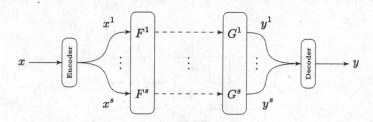

Fig. 3. Schematic illustration of a threshold implementation assuming an equal number of input and output shares [11].

Let $\bar{F}$ be a layer in the threshold implementation corresponding to a part of the circuit $F : \mathbb{F}_2^n \to \mathbb{F}_2^m$. For example, F might be the linear layer of a block cipher. The function $\bar{F} : \mathbb{F}_2^{ns_x} \to \mathbb{F}_2^{ms_y}$, where we assume s_x shares per input bit and s_y shares per output bit, will be called a *masking* of F. The ith share of the function $\bar{F}$ is denoted by $F^i : \mathbb{F}_2^{ns_x} \to \mathbb{F}_2^m$, for $i \in \{1, .., s_y\}$. Maskings can have a number of properties that are relevant in the security argument for a threshold implementation; these properties are summarized in Definition 1.

Definition 1 (Properties of threshold implementations [21]). *Let $F : \mathbb{F}_2^n \to \mathbb{F}_2^m$ be a function and $\bar{F} : \mathbb{F}_2^{ns_x} \to \mathbb{F}_2^{ms_y}$ be a masking of F. The masking $\bar{F}$ is said to be*

correct

    if $\sum_{i=1}^{s_y} F^i(x^1, \ldots, x^{s_x}) = F\left(\sum_{i=1}^{s_x} x^i\right)$ for all shares $x^1, \ldots, x^{s_x} \in \mathbb{F}_2^n$,

non-complete

    if any function F^i, measured between register stages, depends on at most $s_x - 1$ input shares,

uniform

    if $\bar{F}$ maps a uniform random masking of any $x \in \mathbb{F}_2^n$ to a uniform random masking of $F(x) \in \mathbb{F}_2^m$.

Considering that, in a threshold implementation, all input/outputs of the functions are stored in registers, placing a glitch-extended probe in a layer of a threshold implementation returns all inputs of the probed masked Boolean function. If all layers of a threshold implementation are non-complete and uniform, the resulting masked circuit can be proven secure in the first-order probing model with glitches [11].

3   Changing of the Guards with Randomness

The changing of the guards method proposed by Daemen [7] is a technique that
transforms a non-complete masking into a uniform and non-complete masking.
The technique works by embedding the masking into a Feistel-like structure. In
this paper, we slightly generalize the method by considering a first-order probing
secure masking. Such a masking potentially requires multiple register stages and
extra randomness to guarantee its security. The adapted changing of the guards
method still ensures uniformity while allowing the re-use of the randomness. An
example of the method with two shares is shown in Fig. 4.

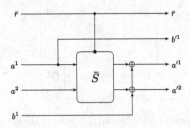

Fig. 4. Changing of the guards method with two shares where the masked S-box $\bar{S}$
uses the randomness $\bar{r}$.

With the original changing of the guards method, the function $\bar{S}$ was non-
complete. Meaning that $\bar{S}$ typically did not use fresh randomness and was computed in one cycle. Instead, with the generalized method, $\bar{S}$ can use fresh randomness (which can be recycled and used for different S-boxes) and be computed
in multiple stages (in this work the whole tower-field decomposition of the AES
S-box).

We give the changing of the guards method formally in Definition 2.

Definition 2. *The generalized changing of the guards method applied to a
masked map $\bar{S}$ given inputs $(a^1, ..., a^s)$, $(b^1, ..., b^{s-1})$, and randomness $\bar{r}$ is calculated as follows*

$$\bar{r}' = \bar{r} \tag{1}$$

$$a'^1 = S^1(a^1, ..., a^s, \bar{r}) \oplus b^1, \quad ... \quad , \quad a'^{s-1} = S^{s-1}(a^1, ..., a^s, \bar{r}) \oplus b^{s-1}, \tag{2}$$

$$a'^s = S^s(a^1, ..., a^s, \bar{r}) \oplus b^1 \oplus ... \oplus b^{s-1} \tag{3}$$

$$b'^1 = a^1, \quad ... \quad , \quad b'^{s-1} = a^{s-1}. \tag{4}$$

In general, we refer to the $(b^1, ..., b^{s-1})$ as the *guards* of the masked S-box $\bar{S}$.
We show that the changing of the guards construction with randomness retains
the correctness and probing security properties from $\bar{S}$ and makes the masking
uniform.

Theorem 1. *The method from Definition 2 is correct, first-order probing secure,
and uniform.*

Proof. The correctness of the construction follows from the correctness of $\bar{S}$ and the fact that each share b^i is added to two different output shares in Eq. (2)–(3).

First-order probing security of the construction, assuming a joint uniform input, follows from the first-order probing security of $\bar{S}$ and the facts that the share b^s is not used in the construction and that each share b'^i is calculated using only one share a^i using Eq. (2)–(3).

For the proof of uniformity, we first take an arbitrary input secret a. We show that the above construction is invertible. In other words, given the secret a and the outputs $(a'^1, ..., a'^s)$, $(b'^1, ..., b'^{s-1})$, $\bar{r}'$, we show it is possible to construct the inputs $(a^1, ..., a^s)$, $(b^1, ..., b^{s-1})$, $\bar{r}$.

Since the input secret a is given, we can construct the inputs $(a^1, ..., a^s)$ from $(b'^1, ..., b'^{s-1})$ using Eq. (4). From $\bar{r}'$ we can evidently construct $\bar{r}$ since the two are equal (see Eq. (1)). By running $(a^1, ..., a^s)$ and $\bar{r}$ through $\bar{S}$ and XORing the output $(a'^1, ..., a'^s)$, we can also construct $(b^1, ..., b^s)$ (see Eq. (2)–(3)) which concludes the proof. □

Thus, the changing of the guards method allows for the transformation of any first-order probing secure masking into a uniform one which allows the re-use of the randomness used in the S-box.

4   Maskings of the S-Box

This section describes the three first-order glitch-extended probing secure S-box designs of the AES. We first go over the components which are masked between all designs and then provide the three designs themselves.

4.1   Overarching Components

We quickly review the functions used in the tower-field decomposition of the S-box.

Input/Output Isomorphism. The first operation occurring in the decomposed S-box performs a change of basis through a linear map. This mapping is implemented in combinational logic, and it maps the 8-bit input $(a_1^i, ..., a_8^i)$ to the 8-bit output $(y_1^i, ..., y_8^i)$ for each share $i \in \{1, 2\}$ as follows:

$$y_8^i = a_8^i \oplus a_7^i \oplus a_6^i \oplus a_3^i \oplus a_2^i \oplus a_1^i \qquad y_4^i = a_8^i \oplus a_5^i \oplus a_4^i \oplus a_2^i \oplus a_1^i$$
$$y_7^i = a_7^i \oplus a_6^i \oplus a_5^i \oplus a_1^i \qquad y_3^i = a_1^i$$
$$y_6^i = a_7^i \oplus a_6^i \oplus a_2^i \oplus a_1^i \qquad y_2^i = a_7^i \oplus a_6^i \oplus a_1^i$$
$$y_5^i = a_8^i \oplus a_7^i \oplus a_6^i \oplus a_1^i \qquad y_1^i = a_7^i \oplus a_4^i \oplus a_3^i \oplus a_2^i \oplus a_1^i$$

The inverse linear map (including the AES affine transformation) maps the 8-bit input $(u_1^i, ..., u_8^i)$ to the 8-bit output $(y_1^i, ..., y_8^i)$ for each share i as follows:

$$y_8^i = u_6^i \oplus u_4^i \qquad\qquad y_4^i = u_8^i \oplus u_7^i \oplus u_6^i \oplus u_5^i \oplus u_4^i$$
$$y_7^i = u_8^i \oplus u_4^i \qquad\qquad y_3^i = u_7^i \oplus u_6^i \oplus u_4^i \oplus u_3^i \oplus u_1^i$$
$$y_6^i = u_7^i \oplus u_1^i \qquad\qquad y_2^i = u_6^i \oplus u_5^i \oplus u_2^i$$
$$y_5^i = u_8^i \oplus u_6^i \oplus u_4^i \qquad\qquad y_1^i = u_7^i \oplus u_5^i \oplus u_2^i$$

The constant 0x63 is then added to the first share $(y_8^1, ..., y_1^1)$.

Finite Field Multipliers. In the computation of the masking of the S-box, we make use of multipliers over $\mathbb{F}_{2^2}$ and $\mathbb{F}_{2^4}$. We note that these equations only hold for the multipliers in the masked S-box and not for any other operation in the AES like the MixColumns. We recall the equations for the multiplication over $\mathbb{F}_{2^2}$. These map the two 2-bit inputs $(a_1, a_2), (b_1, b_2)$ to the 2-bit output (c_1, c_2) as follows:

$$c_1 = (a_2 \oplus a_1)(b_2 \oplus b_1) \oplus a_1 b_1 \qquad c_2 = (a_2 \oplus a_1)(b_2 \oplus b_1) \oplus a_2 b_2$$

We then recall the equations for the multiplication over $\mathbb{F}_{2^4}$. This maps the two 4-bit inputs $(a_1, a_2, a_3, a_4), (b_1, b_2, b_3, b_4)$ to the 4-bit output (c_1, c_2, c_3, c_4) as follows:

$$c_1 = a_4 b_4 \oplus a_2 b_4 \oplus a_3 b_3 \oplus a_1 b_3 \oplus a_4 b_2 \oplus a_1 b_2 \oplus a_3 b_1 \oplus a_2 b_1 \oplus a_1 b_1$$
$$c_2 = a_4 b_4 \oplus a_3 b_4 \oplus a_2 b_4 \oplus a_1 b_4 \oplus a_4 b_3 \oplus a_2 b_3 \oplus a_4 b_2 \oplus a_3 b_2 \oplus a_2 b_2 \oplus a_4 b_1 \oplus a_1 b_1$$
$$c_3 = a_3 b_4 \oplus a_2 b_4 \oplus a_4 b_3 \oplus a_3 b_3 \oplus a_1 b_3 \oplus a_4 b_2 \oplus a_2 b_2 \oplus a_3 b_1 \oplus a_1 b_1$$
$$c_4 = a_4 b_4 \oplus a_2 b_4 \oplus a_1 b_4 \oplus a_3 b_3 \oplus a_2 b_3 \oplus a_4 b_2 \oplus a_3 b_2 \oplus a_2 b_2 \oplus a_1 b_2 \oplus a_4 b_1 \oplus a_2 b_1$$

Scaling Functions. In the decomposed S-box, there are two scaling functions, a "square scale function" and a "scale" function. The linear square scaling $SqSc$ maps the 4-bit input $(x_1, ..., x_4)$ to the 4-bit output $(y_1, ..., y_4) = (x_1, x_1 \oplus x_2, x_2 \oplus x_4, x_1 \oplus x_3)$. The scaling operation $Scale$ maps a 2-bit input (x_1, x_2) to the 2-bit output $(y_1, y_2) = (x_1 \oplus x_2, x_2)$.

Two-Bit Inverter. The inversion Inv in $\mathbb{F}_{2^2}$ is linear and is implemented by swapping the bits. The inversion operation maps a 2-bit input (x_1, x_2) to the 2-bit output $(y_1, y_2) = (x_2, x_1)$.

4.2   Design I: Two-Share S-Box

The first design uses two shares and is divided into five cycles. The S-box uses a total of 54 random bits which can be re-used over all the S-boxes. The design of the masking is given in Fig. 5.

The masked multipliers are calculated by first computing all cross products of the input shares and then re-masking them with a random masking of zero before compressing the cross products back to the input number of shares. More specifically, the two-shared multiplier is given by

$$
\begin{aligned}
a^0 &\to a^0 b^0 + r^0 \\
a^1 &\to a^0 b^1 + r^1 \to x^0 = a^0 b + r^0 + r^1 \\
b^0 &\to a^1 b^0 + r^2 \to x^1 = a^1 b + r^2 + r^3 \\
b^1 &\to a^1 b^1 + r^3
\end{aligned}
\tag{5}
$$

where $r^3 = r^0 + r^1 + r^2$. Note that the above multiplier is non-complete and uniform in both stages when fresh randomness r^i is used. More specifically, the first stage expands the two input shares to a four-sharing and the second stage compresses the uniform four shares back to two output shares.

This masking is the most efficient in terms of area considering the three designs in this paper. However, the decrease in area is traded for by an increase in latency.

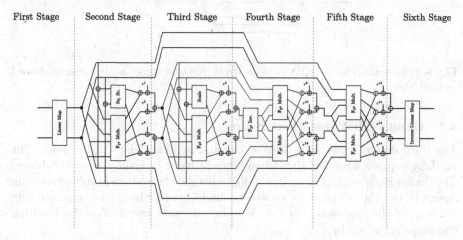

Fig. 5. Representation of the S-box of design I. Register stages are denoted by dashed vertical lines.

4.3 Design II: Three-Share S-Box

The second design uses three shares and is divided into four cycles. The S-box uses a total of 36 random bits which can be re-used in every S-box. The masking is shown in Fig. 6.

The masked multipliers are calculated by combining the cross-products in a non-complete way and re-masking them with a random masking of zero. More specifically, the three-shared multiplier is given by

$$
\begin{aligned}
a^0, b^0 &\to a^0 b^0 + a^0 b^1 + a^1 b^0 + r^0 \\
a^1, b^1 &\to a^1 b^1 + a^1 b^2 + a^2 b^1 + r^1 \\
a^2, b^2 &\to a^2 b^2 + a^2 b^0 + a^0 b^2 + r^2
\end{aligned}
\tag{6}
$$

where $r^2 = r^0 + r^1$.

Compared to the previous design in Sect. 4.2, this S-box is larger in terms of area but has a reduced latency of four cycles compared to five of the previous design. This design works well with both a serialized and a round-based architecture.

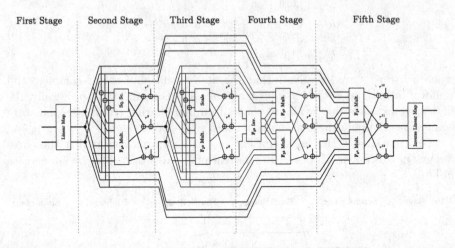

Fig. 6. Representation of the S-box of design II. Register stages are denoted by dashed vertical lines.

4.4 Design III: Three-Share S-Box

The third design works over three shares and is divided into three cycles. The masking requires a total of 40 random bits which can be re-used over all S-boxes. The design is shown in Fig. 7. The design uses the three-shared multipliers from design II, but the design switches to four shares in order to make a non-complete masking of the inversion over $\mathbb{F}_{2^4}$. For the inversion Inv in $\mathbb{F}_{2^4}$, the resulting equations are given by:

$$y_1 = x_1 x_3 \oplus x_1 x_4 \oplus x_2 x_3 x_4 \oplus x_2 x_4 \oplus x_3$$
$$y_2 = x_1 x_3 x_4 \oplus x_1 x_4 \oplus x_2 x_4 \oplus x_3 \oplus x_4$$
$$y_3 = x_1 x_2 x_4 \oplus x_1 x_3 \oplus x_1 \oplus x_2 x_3 \oplus x_2 x_4$$
$$y_4 = x_1 x_2 x_3 \oplus x_1 \oplus x_2 x_3 \oplus x_2 x_4 \oplus x_2$$

For each cubic term $x_a x_b x_c$, the ith share for $i \in \{1, 2, 3, 4\}$ is calculated as follows, where the convention is used that the superscripts wrap around at four.

$$x_a x_b x_c^i = x_a^i x_b^i x_c^i \oplus x_a^i x_b^i x_c^{i+1} \oplus x_a^i x_b^i x_c^{i+1} \oplus x_a^i x_b^{i+1} x_c^{i+1} \oplus x_a^i x_b^{i+1} x_c^{i+2}$$
$$\oplus x_a^i x_b^{i+2} x_c^{i+1} \oplus x_a^i x_b^i x_c^{i+2} \oplus x_a^i x_b^{i+2} x_c^i \oplus x_a^i x_b^{i+2} x_c^{i+2} \oplus x_a^{i+2} x_b^{i+1} x_c^i$$
$$\oplus x_a^{i+1} x_b^{i+2} x_c^i \oplus x_a^{i+2} x_b^i x_c^{i+1} \oplus x_a^{i+1} x_b^i x_c^{i+2} \oplus x_a^{i+2} x_b^{i+1} x_c^{i+1}$$
$$\oplus x_a^{i+2} x_b^{i+2} x_c^{i+1} \oplus x_a^{i+2} x_b^{i+1} x_c^{i+2}$$

The ith share for $i \in \{1, 2, 3, 4\}$ of a quadratic term $x_a x_b$ is calculated as follows:

$$x_a x_b^i = x_a^i x_b^i \oplus x_a^i x_b^{i+1} \oplus x_a^i x_b^{i+2} \oplus x_a^{i+2} x_b^{i+1}$$

The linear terms are added share-wise. The output of the inversion is then refreshed with a zero-masking.

Compared to the previous designs, this one reduces the latency to three cycles, but it trades off a larger area cost. This design is better suited for a round-based architecture.

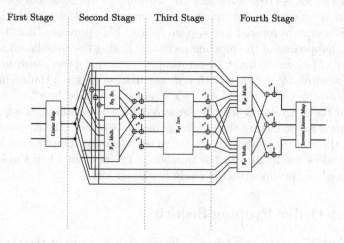

Fig. 7. Representation of the S-box of design III. Register stages are denoted by dashed vertical lines.

5   Architecture

Our designs of the masked AES-128 encryption are implemented in both a serialized architecture and a round-based architecture.

In the round-based AES-128 architecture, 20 S-boxes are instantiated in which 16 S-boxes process `SubBytes` for the state function and 4 for `SubWord` in the key expansion. The bus width of `MixColumns` and `ShiftRows` is $n\_shares \times$ 128 bits. After each round of the AES encryption, the state of the cipher is stored in a register array. Thus, the whole encryption needs 10 rounds of $sbox\_latency+1$ cycles plus the latency by the control logic. The changing of the guards method is applied to the bricklayer of S-boxes following Definition 1 by Daemen [6].

The implementations that are evaluated in Sect. 7.2 use the serialized architecture where there is one S-box for both the key expansion and the state. The bus width is thus $n\_shares \times 8$ bits. The state registers and key registers can be viewed as 4×4 arrays, similar to the serialized encryption modules by Shahmirzadi and Moradi [24] and De Meyer et al. [10]. Each masked byte after

AddRoundKey, or after RotWord in the key expansion, is fed into the masked S-box. Thus, the serialized architecture needs at least 20 cycles per round. If the latency of the S-box is 4 cycles, the stages of the S-box are pipelined without wasting a cycle. If the latency is larger than 4 cycles, waiting cycles are inserted at the end of each round. If the latency is less than 4 cycles, the round-based architecture is preferred for lower latency applications.

The guard shares are initialized with randomness and the subsequent guards are taken from the input of the S-box. Shift registers are used to store these shares in such a way that they are delayed by one more cycle than the S-box delay before they are applied to the output. This is depicted in Fig. 8. A total of $8 \times (n\_shares - 1)$ random bits are required to initialize the guard shares. These can be static throughout the AES encryption. The random bits in the masked S-boxes are rotated per applied S-box. This prevents transitional leakages from happening in the pipeline registers. Taking the two-shared multiplication of Eq. (5) as an example, by rotating (r^0, r^1, r^2, r^3), two consecutive executions (first with a, b and then with c, d) calculating x^0 gives Hamming distance leakage $HD(a^0 b + r^0 + r^1, c^0 d + r^1 + r^2)$ which is masked by r^0, r^2.

Before the AES starts for the serialized implementations, the masked key and plaintext are loaded to the register array in 16 cycles. The ciphertext is read out from the state register array in 16 cycles when the computation is finished. Table 1 depicts the latency in the number of cycles from when the load signal arrives to when the ciphertext is ready at the output.

6   First-Order Probing Security

In this section, we argue the first-order probing security of the three designs. We show this by arguing that the designs are threshold implementations, see Definition 1. We refer to Dhooghe et al. [11] for the proof that a threshold implementation is first-order robust probing secure.

The masked S-boxes are first-order probing secure due to the extra randomness added to each register stage. This refreshing works as follows. Given an input $(a^1, ..., a^s)$, an arbitrary masked map $\bar{F}$, and randomness $r^1, ..., r^{s-1}$, refreshing is done as follows, for $i \in \{1, ..., s-1\}$

$$a'^i = F^i(a^1, ..., a^s) \oplus r^i, \qquad a'^s = F^s(a^1, ..., a^s) \oplus r^1 \oplus ... \oplus r^{s-1}, \qquad (7)$$

where each F^i is non-complete.

Lemma 1. *The refreshing following Eq. (7) gives a uniform output.*

Proof. We show that the function, taking $(a^1, ..., a^s)$ and $(r^1, ..., r^{s-1})$ as input and $(a^1, ..., a^s)$, $F^i(a^1, ..., a^s) \oplus r^i$ for $i \in \{1, ..., s-1\}$, and $F^s(a^1, ..., a^s) \oplus r^1 \oplus ... \oplus r^{s-1}$ as output, is invertible. Removing the $(a^1, ..., a^s)$ then gives a balanced (or uniform) output of the refreshing detailed in Eq. (7).

The derivation is straightforward. Since $(a^1, ..., a^s)$ is given in the output, one can calculate $F^i(a^1, ..., a^s)$ for $i \in \{1, ..., s\}$. Subtracting this from the output $(a^1, ..., a^{s-1})$ then gives $(r^1, ..., r^{s-1})$ showing the map is invertible.   □

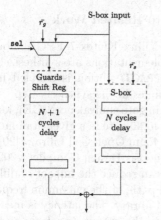

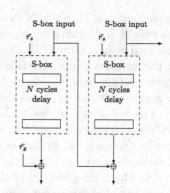

(a) Application of the guards in the serialized architecture.

(b) Application of the guards in the round-based architecture.

Fig. 8. Application of changing of the guards in our implementations.

Theorem 2. *Designs I, II, and III from Sects. 4.2, 4.3, and 4.4 are threshold implementations as given by Definition 1.*

Proof. First, each design uses a changing of the guards method. We refer to Theorem 1 for the proof that the masked input and output of each masked S-box is uniform.

Since firstly, the linear layers of the construction are evidently non-complete; secondly, they work share-wise and uniform; and thirdly, the unmasked linear functions are permutations, these comply with the properties from Definition 1.

We then show that the masked S-box from designs I, II, and III are first-order probing secure. Since a probe in the designs can only view one masked S-box, it suffices to show that each stage in the masked S-box complies with the threshold implementation properties.

Each stage in the masked S-box either maps a part of the input to the output (such as in the outer wires of Fig. 5) or the output is masked using the randomness $\bar{r}$. Due to the changing of the guards structure this randomness $\bar{r}$ is joint uniform with the input of the masked S-box. From Lemma 1, we find that each stage of the masked S-box of design I, II, or III is uniform.

Finally, each stage in the masked S-box is also non-complete. As a result, the masked S-boxes from designs I, II, and III are itself threshold implementations and thus first-order robust probing secure.   □

7   Implementations and Physical Evaluations

In this section, we explain the hardware implementations of the three AES designs and their side-channel analysis security.

7.1 Implementations and Comparison to Related Work

We implement our three AES designs for a Xilinx Kintex-7 FPGA mounted on a Sakura-X [19] evaluation board. The implementations are synthesized and programmed using Xilinx ISE. The KEEP_HIERACHY option is enabled in the Xilinx ISE to prevent optimization across modules in the synthesis step.

The area of the masked ciphers is measured in gate equivalences (GE), i.e., the cipher area normalized to the area of a 2-input NAND gate in a given standard cell library. The cell library we use is the NANGATE 45nm Open Cell Library [20], and the synthesis results are obtained with the Synopsis Design Compiler v2021.06. Comparing the area cost in gate equivalences can reduce the impact of different cell libraries. However, the delay on the critical path, or the maximum frequency, depends on the timing metrics of the used cell library. The latency is measured in the number of cycles to get from the input to the output (*e.g.* from an S-box or a cipher).

Table 1 and Table 2 show the results of our implementation. Table 1 contains the serialized AES implementations and their comparison to other implementations in the literature. Design I gives a 2-share masked AES with the changing of the guards techniques and opts for a low area cost and a moderate latency. Compared to the 4-share AES by Wegener and Moradi [26] and the 3-share AES by Sugawara [25] which also uses the changing of the guards technique, this implementation costs 91.37% fewer cycles and 52.46% less area, respectively. Design II reduces 10 cycles of latency in the AES encryption with the trade-off on a 56.75% larger area and a 37.11% lower maximum frequency. Shahmirzadi and Moradi [24] present a randomness-free 2-share AES implementation with a 5.2% smaller area and a design using one bit per S-box with a 12.2% smaller area. Both designs require a lower maximum frequency compared to design I, although the UMC 180 standard cell library was used. However, we wish to emphasize that both designs are based on using a non-uniform masked S-box whereas our S-box is uniform allowing for a stronger security argument. Moreover, in the work of Shahmirzadi and Moradi, it is mentioned that "the application of changing of the guards on 2-share implementations to nullify the required fresh randomness does not seem trivial (or even possible)". As a result, our work indicates that this is possible. Table 2 covers the round-based implementations. The low-latency S-box reduces the latency of the whole encryption down to 42 cycles using design III. Although the same S-boxes are used in the round-based AES compared to the serialized implementations in Table 1, each estimated f_{max} is 3% to 20% lower than its serialized counterpart. The reason of this reduction is the wider bus in the round-based architecture making the fan-out in control logic cost an extra delay on the critical path. Design III costs 26.52% less area compared to the work by Sasdrich *et al.* [23]. Note that Sasdrich *et al.* used LUT-based Masked Dual-Rail with Pre-charge Logic (LMDPL) and included the mask-table generation circuit in the implementation.

Table 1. Implementation cost of the serialized AES.

Design	Area (S-box) [kGE]	Area (AES) [kGE]	Latency (S-box) [cc]	Latency (AES) [cc]	Random[a] [bpc]	f_{max} [MHz][b]
Design I: Sect. 4.2	2.71	8.13	6	242	0	820
Design II: Sect. 4.3	4.33	12.75	5	232	0	515
Design III: Sect. 4.4	5.83	13.66	4	232	0	534
Bilgin et al. [3][d]	2.84	8.12	4	246	32	-
De Cnudde et al. [9]	1.98	6.68	6	276	54	−
Wegener-Moradi [26]	4.20	7.60	16	2804	0	−
Sugawara [25]	3.50	17.10	4	266	0	-
Shahmirzadi-Moradi [24][c]	−	7.14	6	246	1	160
Shahmirzadi-Moradi [24][c]	−	7.71	6	246	0	160

a. Cost of online fresh random bits per cycle (bpc)
b. Depending on different standard cell libraries
c. With a non-uniform masked S-box
d. With an additional register stage in the S-box to store the state

7.2 Evaluation

In this section, we describe the practical evaluation of the three masked designs. We collect the amplified power traces from the measurement point on the Sakura-X board. The power measurements are amplified by a 30 dB SMA pre-amplifier [13]. The traces are captured by an oscilloscope at a sample rate of 500 MS/s while the FPGA is clocked at 6.144 MHz.

The non-specific leakage detection test from Goodwill et al. [14] is performed which verifies that our implementations do not show first-order leakage. The measured power traces are partitioned into two sets, where the first set S_0 receives fixed plaintexts and the second set S_1 contains random plaintexts. The two sets of measurements are compared using the t-test statistic:

$$t = \frac{\mu(S_0) - \mu(S_1)}{\sqrt{\frac{\sigma^2(S_0)}{|S_0|} + \frac{\sigma^2(S_1)}{|S_1|}}} \tag{8}$$

The t-test verifies whether the two sets have the same first-order moment. Its null hypothesis H_0 states that "the sets S_0 and S_1 are drawn from populations with the same mean." Large absolute values of this t-statistic indicate that the null hypothesis can be rejected with a high degree of confidence. The threshold value of the t-test commonly used by the side-channel research community is 4.5. If the t-test value of the measured power trace grows over 4.5, the implementation under test is considered as insecure.

Figure 9, 11, and 12 illustrate the TVLA result of the three serialized masked AES implementations. Figure 10 shows the maximum univariate t-test value changing over time which is denoted by the number of traces. The first-order

Table 2. Implementation cost of the round-based AES.

Design	Area (S-box) [kGE]	Area (AES) [kGE]	Latency (S-box) [cc]	Latency (AES) [cc]	Random[a] [bpc]	f_{max} [MHz][b]
Design I: Sect. 4.2	2.71	63.91	6	62	0	800
Design II: Sect. 4.3	4.33	102.44	5	52	0	505
Design III: Sect. 4.4	5.84	115.73	4	42	0	426
Gross et al. [16]	60.76	–	1	–	2048	356
Gross et al. [16]	6.74	–	2	–	416	584
Sasdrich et al. [23]	3.48	157.50	1	10	720	400

a. Cost of online fresh random bits per cycle (*bpc*)
b. Depending on different standard cell libraries

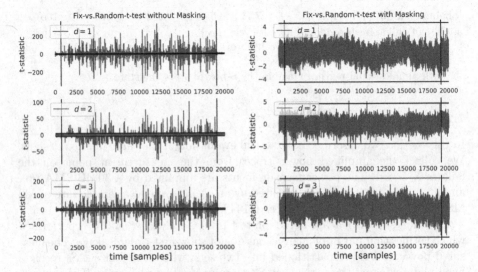

Fig. 9. Univariate fixed-vs.-random t-test results for the serialized AES-128 encryption for design I. The right three plots show the results for the first to the third order statistical moments of the masked implementations at 10^8 traces. The results in the left column are for the unmasked implementations at 10^4 traces. The regions of interest are between vertical red lines, which indicate the start and end of the AES encryption. (Color figure online)

probing security is verified physically for the implementations. The second-order leakages shows as anticipated. The exception is the third order leakage for design I which does not leak due to the noise amplification for third-order moments.

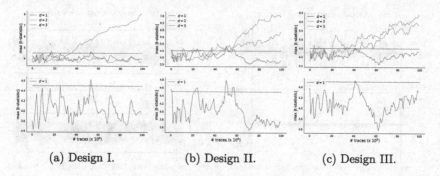

(a) Design I.   (b) Design II.   (c) Design III.

Fig. 10. The Y-axes of each subfigure is the maximum value of the univariate t-test results for the serialized AES-128 designs in this paper. The X-axes are the numbers of traces. The first two figures show the maximum t-test values of the first to the third statistical moments. The last one shows the first-order maximum t-test values only.

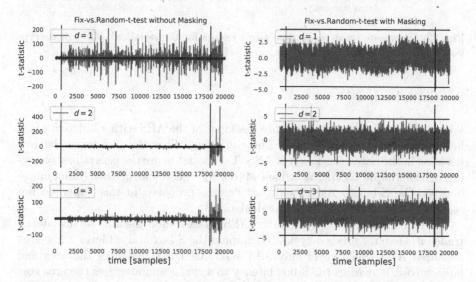

Fig. 11. Univariate fixed-vs.-random t-test results for the serialized AES-128 encryption for design II. The layout is the same as in Fig. 9.

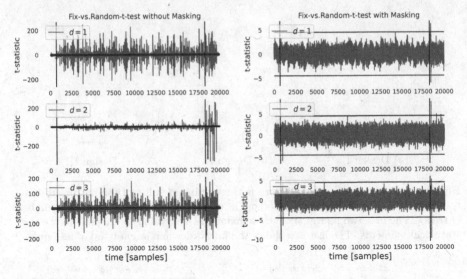

Fig. 12. Univariate fixed-vs.-random t-test results for the serialized AES-128 encryption for design III. The layout is the same as in Fig. 9.

8   Conclusion

We proposed three first-order secure maskings of the AES with a different number of shares and register stages. These maskings allow the area versus latency trade-off in the hardware design of AES. The maskings use the generalized changing of the guards technique, which allows for the re-use of their randomness between the S-boxes. As a result, all designs proposed in this paper do not require fresh randomness in their calculation.

The three variations of the masked AES S-boxes show the area versus latency trade-off. Design I needs 5 cycles to compute the S-box and achieves a low-area serialized AES which costs only 8.13 kGE. Design II balances the area and latency cost. It reduces the S-box latency to 4 cycles and increases the area cost to 12.75 kGE. Design III is made for low latency applications, which requires 3 cycles for the S-box and costs 13.66 kGE for the serialized AES encryption. The designs do not require online randomness and are proven to be probing secure. The maskings are also implemented on FPGA and tested in practice. The TVLA results verify that the designs are first-order secure.

Acknowledgments. This work was supported by CyberSecurity Research Flanders with reference number VR20192203. Siemen Dhooghe is supported by a PhD Fellowship from the Research Foundation - Flanders (FWO). Zhenda Zhang is funded by a research grant from KU Leuven.

References

1. Beyne, T., Dhooghe, S., Zhang, Z.: Cryptanalysis of masked ciphers: a not so random idea. In: Moriai, S., Wang, H. (eds.) ASIACRYPT 2020. LNCS, vol. 12491, pp. 817–850. Springer, Cham (2020). https://doi.org/10.1007/978-3-030-64837-4_27
2. Bilgin, B., Gierlichs, B., Nikova, S., Nikov, V., Rijmen, V.: Higher-order threshold implementations. In: Sarkar, P., Iwata, T. (eds.) ASIACRYPT 2014. LNCS, vol. 8874, pp. 326–343. Springer, Heidelberg (2014). https://doi.org/10.1007/978-3-662-45608-8_18
3. Bilgin, B., Gierlichs, B., Nikova, S., Nikov, V., Rijmen, V.: Trade-offs for threshold implementations illustrated on AES. IEEE Trans. CAD ICs Syst. **34**(7), 1188–1200 (2015). https://doi.org/10.1109/TCAD.2015.2419623
4. Canright, D.: A very compact s-box for AES. In: Rao, J.R., Sunar, B. (eds.) CHES 2005. LNCS, vol. 3659, pp. 441–455. Springer, Heidelberg (2005). https://doi.org/10.1007/11545262_32
5. Chari, S., Jutla, C.S., Rao, J.R., Rohatgi, P.: Towards sound approaches to counteract power-analysis attacks. In: Wiener, M. (ed.) CRYPTO 1999. LNCS, vol. 1666, pp. 398–412. Springer, Heidelberg (1999). https://doi.org/10.1007/3-540-48405-1_26
6. Daemen, J.: Changing of the guards: a simple and efficient method for achieving uniformity in threshold sharing. Cryptology ePrint Archive, Report 2016/1061 (2016). https://eprint.iacr.org/2016/1061
7. Daemen, J.: Changing of the guards: a simple and efficient method for achieving uniformity in threshold sharing. In: Fischer, W., Homma, N. (eds.) CHES 2017. LNCS, vol. 10529, pp. 137–153. Springer, Cham (2017). https://doi.org/10.1007/978-3-319-66787-4_7
8. Daemen, J., Rijmen, V.: Advanced Encryption Standard (AES). National Institute of Standards and Technology (NIST), FIPS PUB 197, U.S. Department of Commerce (2001)
9. De Cnudde, T., Reparaz, O., Bilgin, B., Nikova, S., Nikov, V., Rijmen, V.: Masking AES with $d+1$ shares in hardware. In: Gierlichs, B., Poschmann, A.Y. (eds.) CHES 2016. LNCS, vol. 9813, pp. 194–212. Springer, Heidelberg (2016). https://doi.org/10.1007/978-3-662-53140-2_10
10. De Meyer, L., Reparaz, O., Bilgin, B.: Multiplicative masking for AES in hardware. IACR TCHES **2018**(3), 431–468 (2018). https://doi.org/10.13154/tches.v2018.i3.431-468
11. Dhooghe, S., Nikova, S., Rijmen, V.: Threshold implementations in the robust probing model. In: Bilgin, B., Petkova-Nikova, S., Rijmen, V., (eds.) Proceedings of ACM Workshop on Theory of Implementation Security, TIS@CCS 2019, London, UK, 11 November 2019, pp. 30–37. ACM (2019). https://doi.org/10.1145/3338467.3358949
12. Faust, S., Grosso, V., Pozo, S.M.D., Paglialonga, C., Standaert, F.X.: Composable masking schemes in the presence of physical defaults & the robust probing model. IACR TCHES **2018**(3), 89–120 (2018). https://doi.org/10.13154/tches.v2018.i3.89-120
13. GmbH, L.E.T.: Langer EMV - pa 303 SMA, preamplifier 100 kHz up to 3 GHz
14. Goodwill, G., Jun, B., Jaffe, J., Rohatgi, P.: A testing methodology for side-channel resistance validation (2011)
15. Goubin, L., Patarin, J.: DES and differential power analysis the "Duplication" method. In: Koç, Ç.K., Paar, C. (eds.) CHES 1999. LNCS, vol. 1717, pp. 158–172. Springer, Heidelberg (1999). https://doi.org/10.1007/3-540-48059-5_15

16. Gross, H., Iusupov, R., Bloem, R.: Generic low-latency masking in hardware. IACR TCHES **2018**(2), 1–21 (2018). https://doi.org/10.13154/tches.v2018.i2.1-21

17. Ishai, Y., Sahai, A., Wagner, D.: Private circuits: securing hardware against probing attacks. In: Boneh, D. (ed.) CRYPTO 2003. LNCS, vol. 2729, pp. 463–481. Springer, Heidelberg (2003). https://doi.org/10.1007/978-3-540-45146-4_27

18. Kocher, P., Jaffe, J., Jun, B.: Differential power analysis. In: Wiener, M. (ed.) CRYPTO 1999. LNCS, vol. 1666, pp. 388–397. Springer, Heidelberg (1999). https://doi.org/10.1007/3-540-48405-1_25

19. Lab./UEC, S.: Sakura (sasebo-giii) (2014). https://satoh.cs.uec.ac.jp/SAKURA/hardware/SAKURA-X.html

20. NANGATE: The NanGate 45nm Open Cell Library, version: PDKv1.3_v2010_12.Apache.CCL. https://github.com/The-OpenROAD-Project/OpenROAD-flow-scripts/tree/master/flow/platforms/nangate45

21. Nikova, S., Rechberger, C., Rijmen, V.: Threshold implementations against side-channel attacks and glitches. In: Ning, P., Qing, S., Li, N. (eds.) ICICS 2006. LNCS, vol. 4307, pp. 529–545. Springer, Heidelberg (2006). https://doi.org/10.1007/11935308_38

22. Reparaz, O., Bilgin, B., Nikova, S., Gierlichs, B., Verbauwhede, I.: Consolidating masking schemes. In: Gennaro, R., Robshaw, M. (eds.) CRYPTO 2015. LNCS, vol. 9215, pp. 764–783. Springer, Heidelberg (2015). https://doi.org/10.1007/978-3-662-47989-6_37

23. Sasdrich, P., Bilgin, B., Hutter, M., Marson, M.E.: Low-latency hardware masking with application to AES. IACR TCHES **2020**(2), 300–326 (2020). https://doi.org/10.13154/tches.v2020.i2.300-326

24. Shahmirzadi, A.R., Moradi, A.: Re-consolidating first-order masking schemes. IACR TCHES **2021**(1), 305–342 (2021). https://doi.org/10.46586/tches.v2021.i1.305-342

25. Sugawara, T.: 3-share threshold implementation of AES s-box without fresh randomness. IACR TCHES **2019**(1), 123–145 (2018). https://doi.org/10.13154/tches.v2019.i1.123-145

26. Wegener, F., Moradi, A.: A first-order SCA resistant AES without fresh randomness. In: Fan, J., Gierlichs, B. (eds.) COSADE 2018. LNCS, vol. 10815, pp. 245–262. Springer, Cham (2018). https://doi.org/10.1007/978-3-319-89641-0_14

Rivain-Prouff on Steroids: Faster and Stronger Masking of the AES

Luan Cardoso dos Santos, François Gérard, Johann Großschädl[✉],
and Lorenzo Spignoli

DCS and SnT, University of Luxembourg, 6, Avenue de la Fonte,
4364 Esch-sur-Alzette, Luxembourg
{luan.cardoso,francois.gerard,johann.groszschaedl,
lorenzo.spignoli}@uni.lu

Abstract. At CHES 2010, Rivain and Prouff (RP) introduced an elegant masking technique to protect the Advanced Encryption Standard (AES) against power analysis attacks. RP masking is provable secure in the probing model, but this solid theoretical underpinning comes at the cost of a massive increase in execution time. In this paper, we describe software optimization methods to accelerate the low-level arithmetic in the field $\mathbb{F}_{2^8}$, which has a significant impact on the overall performance of a masked implementation of the AES. Among these optimizations is an improved technique for table-based multiplication in $\mathbb{F}_{2^8}$ that allows one to avoid the special treatment of 0-values, thereby speeding up the multiplication of masked operands. Furthermore, we introduce a novel exponentiation-based algorithm for inversion in $\mathbb{F}_{2^8}$, which reduces the overall number of table look-ups and the amount of randomness needed for the refreshing of masks compared to the original RP inversion. This new inversion provides some advanced (theoretical) security properties for the composition of gadgets, e.g. Strong Non-Interference (SNI) and Probe Isolating Non-Interference (PINI). We also describe a prototype implementation of a first-order masked inversion and AES encryption in ARMv7-M Assembly language. According to our simulation results, the first-order masked AES has an execution time of about 25k clock cycles per block when using a generic Cortex-M3 as target platform, which is roughly twice as fast as the RP-masked AES Assembly implementation presented at EUROCRYPT 2017 by Goudarzi and Rivain.

1 Introduction

Some 20 years ago, the National Institute of Standards and Technology (NIST) selected the block cipher Rijndael to become the AES and replace the DES as most-widely used symmetric cryptosystem [8]. Since then, much effort has been devoted to the development of efficient and secure implementations on various platforms [4,21]. A common attack scenario against concrete implementations is the exploitation of physical leakages of the device executing a cryptographic algorithm. This so-called *Side-Channel Analysis (SCA)*, along with techniques

I. Buhan and T. Schneider (Eds.): CARDIS 2022, LNCS 13820, pp. 123–145, 2023.
https://doi.org/10.1007/978-3-031-25319-5_7

to defend against it (i.e. SCA countermeasures), has received a lot of attention from cryptographers in both academia and industry. SCA is a very broad field covering a variety of techniques that can be used to extract secret information from a device. A well-known example is the statistical analysis of the device's power consumption (resp. its electromagnetic emanations) to find a correlation between these physical leakages and the secret value that is manipulated when executing a cryptographic operation (e.g. an encryption), commonly referred to as *Differential Power Analysis (DPA)* [18]. Unprotected (or insufficiently protected) implementations of the AES and other block ciphers are easily broken through DPA, as has been demonstrated in dozens of papers, e.g. [3].

Masking is a common and well-studied mitigation against DPA that aims to break the link between the internal intermediate values and measured leakages by applying secret-sharing techniques throughout the execution. Any sensitive variable s has to be split up into two or more statistically independent shares $(s_0, s_1, \ldots, s_d)$ for which the relation $s = s_0 \odot s_1 \odot \cdots \odot s_d$ holds for a certain operation $\odot$ (normally XOR) [18,20]. We call d the *masking order*. The main challenge is to find an efficient algorithm that takes values split into shares as input and computes correctly the results of the cryptographic transformations without ever recombining any of the shares. Masked versions of the four round transformations of the AES have been extensively studied in the literature, see e.g. [1,11]. Three of those are easy to mask since they are linear. Indeed, when $f(s_0) \odot f(s_1) \odot \cdots \odot f(s_d) = f(s_0 \odot s_1 \odot \cdots \odot s_d)$ holds for a cryptographic function f, the non-masked version of this function can simply be applied to each share individually. Hence, the majority of research focuses on devising an efficient technique to carry out the only non-linear transformation of the AES (i.e. SubBytes) with shared inputs. Algebraically, this operation is an inversion in the finite field $\mathbb{F}_{2^8}$ (followed by an affine transformation, which is trivial to mask), with the caveat that 0 is a valid input and mapped to itself.

At CHES 2010, Rivain and Prouff [20] introduced a new masking technique to protect the $\mathbb{F}_{2^8}$-inversion against DPA, which has been widely analyzed and initiated lots of follow-up research, e.g. [10,17]. This so-called RP masking has some attractive features that distinguish it from (most) other masking schemes for the AES, namely (i) it comes with strong (theoretical) security guarantees since it is provably secure in the well-known probing model of Ishai, Sahai, and Wagner (ISW) [16], and (ii) it can be relatively straightforwardly extended to any masking order $d \geq 2$. RP masking was inspired by ISW's secure AND gate ("gadget") from [16], which consists of $(d+1)^2$ ordinary AND gates, $2d(d+1)$ conventional XOR gates, and consumes $d(d+1)/2$ random bits every time it is evaluated. Rivain and Prouff observed that ISW's method to securely compute a logical AND (which is nothing else than a multiplication in $\mathbb{F}_2$) can be generalized to secure multiplication in $\mathbb{F}_{2^8}$ when the $(d+1)^2$ binary multiplications (i.e. 1-bit AND gates) are replaced by ordinary multiplications in $\mathbb{F}_{2^8}$ and the $2d(d+1)$ binary additions (i.e. 1-bit XOR gates) by conventional additions in $\mathbb{F}_{2^8}$ (i.e. 8-bit XORs). They also introduced a secure inversion in $\mathbb{F}_{2^8}$ based on exponentiation by 254 (according to Fermat's Little Theorem), which requires four secure multiplications and seven secure squarings in $\mathbb{F}_{2^8}$.

Even though RP masking has some attractive features, especially regarding its (theoretical) security, it turned out to be highly challenging to implement in practice. Goudarzi and Rivain [13,14] studied various implementation options and optimization techniques for masking of the AES and reported the to-date best execution times for RP masking on a 32-bit ARM processor. According to the results given in [14, Table 3], a full AES128 encryption with (non-bitsliced) first-order RP masking has an execution time of (at least) $49,329$ clock cycles per block on the ARM v7 architecture[1]. For comparison, an unprotected ARM implementation that performs SubBytes using a 256-byte look-up table has an execution time of roughly $1,700$ cycles when the round keys are pre-computed (see e.g. [4]). Consequently, the currently best implementation of RP masking introduces a 29-fold execution-time overhead for first-order DPA protection. In addition, Beckers et al. [3] recently showed that virtually all publicly available C and ARM Assembly implementations of masked AES (including RP-masked variants, but also others based on e.g. bit-slicing [21] and fix-slicing [1]) can be easily broken by first-order DPA with a few thousand traces.

In this paper, we introduce some simple yet effective software optimization techniques to significantly reduce the execution time of RP-masked AES. The first optimization targets the (conventional) multiplication in $\mathbb{F}_{2^8}$, which is the most performance-critical operation of a masked multiplication in $\mathbb{F}_{2^8}$ (i.e. the SecMult gadget of Rivain and Prouff). A (conventional) multiplication in $\mathbb{F}_{2^8}$ is commonly performed with the help of Log and AntiLog tables, which map the elements to their orders in the multiplicative group of the field $\mathbb{F}_{2^8}$ (and vice versa) [12]. However, in such a *table-based multiplication*, operands that are 0 have to be treated as a special case, which slows down the multiplication. We found an efficient way to get rid of this special treatment of 0-operands at the expense of larger table sizes. Our second contribution is an alternate sequence of multiplications/squarings to compute an exponentiation by 254 that reduces not only the number of look-ups into the Log and AntiLog tables, but also the amount of randomness to be secure in the ISW probing model. We describe an ARMv7-M Assembly implementation of our exponentiation and give detailed results for a first-order masked AES. These results show that our optimizations make RP masking almost twice as fast as the to-date best implementation. We also performed a t-test with $5,000$ traces to confirm that our implementation is not flawed. However, we defer a fully-protected implementation with special countermeasures against micro-architectural leakage to future work.

2   Preliminaries

2.1   Provably-Secure Masking of the AES

Ishai, Sahai, and Wagner introduced in their seminal paper [16] a technique to transform any circuit C composed of logic gates into an equivalent randomized

---

[1] ARM v7 is not a recognized ARM architecture. Rather, it denotes the features that are common to all of the ARMv7-A, ARMv7-R, and ARMv7-M architectures.

Algorithm 1. SecMult: dth-order secure multiplication over $\mathbb{F}_{2^k}$

Input: shares a_i satisfying $a = \bigoplus_{i=0}^{d} a_i$, shares b_i satisfying $b = \bigoplus_{i=0}^{d} b_i$
Output: shares c_i satisfying $c = \bigoplus_{i=0}^{d} c_i = ab$

1: **for** $i = 0$ **to** d **do**
2:     **for** $j = i + 1$ **to** d **do**
3:         $r_{i,j} \leftarrow \mathsf{rand}(k)$
4:         $r_{j,i} \leftarrow (r_{i,j} \oplus a_i b_j) \oplus a_j b_i$
5: **for** $i = 0$ **to** d **do**
6:     $c_i \leftarrow a_i b_i$
7:     **for** $(j = 0$ **to** $d) \wedge (i \neq j)$ **do**
8:         $c_i \leftarrow c_i \oplus r_{i,j}$

---

circuit C' that remains secure against an adversary who is allowed probe up to t wires within a given time interval. Security against such *t-probing attacks* is achieved by mapping each binary input to $d + 1$ binary values and constructing C' so that each wire in C corresponds to $d + 1$ wires in C' carrying an additive $(d + 1)$-out-of-$(d + 1)$ secret sharing of the value on that wire of C. If x is an input to C, the mapped input for C' consists of d values $x_1, x_2, \ldots, x_d$ drawn uniformly at random, and a further value x_0 computed to satisfy the equation $x = x_0 \oplus x_1 \oplus \cdots \oplus x_d$. The circuit C' is obtained by transforming the normal gates into secure variants ("gadgets") as specified in [16] for the AND and the NOT gate. Also given in [16] is a proof for the security of C' against a t-bound adversary for $t = \lfloor d/2 \rfloor$, i.e. the proof holds when the number of wires that are probed in C' is less than half of the $d + 1$ wires of gadget inputs/outputs.

Rivain and Prouff generalized in [20] the construction of a secure AND gate from [16] to secure multiplication $\mathbb{F}_{2^8}$, yielding the SecMult gadget specified in Algorithm 1. In general, for d-th order masking, the inputs of this gadget are two operands $a, b \in \mathbb{F}_{2^8}$, each represented via $d + 1$ shares, and the output is a $(d + 1)$-sharing of the product $c = ab$. Algorithm 1 requires the computation of $(d + 1)^2$ conventional $\mathbb{F}_{2^8}$-multiplications and consumes $d(d + 1)/2$ random bytes. The SecMult gadget features a *tight* security proof because Rivain and Prouff showed that it can withstand an adversary who knows $t = d$ out of the $d + 1$ shares of a sensitive variable (rather than just $t = \lfloor d/2 \rfloor$ shares as in the case of the ISW proof). However, this proof requires the input $(d + 1)$-sharings to be mutually independent. If this is not guaranteed then one of the sharings needs to be refreshed using the RefreshMasks gadget in Algorithm 2.

Algorithm 2. RefreshMasks

Input: shares x_i satisfying $x = \bigoplus_{i=0}^{d} x_i$
Output: shares x_i satisfying $x = \bigoplus_{i=0}^{d} x_i$

1: **for** $i = 1$ **to** d **do**
2:     $tmp \leftarrow \mathsf{rand}(k)$
3:     $x_0 \leftarrow x_0 \oplus tmp$
4:     $x_i \leftarrow x_i \oplus tmp$

Algorithm 3. SecExp254: dth-order secure exponentiation by 254 over $\mathbb{F}_{2^k}$

Input: shares $x_0, x_1, \ldots, x_d$ satisfying $x = \bigoplus_{i=0}^d x_i$
Output: shares $y_0, y_1, \ldots, y_d$ satisfying $y = \bigoplus_{i=0}^d y_i = x^{254}$
1: **for** $i = 0$ **to** d **do** $z_i \leftarrow x_i^2$ $\triangleright \bigoplus_i z_i = x^2$
2: $(x_0, x_1, \ldots, x_d) \leftarrow$ RefreshMasks$(x_0, x_1, \ldots, x_d)$
3: $(y_0, y_1, \ldots, y_d) \leftarrow$ SecMult$((z_0, z_1, \ldots, z_d), (x_0, x_1, \ldots, x_d))$ $\triangleright \bigoplus_i y_i = x^3$
4: **for** $i = 0$ **to** d **do** $w_i \leftarrow y_i^4$ $\triangleright \bigoplus_i w_i = x^{12}$
5: $(w_0, w_1, \ldots, w_d) \leftarrow$ RefreshMasks$(w_0, w_1, \ldots, w_d)$
6: $(y_0, y_1, \ldots, y_d) \leftarrow$ SecMult$((y_0, y_1, \ldots, y_d), (w_0, w_1, \ldots, w_d))$ $\triangleright \bigoplus_i y_i = x^{15}$
7: **for** $i = 0$ **to** d **do** $y_i \leftarrow y_i^{16}$ $\triangleright \bigoplus_i y_i = x^{240}$
8: $(y_0, y_1, \ldots, y_d) \leftarrow$ SecMult$((y_0, y_1, \ldots, y_d), (w_0, w_1, \ldots, w_d))$ $\triangleright \bigoplus_i y_i = x^{252}$
9: $(y_0, y_1, \ldots, y_d) \leftarrow$ SecMult$((y_0, y_1, \ldots, y_d), (z_0, z_1, \ldots, z_d))$ $\triangleright \bigoplus_i y_i = x^{254}$

Algorithm 3 shows the secure exponentiation method for inversion in $\mathbb{F}_{2^8}$ as proposed by Rivain and Prouff [20]. This exponentiation consists of four secure multiplications (i.e. SecMult gadgets) and seven secure squarings, whereby two squarings are combined to an exponentiation by 4 (line 4) and four squarings to an exponentiation by 16 (line 7). The first two SecMult gadgets (line 3 and line 6) require one of their input sharings to be refreshed to overcome a dependency between the two inputs (concretely, one input is a linear function of the other input). Besides the inversion in $\mathbb{F}_{2^8}$, the SubBytes operation of the AES includes also a simple affine transformation, which is linear and can be masked as described in [20]. While Algorithm 3 is provably secure in the probing model for $d = 1$ (i.e. first-order masking), it should be mentioned that a flaw has been discovered (and corrected) for higher orders in [7]. However, since we present in Sect. 4 a new exponentiation technique along with a different proof, this flaw is not relevant for our work.

2.2 Formal Security Notions

When using the circuit protection method of Ishai, Sahai, and Wagner [16], the number of wires in the transformed circuit C' corresponding to a single wire in the original circuit C has to be $2t + 1$ to guarantee security against a t-limited adversary, i.e. $d = 2t$ since, in our notation, the number of transformed wires is $d + 1$. On the other hand, the SecMult gadget (with independent input sharings) from [20] enjoys a tight proof with $t = d$, i.e. the adversary is allowed to probe d out of the $d + 1$ shares of a masked variable. However, the tight proof for the SecMult gadget in the probing model can not be straightforwardly extended to the full exponentiation (despite the claims in Sect. 4.3 of [20]), though there is a correct non-tight ISW-like proof for $d = 2t$. In order to fill this gap, various improvements of the classic probing model have been proposed, including some stronger notions of security, such as *Strong Non-Interference* (SNI) by Barthe et al. [2]. Their work starts by defining a notion of Non-Interference (NI), see Definition 1, which is sufficient for security in the ISW model. In fact, one can view NI as equivalent to the t-threshold probing security proposed in [16].

Definition 1 (t-NI security). *Let G be a gadget[2] taking as input $(x_i)_{0 \leq i \leq d}$ and outputting $(y_i)_{0 \leq i \leq d}$. The gadget G is said t-NI secure if for any set of t_1 intermediate variables and any subset $\mathcal{O}$ of output indices, there exists a subset $\mathcal{S}$ of input indices with $|I| \leq t_1 + |\mathcal{O}|$ such that the t_1 intermediate variables and the output variables $y_{|\mathcal{O}}$ can be perfectly simulated from $x_{|\mathcal{S}}$.*

Starting from this definition, they proposed a stronger version of NI, where the size of the input-shares subset I solely depends on the number of internal probes (i.e. on t_1) and is independent of the output probes $\mathcal{O}$, as formalized in Definition 2. The main goal was to make security proofs in the probing model simpler; the SNI property will guarantee "isolation" between the input and the output shares, allowing probing security with $t + 1$ shares. Furthermore, it is very easy to obtain composability from it. Namely, it is possible to prove the SNI property of a full construction given the SNI property of its components.

Definition 2 (t-SNI security). *Let G be a gadget taking as inputs $(x_i)_{0 \leq i \leq d}$ and $(y_i)_{0 \leq i \leq d}$, and outputting $(z_i)_{0 \leq i \leq d}$. The gadget G is said to be t-SNI secure if for any set of t_1 intermediate variables and any subset $\mathcal{O}$ of output indices such that $t_1 + |O| \leq t$, there exist two subsets $\mathcal{S}_1$ and $\mathcal{S}_2$ of input indices which satisfy $|\mathcal{S}_1|, |\mathcal{S}_2| \leq t_1$ such that the t_1 intermediate variables and the output variables $z_{|\mathcal{O}}$ can be perfectly simulated from $x_{|\mathcal{S}_1}$ and $y_{|\mathcal{S}_2}$.*

Along these definitions, they provided proofs of SNI for an RP-masked AES, in particular for SecMult (see Lemma 1) and RefreshMasks (see Lemma 2).

Lemma 1 (t-SNI of SecMult). *Let $(x_i)_{0 \leq i \leq d}$ and $(y_i)_{0 \leq i \leq d}$ be the input shares of the SecMult operation from [20], and let $(z_i)_{0 \leq i \leq d}$ be the output shares. For any set of t_1 intermediate variables and any subset $\mathcal{O}$ of output shares such that $t_1 + |\mathcal{O}| < n$, there exist two subsets $\mathcal{S}_1$ and $\mathcal{S}_2$ of indices with $|\mathcal{S}_1|, |\mathcal{S}_2| \leq t_1$ such that those t_1 intermediate variables as well as the output shares $z_{|\mathcal{O}}$ can be perfectly simulated from $x_{|\mathcal{S}_1}$ and $y_{|\mathcal{S}_2}$.*

Lemma 2 (t-SNI of RefreshMasks). *Let $(x_i)_{0 \leq i \leq d}$ be the input shares of the RefreshMasks operation from [2], and let $(y_i)_{0 \leq i \leq d}$ be the output shares. For any set of t_1 intermediate variables and any subset $\mathcal{O}$ of output shares such that $t_1 + |\mathcal{O}| < n$, there exists a subset $\mathcal{S}$ of indices with $|\mathcal{S}| \leq t_1$, such that the t_1 intermediate variables as well as the output shares $y_{|\mathcal{O}}$ can be perfectly simulated from $x_{\mathcal{S}}$.*

3   Low-Level Field Arithmetic

In the case of first-order masking (i.e. $d = 1$), the SecMult gadget of Rivain and Prouff [20] performs four ordinary multiplications in $\mathbb{F}_{2^8}$. More generally, the

---

[2] For the sake of clarity, we refer to the set $(x_0, x_1, \ldots, x_d)$ as $(x_i)_{0 \leq i \leq d}$. When given $d + 1$ as number of shares and a vector $(x_i)_{0 \leq i \leq d}$, we denote by $x_{|I} := (x_i)_{i \in I}$ the set of shares x_i with $i \in I$.

number of multiplications in $\mathbb{F}_{2^8}$ grows with the square of the number of shares (which is $d+1$ in our notation), making it a performance-critical operation. In essence, a multiplication of two elements of $\mathbb{F}_{2^8}$ boils down to a multiplication of binary polynomials of a degree of up to 7, resulting in a product-polynomial with a maximum degree of 14, followed by a reduction modulo the irreducible polynomial $p(x) = x^8 + x^4 + x^3 + x + 1$ [8]. Modern x64 processors provide the GF2P8MULB instruction for multiplication in $\mathbb{F}_{2^8}$, but such kind of instruction is lacking on our target platform, the ARM Cortex-M3, and also on virtually all other 8, 16, and 32-bit microcontrollers. As a consequence, implementers have to resort to either the binary method [13], which is relatively slow, or perform the multiplication with the help of look-up tables.

3.1   Table-Based Multiplication

In its most basic form, a table-based multiplication of elements of $\mathbb{F}_{2^8}$ employs two look-up tables, namely a so-called *Log table* that contains the order of the 255 non-0 elements of the field (based on the generator $g(x) = x + 1$) and an *AntiLog table* containing the elements corresponding to the orders in the range of $[0, 254]$. Each of the tables consists of 255 entries, but for efficiency reasons a Log table with 256 entries is used so that the integer representation of a field element can be directly used as an index for the look-up into the table. Note that only the elements of the multiplicative group of the field $\mathbb{F}_{2^8}$ (i.e. only the 255 non-0 elements) actually have an order. However, by assigning 255 to the Log-table entry at index 0 and assigning 0 to the AntiLog-table entry at index 255, we achieve that AntiLog[Log[a]] = a holds for *any* $a(x) \in \mathbb{F}_{2^8}$, and this includes the special case $a(x) = 0$. Each the Log table and the AntiLog table has a size of 256 bytes, which amounts to 512 bytes in total. The computation of the product $c(x) = a(x)b(x)$ consists of three simple steps [12]. At first, the Log table is queried to obtain $u = \mathrm{ord}(a(x))$ and $v = \mathrm{ord}(b(x))$. Thereafter, the modular sum $s = u + v \bmod 255$ is computed using an 8-bit addition followed by a conditional subtraction of 255. Finally, the element $c(x)$ corresponding to the order s is determined with the help of the AntiLog table. Consequently, the multiplication requires three table look-ups and a modular addition [8,12].

It is possible to speed up the multiplication by replacing the modular addition by a normal addition, but this requires a larger AntiLog table [12]. More concretely, when the AntiLog table is "duplicated" in such a way that we have AntiLog[255+i] = AntiLog[i] for $i \in [0, 254]$, then a conditional subtraction of 255 from $u + v$ is not necessary anymore. Thanks to this optimization, the cost of the table-based multiplication is reduced to three table look-ups and an ordinary addition [12]. The size of the AntiLog table doubles, and both tables amount to 768 bytes altogether. Note that, since the tables are static, they can be kept in non-volatile memory (i.e. ROM or flash) and do not occupy precious space in RAM. However, there is a further implementation detail that requires special attention, namely how to treat the case $a(t) = 0$ or $b(t) = 0$, which we ignored until now. Unfortunately, the table-based multiplication as presented above does not give the correct result when one of the operands is 0. A simple mathematical explanation for this problem is that 0 is not an element of the multiplicative group of the finite field $\mathbb{F}_{2^8}$ and, consequently, does also not have a multiplicative order. Software implementations normally treat the occurrence of 0-operands as special case and explicitly set the result to 0 when one of the two operands is 0 [12]. Hence, a table-based multiplication involves besides the three steps already mentioned above a fourth step in which the operands are checked for 0 and the product is corrected if needed. A pseudo-code description of the table-based multiplication can be found at the end of Sect. 7.2 of [12].

Listing 1. ARM Assembler macro for table-based multiplication in $\mathbb{F}_{2^8}$ [13].

```
1:  .macro F2P8MUL res:req, opA:req, opB:req, tmp:req
2:      // step 1: tmp <- Log[opA], res <- Log[opB]
3:      ldrb \tmp, [ptLog, \opA]
4:      ldrb \res, [ptLog, \opB]
5:      // step 2: tmp <- tmp + res
6:      add  \tmp, \tmp, \res
7:      // step 3: res <- ALog[tmp]
8:      ldrb \res, [ptALog, \tmp]
9:      // step 4: treat (opA == 0) or (opB == 0) case
10:     rsb  \tmp, \opA, #0             // tmp <- 0 - opA
11:     and  \tmp, \opB, \tmp, asr #32  // tmp <- opB & (tmp >> 32)
12:     rsb  \tmp, \tmp, #0             // tmp <- 0 - tmp
13:     and  \res, \res, \tmp, asr #32  // res <- res & (tmp >> 32)
14:  .endm
```

Listing 1 contains an optimized implementation of the table-based multiplication in ARMv7M Assembly language (GNU syntax) that can be executed on e.g. Cortex-M3 and Cortex-M4 microcontrollers. This implementation is based on the assembly code provided in Appendix A (i.e. Fig. 24) of [13], which is an extended version of [14]. It follows the four steps outlined above, but differs slightly from the original Assembler code of [13]. For example, our version does not contain the reduction of the sum of the orders modulo 255 because we take advantage of the "duplicated" AntiLog table, which saves an add as well as an and instruction. Furthermore, our source code does not contain the ldr ("load register") instruction at the beginning of the Assembler macro from [13] (used to initialize the register ptLog with the pointer to the Log table) since we assume that ptLog already contains the Log-table address at any time the macro gets executed. The last four instructions do a 0-testing of the operands and set the result in register res to 0 when (at least) one of the two operands is 0. These four ARM instructions are exactly the same as in Subsect. 3.2 of [13,14] and in Fig. 24 of [13]. The result of the rsb ("reverse subtract") instruction at line 10 is either 0 (when register opA was 0) or the two's complement of opA; in the latter case, the result's most-significant bit is 1. Note that the second operand of the subsequent and instruction (i.e. register tmp) is *arithmetically* shifted to the right before the AND is actually carried out. Hence, the second operand is either 0 or the "all-1" word $2^{32} - 1$, i.e. 0xffffffff in hex notation, and the content of register tmp is either 0 or opB. After the execution of the other two rsb and and instructions in line 12 and 13, respectively, the content of res is either 0 or the multiplication result from line 8.

The ldrb ("load register with byte") instruction has a latency of two clock cycles[3] on our target device, which is a VL Discovery development board from STMicroelectronics with an STM32F100RB Cortex-M3 microcontroller that is clocked at a nominal frequency of 24 MHz [22]. All other ARM instructions in Listing 1 have a latency of one cycle. Hence, the F2P8MUL macro has an overall execution time of 11 clock cycles, of which four are spent for the special treatment of 0-operands. These four clock cycles represent a whopping 36.4% of the execution time and have a huge impact on the performance of RP masking in software. Namely, as we will see in Sect. 5 by using first-order masking as case study, the inversion of the SubBytes transformation contributes about 90% to the execution time of an AES round. The performance of the inversion depends heavily on that of the masked multiplication (SecMul gadget), which in turn is dominated by ordinary multiplication in $\mathbb{F}_{2^8}$ (i.e. F2P8MUL macro). Getting rid of the special treatment of 0-operands in Listing 1 would lead to an enormous speed-up for RP-masking. However, this seems to be a non-trivial problem as neither Goudarzi and Rivain [14] nor Rivain and Prouff [20] found a solution.

Eliminating the Special Treatment of 0-Operands. In the following, we demonstrate that it is possible to completely avoid a special treatment of 0-elements and thereby improve the efficiency of RP masking considerably, at the expense of an increased size of both the Log and the AntiLog table. Though our idea is surprisingly simple, did not see it published anywhere, neither in the landmark paper of Rivain and Prouff [20], nor any subsequent publication on optimized implementation of RP masking like [14]. Recall that the very first entry of the Log table (i.e. the entry with index 0) is 255. Furthermore, as also mentioned before, the entry with index 255 of the original (unduplicated) AntiLog table is 0 to ensure AntiLog[Log[0]] = 0. Our idea is based on the fact that *the sum of the order of two non-0 elements of* $\mathbb{F}_{2^8}$ *is strictly less than 509.* Hence, when we modify the Log table by setting the very first entry to 509 (which could be interpreted as assigning the field-element 0 the order 509, although this makes no sense from a mathematical viewpoint), the special treatment of 0-operands can be simply avoided if the AntiLog table is modified accordingly. This means we have to "extend" the AntiLog table from 512 to 1024 entries, whereby the upper half of this extended table (i.e. all entries with index 509 to index 1023) contain 0. Note that these modifications do not impact a multiplication of two non-0 elements since it still works in exactly the same way as before and yields the correct result. However, when one of the two operands is 0, the sum of the two orders is at least 509, which implies that an entry in the upper part of the AntiLog table is accessed and the result of the multiplication is 0.

---

[3] On certain Cortex-M models, neighboring load and store instructions can pipeline their address and data phases, enabling them to complete in one execution cycle.

These small modifications of the Log and AntiLog table make it possible to remove the last four instructions of Listing 1, which reduces the execution time of the F2P8MUL macro by 36.4%. However, this speed-up comes at the expense of larger look-up tables. The Log table still consists of 256 entries, but needs to be able to accommodate the 9-bit integer value 509, which means it has to be of 16-bit type hword instead of 8-bit type byte. In addition, the AntiLog table contains now 1024 entries altogether instead of 512, whereby all entries in the upper half (i.e. index 509 and above) are 0. Therefore, the two tables become twice as large and the size of both tables amounts to 1536 bytes, which is still relatively small compared to the non-volatile memory (i.e. flash) consumption of most masked AES implementations in the literature (see Sect. 5).

Reducing the Number of Table Look-Ups. The exponentiation-based inversion of Rivain and Prouff [20], reproduced in Algorithm 3, consists of four masked multiplications (i.e. SecMult gadgets), two mask refreshings (i.e. RefreshMasks gadgets), and three other masked operations in $\mathbb{F}_{2^8}$: squaring, 4th-power, and 16th-power[4]. However, the latter three arithmetic operations are all linear in $\mathbb{F}_{2^8}$ and can, therefore, be performed separately on each of the $d + 1$ shares. In addition, the three operations have in common that they can simply be carried out with a 256-byte look-up table. For example, the squaring is a conventional table look-up, which, given $a \in \mathbb{F}_{2^8}$, simply returns the field-element $b = a^2$ as result, i.e. it is not necessary to obtain the order of a first as is the case for the multiplication. When we ignore the mask refreshings, the first two steps of the inversion in Algorithm 3 are a squaring followed by a multiplication[5], and this operation sequence (i.e. *a multiplication preceded by a linear operation* that is nothing else than an simple table look-up) appears two more times, namely in lines 4 and 6, and in lines 7 and 8. Based on this observation, it is possible to reduce the number of table look-ups during an exponentiation by adapting the squaring table (resp. 4th-power, 16th-power table) to contain *the order of the square* ord (b) *instead of the actual square* b. This modification does not change the number of entries the linear tables contain, but allows one to accelerate the exponentiation by using the order read from the linear table as operand in the subsequent multiplication, which saves a look-up into the Log table. In the ideal case (i.e. both operands are given through their orders), the multiplication only consists of only an addition and a single look-up into the AntiLog table.

This optimization to reduce the number of table look-ups is not applicable in a straightforward way without the first optimization (i.e. assigning the field-element 0 the order 509) since operands that are 0 would still require a special treatment. The proposed reduction of the look-ups is also not particularly well suited for the original masked exponentiation (Algorithm 3) because of the two

---

[4] As mentioned in Sect. 2, the 4th-power function $x \mapsto x^4$ and 16th-power function $x \mapsto x^{16}$ can be performed via squarings, in which case seven squarings have to be carried out in the course of an inversion.

[5] To be more precise, these two steps are a *masked* squaring and a *masked* multiplication. However, for the explanation of our software optimization methods, it does not matter whether these operations are masked or not.

mask refreshings between the squaring and multiplication at line 2 and line 5 (which we ignored in the description of our idea). However, in Sect. 4, we will introduce a modified variant of the secure exponentiation that is better suited for our optimization and, therefore, faster than the RP-exponentiation.

3.2 Basic First-Order Gadgets

As mentioned in Sect. 2, the RP masking is generic in the sense that it can be used to achieve arbitrary masking orders, thereby enabling different trade-offs between execution time and security, i.e. resistance against DPA. However, in the rest of this section, we focus on first-order masking, i.e. we always assume $d + 1 = 2$ shares to simplify the description of how we implemented the basic gadgets for masked operations in $\mathbb{F}_{2^8}$. But we emphasize that all optimizations we discuss in the sequel translate over to higher orders in a natural fashion.

In the case of first-order masking (i.e. $d = 1$), all sensitive variables have to be split into $d + 1 = 2$ shares. Therefore, the two operands $a, b \in \mathbb{F}_{2^8}$ that are input to the secure (i.e. masked) multiplication given in Algorithm 1 have the form $a = a_0 \oplus a_1$ and $b = b_0 \oplus b_1$, respectively. If $d = 1$, the computation of the two shares c_0 and c_1 of the product $c = ab = c_0 \oplus c_1$ involves four conventional multiplications in $\mathbb{F}_{2^8}$ as follows.

$$r_{0,1} = \mathsf{rand}(8), \quad r_{1,0} = (r_{0,1} \oplus a_0 b_1) \oplus a_1 b_0$$
$$c_0 = a_0 b_0 \oplus r_{0,1}, \quad c_1 = a_1 b_1 \oplus r_{1,0} \tag{1}$$

As explained earlier, it is possible to have a Log table and an AntiLog table so that `AntiLog[Log[a]]` = a holds for any $a \in \mathbb{F}_{2^8}$ (including 0) and not just the non-0 elements. Therefore, it is possible to define a bijective mapping between the elements of $\mathbb{F}_{2^8}$ and their orders, although, strictly speaking, the element 0 does not have an order in a mathematical sense. This means that it is possible to represent any element of the field $\mathbb{F}_{2^8}$ by its order and vice versa, and we can always easily convert between the two representations, which we refer to as the *normal domain* and the *order domain*, with the help of look-up tables.

Listing 2 shows an ARM Assembly macro that implements a SecMult gadget for the first-order case (i.e. $d = 1$) based on Eq. (1). This implementation incorporates the two optimizations described in the last subsection, i.e. it does not treat 0-operands in a special way and assumes the two operands (or, more precisely, the shares of these operands) to be in the order domain, whereas the shares of the result are in the normal domain to minimize the number of table look-ups in a masked inversion. Since this macro follows Eq. (1), it performs four multiplications in $\mathbb{F}_{2^8}$ and four XOR operations (`eor` instructions). However, since the operands are given in the order domain, the four multiplications to produce $a_0 b_0$, $a_1 b_0$, $a_0 b_1$, and $a_1 b_1$ are just additions of the orders (i.e. the `add` instructions at lines 3 to 6). The subsequent `ldrb` instructions convert the four products from the order domain to the normal domain through look-ups into the AntiLog table (similar to Listing 1, the register `ptALog` contains the start address of the table). Finally, the four `eor` instructions compute the two

Listing 2. ARM Assembler macro of a two-share SecMult gadget (the shares a_0, a_1 and b_0, b_1 of the operands a and b are in the order domain, whereas the shares c_0, c_1 of the product $c = ab$ are in the normal domain).

```
 1:    .macro SECMULT c0:req,c1:req, a0:req,a1:req, b0:req,b1:req, rn:req
 2:        // products in order domain
 3:        add     \c0, \a0, \b0
 4:        add     \c1, \a1, \b1
 5:        add     \a0, \a0, \b1
 6:        add     \a1, \a1, \b0
 7:        // products in normal domain
 8:        ldrb    \c0, [ptALog, \c0]
 9:        ldrb    \c1, [ptALog, \c1]
10:        ldrb    \a0, [ptALog, \a0]
11:        ldrb    \a1, [ptALog, \a1]
12:        // compute masked result c0,c1
13:        eor     \c0, \c0, \rn
14:        eor     \rn, \a0, \rn
15:        eor     \rn, \a1, \rn
16:        eor     \c1, \c1, \rn
17:    .endm
```

shares c_0, c_1 of the result c in the normal domain (rn contains a random byte corresponding to the random value $r_{0,1}$ in Algorithm 1).

On a Cortex-M3 microcontroller, the 12 instructions of the SECMULT macro have an execution time of 16 clock cycles (we assume here that each ldrb has a latency of two cycles). On the other hand, a first-order SecMult gadget implemented as described in [13] would carry out the multiplication in $\mathbb{F}_{2^8}$ with the help of the F2P8MUL macro provided in Listing 1, which has an execution time of 11 cycles. Four executions of F2P8MUL together with four eor instructions results in an overall execution time of 48 cycles. Hence, the two optimizations we propose in this section, which eliminate the special treatment of 0-operands and reduce the number of look-ups, make the first-order SecMult gadget three times faster then the best implementation in the literature.

In addition to SecMult, a masked inversion in $\mathbb{F}_{2^8}$ requires a few more gadgets, all of which perform linear operations and are, therefore, relatively easy to implement. The most important ones[6] are those for the masked squaring and 16th-power function, which we call SecSquare and SecPow16, respectively. Both have in common that the shares of the input are in the normal domain, while the shares of the output are in the order domain and can, thus, be directly fed into a subsequent SecMult gadget. The underlying squaring table shall contain the order of the squares instead of the actual squares, and this also applies to the 16th-power table. Since these operations are linear, we can simply perform

---

[6] An efficient implementation of the original exponentiation (Algorithm 3) also needs a gadget for the 4th-power function, but this is not the case for the exponentiation algorithm we introduce in the next section.

Algorithm 4. New SecExp254 computation

Input: Shares $x_0, x_1, \ldots, x_d$ satisfying $x = \bigoplus_{i=0}^{d} x_i$
Output: Shares $e_0, e_1, \ldots, e_d$ satisfying $e = \bigoplus_{i=0}^{d} e_i = x^{254}$

1: $(z_i)_{0 \leq i \leq d} \leftarrow \mathsf{SecSquare}((x_i)_{0 \leq i \leq d})$ $\qquad \triangleright \bigoplus_i z_i = x^2$
2: $(z_i)_{0 \leq i \leq d} \leftarrow \mathsf{RefreshMasks}((z_i)_{0 \leq i \leq d})$
3: $(y_i)_{0 \leq i \leq d} \leftarrow \mathsf{SecMult}((z_i)_{0 \leq i \leq d}, (x_i)_{0 \leq i \leq d})$ $\qquad \triangleright \bigoplus_i y_i = x^2 \cdot x = x^3$
4: $(z_i)_{0 \leq i \leq d} \leftarrow \mathsf{SecSquare}((y_i)_{0 \leq i \leq d})$ $\qquad \triangleright \bigoplus_i z_i = (x^3)^2 = x^6$
5: $(y_i)_{0 \leq i \leq d} \leftarrow \mathsf{SecMult}((z_i)_{0 \leq i \leq d}, (x_i)_{0 \leq i \leq d})$ $\qquad \triangleright \bigoplus_i y_i = x^6 \cdot x = x^7$
6: $(z_i)_{0 \leq i \leq d} \leftarrow \mathsf{SecSquare}((y_i)_{0 \leq i \leq d})$ $\qquad \triangleright \bigoplus_i z_i = (x^7)^2 = x^{14}$
7: $(y_i)_{0 \leq i \leq d} \leftarrow \mathsf{SecMult}((z_i)_{0 \leq i \leq d}, (x_i)_{0 \leq i \leq d})$ $\qquad \triangleright \bigoplus_i y_i = x^{14} \cdot x = x^{15}$
8: $(y_i)_{0 \leq i \leq d} \leftarrow \mathsf{SecPow16}((y_i)_{0 \leq i \leq d})$ $\qquad \triangleright \bigoplus_i y_i = (x^{15})^{16} = x^{240}$
9: $(e_i)_{0 \leq i \leq d} \leftarrow \mathsf{SecMult}((y_i)_{0 \leq i \leq d}, (z_i)_{0 \leq i \leq d})$ $\qquad \triangleright \bigoplus_i e_i = x^{240} \cdot x^{14} = x^{254}$

them on each share individually, i.e. the corresponding Assembly code consists of just two load instructions. A further gadget needed by the exponentiation is RefreshMasks, which only injects fresh randomness to the masked representation of a sensitive variable. Finally, we also implemented a couple of auxiliary gadgets for such functions like the conversion of operands between the normal domain and the order domain.

4   New Exponentiation-Based Inversion

The exponentiation $x \mapsto x^{254}$ is the core operation of our masking scheme. It is crucial to find the most efficient way to compute it using the $\mathbb{F}_{2^8}$ multiplication technique described in the previous section, and this requires to minimize the number of table look-ups by minimizing the number of conversions between the normal domain and the order domain. There are four main operations ("gadgets") used in a masked exponentiation: SecMult, SecSquare, SecPow16, as well as RefreshMasks. The ideal representation for these operations is as follows:

1. RefreshMasks: As state before, this operation injects fresh randomness into the shares. Since there is no trivial way to do it from the order domain, the inputs and the outputs should be in the normal domain.
2. SecMult: Since the first operations of our masked multiplication are normal multiplications in $\mathbb{F}_{2^8}$, the order domain is preferred for the inputs. On the other hand, the last step is somewhat similar to a mask refreshing and, as a consequence, the outputs are in the normal domain.
3. SecSquare and SecPow16: These are the most flexible operations. Since the tables are pre-computed and the look-ups are done share by share, there is no efficiency difference between the two representations. An implementer is free to chose them for inputs and outputs, but of course the look-up tables have to be generated accordingly.

In the original RP exponentiation (Algorithm 3), the order of operations is such that, when using the gadgets we propose, some extra conversions between representations would be needed. For example, at the end of Algorithm 3, the two SecMult operations are sequential, which means a conversion from normal to order domain is required in between. To improve the efficiency, we propose a novel exponentiation technique in Algorithm 4. In this algorithm, we can see that the four SecMult operations are separated by SecSquare (resp. SecPow16) operations, so that, when the squaring and 16th-power table were generated to output the result in the order domain, no extra conversions are needed except after the RefreshMasks. Furthermore, one less RefreshMasks operation has to be carried out compared to the original RP-exponentiation, which saves one byte of randomness and a number of XORs. However, we point out that this single remaining RefreshMasks is strictly required, which will become more clear from the proof below. In short, as explained in Sect. 2, the SecMult gadget requires mutually independent input shares to perform the multiplication securely. The first SecMult gadget (line 3 of Algorithm 4) takes (x_i) and (x_i^2) as input, where clearly (x_i^2) depends on (x_i). Therefore, (x_i^2) needs to be refreshed, and once RefreshMasks has been applied on the shares $(z_i)_{0 \leq i \leq d}$, any pair of inputs given to the remaining three SecMult gadgets is independent. With this in mind, we can now prove the t-SNI property of our new exponentiation algorithm.

Lemma 3 (t-**SNI of** SecExp254)**.** *Let $(x_i)_{1 \leq i \leq t+1}$ be the input shares of the x^{254} operation, and let $(y_i)_{1 \leq i \leq t+1}$ be the output shares. For any set of t_1 intermediate variables and any subset $\mathcal{O}$ of output shares such that $t_1 + |\mathcal{O}| \leq t$, there exists a subset $\mathcal{I}$ of indexes with $|\mathcal{I}| \leq t_1$, such that those t_1 intermediate variables as well as the output shares $y_{|\mathcal{O}}$ can be perfectly simulated from $x_{|\mathcal{I}}$.*

Proof The proof makes use of the SNI property of SecMult (see Lemma 1) and RefreshMasks (see Lemma 2), as well as the linearity of the squaring operation (note that an exponentiation by 16 consists of four subsequent squarings).

Let $\mathcal{I}$ be the set of indices corresponding to the internal variables observed by the adversary, so that $|\mathcal{I}| \leq t_1$. Let $\mathcal{O}$ be the set of indices corresponding to the output probes observed by the adversary, so that $t_1 + |\mathcal{O}| \leq t$. Consider an partition of indices $\mathcal{I} = \bigcup_{i=1}^{9} \mathcal{I}^i \leq t_1$ depending on the subgadget in which the observation occurs (see Fig. 1).

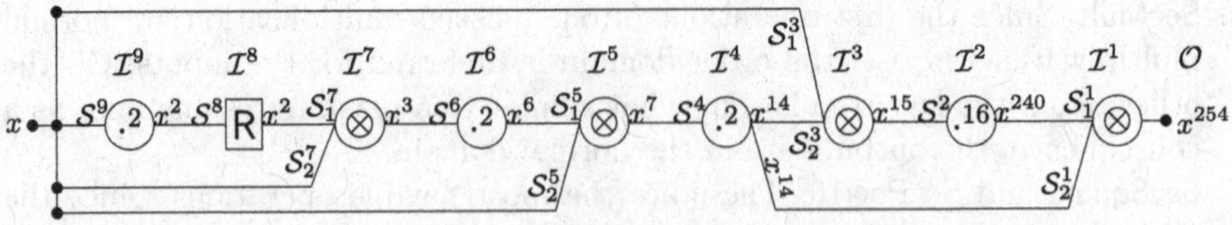

Fig. 1. x^{254} as a composition of squaring $(\cdot^2)$, multiplication $(\otimes)$ and mask refreshing (R) gadgets.

Thus, to prove the SNI property of Algorithm 4, by Definition 2 there has to exist a subset of input indices $\mathcal{S}$ with $|\mathcal{S}| \leq t_1$, such that the t_1 intermediate variables along with the output variables corresponding to $\mathcal{O}$ can be perfectly simulated from $x_{|S}$. In the following, we will build such a set $\mathcal{S}$, starting from the output probes and continuing backwards subgadget by subgadget, until the input shares. Formally:

Gadget 1. By assumption, $|\mathcal{I}^1 \cup \mathcal{O}| \leq t_1 + |\mathcal{O}| \leq t$. This enables us to use the SNI property of the multiplication gadget $\otimes$. Namely, there exist due to Lemma 1 two sets of indices $\mathcal{S}_1^1$, $\mathcal{S}_2^1$ so that $|\mathcal{S}_1^1|, |\mathcal{S}_2^1| \leq |\mathcal{I}^1|$. In particular, it is possible to simulate the gadget probes along with the observed variables $\mathcal{O}$ from its input shares corresponding to indices in $\mathcal{S}_1^1$ and $\mathcal{S}_2^1$.

Gadget 2. As a composition of squarings over a characteristic-2 field, the gadget $\cdot^{16}$ is linear. Hence, by its NI property, there exists a set of indices $\mathcal{S}^2$ such that $|\mathcal{S}^2| \leq |\mathcal{I}^2| + |\mathcal{S}_1^1|$ and every probe in the last two gadgets (included the output probes) can be simulated from its input shares corresponding to $\mathcal{S}^2$ and $\mathcal{S}_2^1$. Note that, $|\mathcal{S}^2| \leq |\mathcal{I}^2| + |\mathcal{S}_1^1| \leq |\mathcal{I}^2| + |\mathcal{I}^1|$.

Gadget 3. From the previous steps, $|\mathcal{I}^3 \cup \mathcal{S}^2| \leq |\mathcal{I}^3| + |\mathcal{I}^2| + |\mathcal{I}^1| \leq t$. Thus we can apply the SNI property of gadget $\otimes$ and claim that there exist two sets of indices $\mathcal{S}_1^3$, $\mathcal{S}_2^3$ so that $|\mathcal{S}_1^3|, |\mathcal{S}_2^3| \leq |\mathcal{I}^3|$ and every probe in previous gadgets can be simulated from the input shares corresponding to $\mathcal{S}_1^3$, $\mathcal{S}_2^3$, and $\mathcal{S}_2^1$.

Gadget 4. By previous steps, $|\mathcal{I}^4 \cup \mathcal{S}_2^3 \cup \mathcal{S}_2^1| \leq |\mathcal{I}^4| + |\mathcal{I}^3| + |\mathcal{I}^1| \leq t$. This together with the NI property of $\cdot^2$, allows the existence of a set of indices $\mathcal{S}^4$ such that $|\mathcal{S}^4| \leq |\mathcal{I}^4| + |\mathcal{I}^3| + |\mathcal{I}^1|$ and every probe in the previous gadgets can be simulated from the input shares corresponding to $\mathcal{S}^4$, and $\mathcal{S}_1^3$.

Gadget 5. From the above, $|\mathcal{I}^5 \cup \mathcal{S}^4| \leq |\mathcal{I}^5| + |\mathcal{I}^4| + |\mathcal{I}^3| + |\mathcal{I}^1| \leq t$. This means we can use the SNI property of $\otimes$ and define two sets of indices, $\mathcal{S}_1^5$ and $\mathcal{S}_2^5$, such that $|\mathcal{S}_1^5|, |\mathcal{S}_2^5| \leq |\mathcal{I}^5|$ and every probe in the previous gadgets can be simulated from the input shares corresponding to $\mathcal{S}_1^5$, $\mathcal{S}_2^5$ and $\mathcal{S}_1^3$.

Gadget 6. From the previous steps, $|\mathcal{I}^6 \cup \mathcal{S}_1^5| \leq |\mathcal{I}^6| + |\mathcal{I}^5| \leq t$. Hence, due to its NI property, there exists a set of indices $\mathcal{S}^6$ such that $|\mathcal{S}^6| \leq |\mathcal{I}^6| + |\mathcal{I}^5|$ and every probe in the previous gadgets can be simulated from the input shares corresponding to $\mathcal{S}^6$, $\mathcal{S}_2^5$ and $\mathcal{S}_1^3$.

Gadget 7. From the previous steps, $|\mathcal{I}^7 \cup \mathcal{S}^6| \leq |\mathcal{I}^7| + |\mathcal{I}^6| + |\mathcal{I}^5| \leq t$. So we can apply the SNI property of $\otimes$ and claim there exist two sets of indices $\mathcal{S}_1^7$, $\mathcal{S}_2^7$ such that $|\mathcal{S}_1^7|, |\mathcal{S}_2^7| \leq |\mathcal{I}^7|$ and every probe in the previous gadgets can be simulated from the input shares corresponding to $\mathcal{S}_1^7$, $\mathcal{S}_2^7$, $\mathcal{S}_2^5$ and $\mathcal{S}_1^3$.

Gadget 8. From the previous steps, $|\mathcal{I}^8 \cup \mathcal{S}_1^7| \leq |\mathcal{I}^8| + |\mathcal{I}^7| \leq t$. By Lemma 2 also the mask refreshing R is proven to be SNI. Thus there exists a set of indices $\mathcal{S}^8$ so that $|\mathcal{S}^8| \leq |\mathcal{I}^8|$ and every probe in the previous gadgets can be simulated from the its input shares corresponding to $\mathcal{S}^8$, $\mathcal{S}_2^7$, $\mathcal{S}_2^5$ and $\mathcal{S}_1^3$.

Gadget 9. Lastly, $|\mathcal{I}^9 \cup \mathcal{S}^8| \leq |\mathcal{I}^9| + |\mathcal{I}^8| \leq t$. Hence we can again apply the NI property of $\cdot^2$ and define a set of indices $\mathcal{S}^9$ such that $|\mathcal{S}^9| \leq |\mathcal{I}^9| + |\mathcal{I}^8|$ and every probe in the previous gadgets can be simulated from the its input shares corresponding to $\mathcal{S}^9$, $\mathcal{S}_2^7$, $\mathcal{S}_2^5$ and $\mathcal{S}_1^3$.

In conclusion, due to all above statements, we can guarantee the existence of a simulator for each gadget. Furthermore, let $S = S^9 \cup S_2^7 \cup S_2^5 \cup S_1^3$, and we can compose the simulators to perfectly simulate the computation of x^{254} from $x_{|S}$. To conclude the proof we show that the cardinality of such a set of indices has to be less or equal to the number of intermediate probed variables t_1. From Step 3, Step 5, Step 7, Step 9, and $\mathcal{I} = \bigcup_{i=1}^9 \mathcal{I}^i \leq t_1$ we have:

$$|S| \leq |S^9| + |S_2^7| + |S_2^5| + |S_1^3| \leq |\mathcal{I}^9| + |\mathcal{I}^8| + |\mathcal{I}^7| + |\mathcal{I}^5| + |\mathcal{I}^3| \leq t_1 \quad (2)$$

Remark 1 One could suggest to use the PINI security notion [6] for the above exponentiation gadget with the argument that it is normally seen stronger than its SNI counterpart. While this is usually true, when the comparison is limited to 1-input gadgets, the SNI definition seems stronger. In a 2-input gadget, the SNI allows two different sets $\mathcal{I}$ and $\mathcal{J}$ with cardinality less than or equal to the number of internal probes t_1 to simulate every corrupted wire from the shares of the first input, whose indices are in $\mathcal{I}$, and the shares of the second input, whose indexes are in $\mathcal{J}$. On the other hand, the PINI security requires a unique set of indexes $\mathcal{I}$ with a cardinality less or equal to the total number of probes t, where every probe needs to be simulated from the input shares of both the inputs, whose indexes are in $\mathcal{I}$. Consequently, the intrinsic isolation guaranteed by a unique set is what makes PINI security stronger. But when applied to a 1-input gadget, the two definitions are almost identical since they both require a unique set for a unique input. Actually, SNI requires even fewer input shares (since $t_1 \leq t$ by definition), which allowed us to claim that, in this specific case of 1-input gadget, SNI implies PINI: the simulator S from above can be used to satisfy the PINI definition too. To conclude, Algorithm 4 is indeed PINI, but we decided to stay with SNI since, in our case, it gives better guarantees.

5   Masked AES

In this section we provide a brief description of how we implemented the first-order masked round transformations of AES128 and report the execution time of each transformation on a Cortex-M3 microcontroller. Thereafter, we compare our results with that of a couple of other masked AES implementations for the 32-bit ARM platform.

5.1   Round Transformations

We implemented all four round transformations, including the simple ones like ShiftRows and AddRoundKey, completely in Assembly language to exclude the theoretical possibility that the compiler introduces any kind of leakage. Unlike some other implementations reported in the literature, we did not solely strive for high speed, but tried to achieve a reasonable trade-off between execution time and (binary) code size. Therefore, we refrained from certain optimization techniques like full loop-unrolling, which often produce only modest speed-ups at the expense of an unreasonably large increase of code size.

SubBytes is, from an implementer's perspective, the most challenging round transformation because it has a massive impact on the overall execution time and is particularly leakage-sensitive because of its non-linear nature [18]. The SubBytes transformation consists of 16 S-box operations, which, in turn, are composed of an inversion in $\mathbb{F}_{2^8}$ along with an affine operation [8]. Our masked inversion is based on the exponentiation technique we presented in Sect. 4 and utilizes (variants of) the optimized low-level gadgets introduced in Sect. 3. The proposed exponentiation has three advantages over the original exponentiation method of Rivain and Prouff [20]. First, it allows us to better exploit our optimization to reduce the overall number of table look-ups and is, therefore, much faster. Second, it needs only one RefreshMasks instead of two and so it reduces the number of random bytes from six to five. Third, as proven in the previous section, it achieves SNI, which is a stronger notion of probing security that is not met by the original (i.e. flawed [7]) exponentiation algorithm from [20].

Although our implementation of the masked exponentiation closely follows Algorithm 4, there are two aspects that deserve more discussion. First, for the sake of brevity, Algorithm 4 provides only a high-level description of our exponentiation using the four main gadgets (i.e. SecMult, SecSquare, SecPow16, and RefreshMasks), but omits gadgets for "auxiliary" operations like the conversion of operands between the normal domain and order domain. For example, at the very beginning of the exponentiation, the shares x_i of the input operand x are converted from the normal domain to the order domain, such that they can be fed into the SecMult gadgets at line 3, 5, and 7. In addition, as explained in the previous section, the RefreshMasks gadget operates in the normal domain; its output masks have to be converted from the normal domain to order domain because they serve as input to the first SecMult gadget at line 3. The input to RefreshMasks is the output of the first SecSquare gadget at line 1, which would normally have outputs in the order domain. However, instead of converting the SecSquare output from the order domain to the normal domain, we decided to implement a second SecSquare gadget that operates completely in the normal domain, i.e. both its inputs and outputs are elements of $\mathbb{F}_{2^8}$.

A second aspect we have not discussed yet concerns the actual implementation of the different gadgets. For reasons of efficiency, the SecMult gadget is "duplicated" in the sense that is actually performs two masked multiplications (in an interleaved fashion) instead of just one, which means the SECMULT macro provided in Listing 2 consists of twice as many add, ldrb, and eor instructions (i.e. eight instead of four). The reason why this may be beneficial can be found in the pipeline structure of ARM Cortex microcontrollers. Namely, on certain Cortex-M3 models, neighboring load and store single instructions can pipeline their address and data phases, which enables them to execute in a single cycle instead of two. Thus, it is potentially beneficial to have sequence of eight ldrb instructions instead of four. Besides SecMult, we also duplicated the SecSquare and SecPow16 gadget, which means they always perform two masked squarings and two masked 16th-power operations. This duplication propagates up to the exponentiation, which is computed "pair-wise" on two masked state-bytes.

Table 1. Comparison of masked AES implementations on ARM Cortex-M3/M4. The code size of the implementation in [21] is given for full AES-CTR and not single AES block encryption. There exist two versions of the fixsliced implementation described in [1] with similar performance, but Beckers et al. [3] do not report which one they used for their benchmarking and leakage assessment.

Reference	Implement. details	Exec. time (cycles)	Code size (kB)	Theoret. security	Practical security
Goudarzi [14]	RP-masked	49,329 (ARM)	4.8	Probing	Unknown
Schwabe [21]	Bitsliced	17,500 (M4)	39.9	Probing?	t-test (failed [3])
Gross [15]	Bitsliced	6,800 (M4)	25.2	Probing	t-test (failed [3])
Adomnicai [1]	Fixsliced	6,200 (M4)	22 or 4	Probing	t-test (failed [3])
ANSSI (small)	Affine	53,072 (M4)	5.5	Affine	Attacked ([5,19])
ANSSI (fast)	Affine	29,920 (M4)	25.0	Affine	Attacked ([5,19])
Our work	RP-masked	25,413 (M3)	3.2	SNI/PINI	t-test (5k traces)

Besides the $\mathbb{F}_{2^8}$-inversion, an S-box operation also involves an affine transformation, which is linear and can be performed share-by-share with a 256-byte look-up table. At the highest level, our implementation of a first-order masked SubBytes function consists of a simple loop that is iterated eight times and in each iteration it executes a pair of masked S-box operations (i.e. two inversions followed by two affine transformations). Each first-order masked S-box requires five bytes of randomness (four for the SecMult gadgets and one further for the RefreshMasks gadget), which amounts to 80 random bytes altogether for a full SubBytes. On our target device (a VL Discovery development board equipped with a STM32F100RB Cortex-M3 microcontroller clocked at 24 MHz [22]), the first-order masked SubBytes transformation has an (measured) execution time of 2299 clock cycles. This cycle count includes the loading of the random bytes from an array in RAM, but not the generation of the random bytes itself.

The other three round transformations are rather easy to implement since they consist solely of linear operations and can be executed in a share-by-share way. Our implementation of the MixColumns transformation is inspired by the approach of Bertoni et al. [4], which is very efficient on 32-bit platforms. The execution time (including full function-call overhead) of our first-order masked ShiftRows, MixColumns, and AddRoundKey transformation on Cortex-M3 is 37, 133, and 78 clock cycles, respectively, which is almost negligible in relation to SubBytes. A single round of masked AES executes in 2552 cycles, and the complete masked AES encryption has an execution time of 25413 cycles.

5.2   Results and Comparison

Table 1 compares our work with previous implementations of masked AES on the 32-bit ARM platform. It is difficult to make a direct comparison since, even if the microcontrollers are similar, the techniques used for the masking are quite different and may not offer similar security, neither in theory nor in practice.

Among all implementations presented in Table 1, the one of Goudarzi and Rivain [14] is the most relevant for comparison with our work as both use the RP masking method. The execution time of 49,329 cycles represents the value for "Standard AES (KHL)" in [14, Table 3] and [13, Table 16] for $d = 2$ (note that d in the paper of Goudarzi and Rivain is not the masking order but the number of shares!). This cycle count exceeds the 25,413 cycles of our masked AES by a factor of 1.94. Thus, we conclude that the optimizations described in this paper roughly halve the execution time of RP-masked AES on ARM. The bit-sliced (resp. fix-sliced) implementations introduced in [1,15,21] outperform our masked AES, but are larger in terms of code size and seem to succumb to first-order DPA (see below for details). Most execution times in Table 1 were measured by Beckers et al. [3] and may differ from the results reported by the implementation authors. The execution time of the bit-sliced AES of Schwabe and Stoffelen (17.5k cycles as per [3, Table 2]) is twice the cycle-count given on the second authors' GitHub[7] repository ($3439.5 + 5288.1 \approx 8.75$k cycles). This discrepancy can be explained by the fact that two data blocks are processed in parallel and the author reports the cycles *per block* on GitHub.

The code size of our masked AES is very small (≈ 3.2 kB), mainly because we did not apply size-increasing optimizations like loop unrolling. Not included in these 3.2 kB are the look-up tables, which have a size of 3328 bytes. We use six tables in total: A Log table (512 bytes), an AntiLog table (1024 bytes), two Squaring tables (one with outputs in normal domain for a subsequent Refresh-Masks, one with outputs in order domain for subsequent SecMult, each of size 512 bytes), a 16th-power table (512 bytes), and an Affine-transformation table (256 bytes). For comparison, a fast RP-masked AES like [14] needs 1792 bytes for six (different) tables: $256 + 512 + 256 + 256 + 256 + 256 = 1792$.

Regarding security, it should be noted that the implementations in [15] and [1] reduce the amount of randomness needed for masking to the bare minimum of two bits. The authors used a formal verification tool to assess the security in the probing model. However, such an aggressive approach naturally raises some concerns about the practical security of the masking scheme and does not fit in our theoretical framework. Furthermore, their approach does not generalize to higher orders. More importantly, all bitsliced/fixsliced maskings in Table 1 are actually leaking according to the recent evaluation in [3]. The bit-sliced masked AES from [21] leaks because of an unintended unmasking caused by accidental overwrites of a register (see [3, Sect. 2.4]), which means the implementation is actually flawed. Such flaws can be discovered with a t-test, whereby a few 1000 traces already suffice to spot leakage. We performed a t-test on our RP-masked AES implementation with 5000 traces that we measured on a ChipWhisperer-Lite board, see Appendix A. This t-test did not show any peaks exceeding the ± 4.5 threshold, which makes us relatively confident that our implementation is not flawed. However, even if a masking scheme is implemented correctly, it can still leak sensitive information through certain micro-architectural effects, see e.g. [9]. We will address this micro-architectural leakage in our future work.

---

[7] See https://github.com/Ko-/aes-armcortexm (accessed 2022-06-21).

6   Conclusions

The masking scheme of Rivain and Prouff, introduced some 12 years ago, has attracted a lot of interest in the cryptographic community because it stands on a solid theoretical foundation and supports arbitrary masking orders. Unfortunately, it turned out that these features come at the expense of relatively bad performance; the to-date best implementation of a first-order masked AES has an execution time of about 50k cycles on a 32-bit ARM microcontroller, which is prohibitively expensive for various classes of applications. We showed in this paper that the masking technique proposed by Rivain and Prouff can be made much faster, namely by a factor of almost two in comparison with the, to this date, best implementation in the literature. We achieved this speed-up by (i) an optimization of the $\mathbb{F}_{2^8}$-multiplication that allows one to overcome the special treatment of 0-operands (thereby reducing the execution time by 36.4%) and (ii) a new exponentiation technique that minimizes the overall number of table look-ups. Our masking scheme is not only almost twice as fast as the original Rivain-Prouff method, it also satisfies Strong Non-Interference (SNI), which is a stronger notion of probing security. But since we do not want to rely only on theoretical security properties like SNI, we also performed a test-vector-based leakage evaluation as "sanity test" to confirm that our AES implementation is not flawed. As part of our future work, we plan to develop higher-order masked implementations for the 32-bit RISC-V platform with special countermeasures against micro-architectural leakages.

Acknowledgements. The second and fourth author were supported by the European Research Council (ERC) under the European Union's Horizon 2020 research and innovation programme (grant agreement No. 787390). The third author was supported, in part, by the Fonds National de la Recherche (FNR) Luxembourg under CORE grant C19/IS/13641232.

A   Practical Leakage Assessment

In order to assess the masked AES implementation described in this paper, we utilized the general Test Vector Leakage Assessment (TVLA) test. The general test compares power measurements on a device with fixed key, across datasets with random and fixed plaintext. To evaluate the leakage, the two datasets are compared for significant statistical differences using Welch's t-test.

Welch's t-test to compare the measured traces from two subsets F and R is done by first computing X_F and X_R, which is the average of all traces in each subset. In the next step, one has to compute S_F and S_R, which is the sample standard deviation of all traces in each subset. Given that each measured trace is a vector of measured values across time, the average and standard deviation are also vectors over time. The t-value is then obtained point-wise as

$$\frac{X_F - X_R}{\sqrt{\frac{S_F^2}{N_F} + \frac{S_R^2}{N_R}}} \tag{3}$$

The test returns a threshold, expressed through standard deviations, that corresponds to the confidence that the difference between the sets is not a random event. In the context of leakage evaluation, it is usual to set the confidence value to 4.5 standard deviations (equivalent to 99.999% confidence). Furthermore, the test should be run twice on independent datasets, and if a point in both tests exceeds $\pm 4.5\sigma$, the device is leaking data-related information.

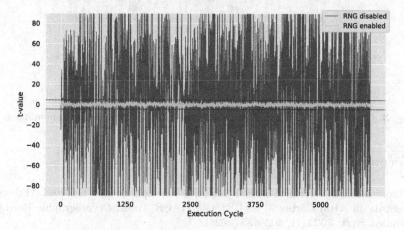

Fig. 2. Graph showing the t-test result of our masked implementation. The normal execution is in light color, showing that the results stay within a $\pm 4.5\sigma$ boundary. In a darker color is the same experiment but with one of the masks set to zero.

Our assessment platform is the ChipWhisperer-Lite 32-Bit, manufactured by NewAE Technology, Inc. This is a single-board solution, fully open source side-channel (i.e. power analysis and glitching) platform. The board houses an STM32F303 32-bit ARM Cortex-M4[8] microcontroller and a 10-bit OpenADC scope. The results of the TVLA can be found in Fig. 2. Our implementation stays within the set boundaries of $\pm 4.5\sigma$, but shows massive leakage when one of the masks is set to zero. Note that the y-axis is cropped and that, at some points, the zero-mask reaches a confidence in excess of 100σ. Figure 3 contains a more detailed plot of t values of the masked AES. This experiment consists of 5000 pairs of power traces and mainly serves as sanity check to confirm the absence of implementation flaws that can cause sensitive-data leakage. Beckers et al. [3] recently demonstrated that a t-test with a few 1000 traces suffices to reveal such flaws.

---

[8] As mentioned in Sect. 5, our actual target device is an STM32F100RB Cortex-M3 microcontroller, but we use a Cortex-M4 for the leakage assessment because we do not have access to an SCA evaluation board with a Cortex-M3.

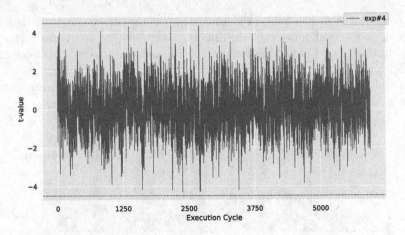

Fig. 3. Graph showing the detailed t-test result of the masked implementation.

References

1. Adomnicai, A., Peyrin, T.: Fixslicing AES-like ciphers: new bitsliced AES speed records on ARM-Cortex M and RISC-V. IACR Trans. Cryptographic Hardware Embed. Syst. **2021**(1), 402–425 (2021)
2. Barthe, G., et al.: Strong non-interference and type-directed higher-order masking. In: Weippl, E.R., Katzenbeisser, S., Kruegel, C., Myers, A.C., Halevi, S. (eds.). Proceedings of the 2016 ACM Conference on Computer and Communications Security (CCS 2016), pp. 116–129. ACM Press (2016)
3. Beckers, A., Wouters, L., Gierlichs, B., Preneel, B., Verbauwhede, I.: Provable secure software masking in the real-world. In: Balasch, J., O'Flynn, C. (eds.) COSADE 2022. LNCS, vol. 13211, pp. 215–235. Springer, Cham (2022). https://doi.org/10.1007/978-3-030-99766-3_10
4. Bertoni, G., Breveglieri, L., Fragneto, P., Macchetti, M., Marchesin, S.: Efficient software implementation of AES on 32-bit platforms. In: Kaliski, B.S., Koç, K., Paar, C. (eds.) CHES 2002. LNCS, vol. 2523, pp. 159–171. Springer, Heidelberg (2003). https://doi.org/10.1007/3-540-36400-5_13
5. Bronchain, O., Standaert, F.-X.: Side-channel countermeasures' dissection and the limits of closed source security evaluations. IACR Trans. Cryptographic Hardware Embed. Syst. **2020**(2), 1–25 (2020)
6. Cassiers, G., Standaert, F.-X.: Trivially and efficiently composing masked gadgets with probe isolating non-interference. IEEE Trans. Inf. Forensics Secur. **15**, 2542–2555 (2020)
7. Coron, J.-S., Prouff, E., Rivain, M., Roche, T.: Higher-order side channel security and mask refreshing. In: Moriai, S. (ed.) FSE 2013. LNCS, vol. 8424, pp. 410–424. Springer, Heidelberg (2014). https://doi.org/10.1007/978-3-662-43933-3_21
8. Daemen, J., Rijmen, V.: The Design of Rijndael: AES - The Advanced Encryption Standard. Springer, Verlag (2002)
9. De Meyer, L., De Mulder, E., Tunstall, M.: On the effect of the (micro)architecture on the development of side-channel resistant software. Cryptology ePrint Archive, Report 2020/1297 (2020). https://ia.cr/2020/1297

10. Faust, S., Paglialonga, C., Schneider, T.: Amortizing randomness complexity in private circuits. In: Takagi, T., Peyrin, T. (eds.) ASIACRYPT 2017. LNCS, vol. 10624, pp. 781–810. Springer, Cham (2017). https://doi.org/10.1007/978-3-319-70694-8_27

11. Fumaroli, G., Martinelli, A., Prouff, E., Rivain, M.: Affine masking against higher-order side channel analysis. In: Biryukov, A., Gong, G., Stinson, D.R. (eds.) SAC 2010. LNCS, vol. 6544, pp. 262–280. Springer, Heidelberg (2011). https://doi.org/10.1007/978-3-642-19574-7_18

12. Gladman, B.R.: A specification for Rijndael, the AES algorithm (2007). Technical report, available for download at https://ccgi.gladman.plus.com/oldsite/cryptography_technology/rijndael/aes.spec.v316.pdf

13. Goudarzi, D., Rivain, M.: How fast can higher-order masking be in software? Cryptology ePrint Archive, Report 2016/264 (2016). https://ia.cr/2016/264

14. Goudarzi, D., Rivain, M.: How fast can higher-order masking be in software? In: Coron, J.-S., Nielsen, J.B. (eds.) EUROCRYPT 2017. LNCS, vol. 10210, pp. 567–597. Springer, Cham (2017). https://doi.org/10.1007/978-3-319-56620-7_20

15. Gross, H., Stoffelen, K., De Meyer, L., Krenn, M., Mangard, S.: First-order masking with only two random bits. In: Bilgin, B., Petkova-Nikova, S., Rijmen, V. (eds). Proceedings of the 1st ACM Workshop on Theory of Implementation Security (TIS@CCS 2019), pp. 10–23. ACM Press (2019)

16. Ishai, Y., Sahai, A., Wagner, D.: Private circuits: securing hardware against probing attacks. In: Boneh, D. (ed.) CRYPTO 2003. LNCS, vol. 2729, pp. 463–481. Springer, Heidelberg (2003). https://doi.org/10.1007/978-3-540-45146-4_27

17. Journault, A., Standaert, F.-X.: Very high order masking: efficient implementation and security evaluation. In: Fischer, W., Homma, N. (eds.) CHES 2017. LNCS, vol. 10529, pp. 623–643. Springer, Cham (2017). https://doi.org/10.1007/978-3-319-66787-4_30

18. Kocher, P., Jaffe, J., Jun, B., Rohatgi, P.: Introduction to differential power analysis. J. Cryptogr. Eng. 1(1), 5–27 (2011). https://doi.org/10.1007/s13389-011-0006-y

19. Masure, L., Strullu, R.: Side channel analysis against the ANSSI's protected AES implementation on ARM. Cryptology ePrint Archive, Report 2021/592 (2021). https://ia.cr/2021/592

20. Rivain, M., Prouff, E.: Provably secure higher-order masking of AES. In: Mangard, S., Standaert, F.-X. (eds.) CHES 2010. LNCS, vol. 6225, pp. 413–427. Springer, Heidelberg (2010). https://doi.org/10.1007/978-3-642-15031-9_28

21. Schwabe, P., Stoffelen, K.: All the AES you need on cortex-M3 and M4. In: Avanzi, R., Heys, H. (eds.) SAC 2016. LNCS, vol. 10532, pp. 180–194. Springer, Cham (2017). https://doi.org/10.1007/978-3-319-69453-5_10

22. STMicroelectronics. STM32F100x4 STM32F100x6 STM32F100x8 STM32F100xB (2016). Data sheet, available for download at https://www.st.com/resource/en/datasheet/stm32f100cb.pdf

Self-timed Masking: Implementing Masked S-Boxes Without Registers

Mateus Simões[1,2(✉)], Lilian Bossuet[1], Nicolas Bruneau[2], Vincent Grosso[1], Patrick Haddad[2], and Thomas Sarno[2]

[1] Laboratoire Hubert Curien, CNRS UMR5516, Saint-Étienne, France
mateus.simoes@univ-st-etienne.fr
[2] STMicroelectronics, Rousset, France

Abstract. Masking is one of the most used side-channel protection techniques. However, a secure masking scheme requires additional implementation costs, e.g. random number, and transistor count. Furthermore, glitches and early evaluation can temporally weaken a masked implementation in hardware, creating a potential source of exploitable leakages. Registers are generally used to mitigate these threats, hence increasing the implementation's area and latency.

In this work, we show how to design glitch-free masking without registers with the help of the dual-rail encoding and asynchronous logic. This methodology is used to implement low-latency masking with arbitrary protection order. Finally, we present a side-channel evaluation of our first and second order masked AES implementations.

Keywords: Side-channel analysis · Masking · Asynchronous circuits

1 Introduction

Different techniques exist to counter side-channel attacks; one of the most studied is the Boolean masking [7,11,14], which splits a sensitive variable into several shares. In this manner, a secret x is d^{th} order masked with $d+1$ shares as shown in Eq. (1), with $(x_0, x_1, \ldots, x_{d-1})$ the random shares and x_d the masked value. The $\oplus$ symbol denotes the XOR operation.

$$x_d = x_0 \oplus x_1 \oplus \ldots \oplus x_{d-1} \oplus x \qquad (1)$$

Thereupon, instead of manipulating the plain data, the circuit performs computation on the shares. This results in a more complex relationship between the side-channel leakages and the sensitive data. At the appropriate moment, the shares can be recombined to uncover the secret data, that is, $x = \bigoplus_{i=0}^{d} x_i$. Note that security comes at the cost of higher implementation complexity, raising the transistor count.

The circuit needs to be transformed to perform the desired computation of the plain data while manipulating the shares. Securely masking a linear function is trivial, since each input share can be manipulated independently and in

© The Author(s), under exclusive license to Springer Nature Switzerland AG 2023
I. Buhan and T. Schneider (Eds.): CARDIS 2022, LNCS 13820, pp. 146–164, 2023.
https://doi.org/10.1007/978-3-031-25319-5_8

parallel. For instance, let $z = f(x, y)$ be a linear Boolean operation, its dth order masking can be expressed as shown in the Eq. (2).

$$z = \bigoplus_{i=0}^{d} z_i = \bigoplus_{i=0}^{d} f_i(x_i, y_i) = f\left(\bigoplus_{i=0}^{d} x_i, \bigoplus_{i=0}^{d} y_i\right) = f(x, y) \qquad (2)$$

On the other hand, the masking of a non-linear function, such as inversion in $\mathbb{F}_{2^n}$, manipulates the sharing in such a way that its intermediate terms require the recombination of several shares of a variable. Moreover, the sharing recombination may be the source of exploitable side-channel leakages, rendering masking of the non-linear operations a critical task for security engineers. In this context, techniques such as threshold implementations (TI) [24] were proposed.

TI limits the share recombination leakages. In this manner, the occurrence of glitches is an important factor to take into account when implementing the masking scheme in hardware. In fact, a glitchy function has an unexpected behavior that may be correlated to the unshared variable [18] — or more broadly to several shares. To guarantee the security of non-linear functions in the presence of glitches, register barriers can be employed to cease the spurious signal propagations [27], thus increasing the overall latency of the masked function.

Theoretical analysis can be used to strengthen confidence in the protection brought by the masked implementation. To evaluate the security of a masking scheme, several methods based on the probing model of Ishai et al. [14] exist. In this security evaluation method, the adversary can place up to d probes on different wires of the circuit in order to obtain their instantaneous logic level, providing clues about a potential dependence between the unshared value and the internal signal states. This model was enhanced to take into account physical defaults such as glitches [9] and composability [2].

Satisfying security in those security models requires additional overhead such as register layers, fresh random bits and higher silicon area. Therefore, reducing the masking costs is a pertinent branch of research in side-channel countermeasures. In this context, this work aims at reducing the number of clock cycles needed to compute masked functions. For that, we present a self-timed masking implementation built upon the Muller c-element [22] latches. We show how to replace registers with those latches, assuring data synchronization among different combinatorial layers. Furthermore, we present our locally asynchronous globally synchronous (LAGS) AES design. Finally, we evaluate the side-channel leakages based on experimental measurements up to the second-order protection.

2   Background

The use of asynchronous methodologies and dual-rail logic to implement low-latency masking was first introduced by Moradi and Schneider in [21]. They designed fully unrolled first-order threshold implementations of PRINCE and Midori based on WDDL gates [30]. Later, Sasdrich et al. [28] used the LUT-based Masked Dual-Rail with Pre-charge Logic (LMDPL) [17] masking scheme to implement a low-latency AES, which is limited to first-order security. More recently, Nagpal et al. [23] presented a low-latency domain-oriented implementation [13] also built upon WDDL gates, but employing Muller c-elements as synchronization modules, whose results have shown to be higher-order secure.

Similarly to these works, we aim at the study of low-latency masking, using dual-rail encoding with pre-charge logic and Muller c-elements to implement masked S-boxes with arbitrary protection order. Relying on the domain-oriented masking (DOM) [13], we focus on presenting and evaluating a generic methodology to replace registers with self-timed latches to design effective single-cycle masked functions.

2.1 Notations

We denote binary random variables in $\mathbb{F}_2$ with lower-case letters, e.g. x. A random variable x is Boolean masked with $d+1$ shares x_i, whose sharing is denoted with calligraphic fonts — e.g., $\mathcal{S} = (x_0, x_1, \ldots, x_d)$ — in such a manner that $x = \bigoplus_{i=0}^{d} x_i$.

We use typewriter fonts to denote binary random variables x, vectors X and signals encoded in the dual-rail protocol with a pair of wires (x.t, x.f). The wire x.t is used for signalling x.t $= x$ while x.f signalizes the complement x.f $= \overline{x}$. A dual-rail token of a variable x is then referred as $\overset{*}{\mathtt{x}} = (\mathtt{x.t}, \mathtt{x.f}) = (x, \overline{x})$.

2.2 The Domain-Oriented Masking

This work relies on the domain-oriented masking (DOM) [13], a known secure arbitrary order masking scheme. Since their gadget has already been formally verified in the original paper, we do not present theoretical proofs of security in this work. In order to satisfy d-glitch-extended probing security [9], their gadget is divided into two register-isolated steps, which we identify, in this work, as processing and compression.

Let us take the 2-share DOM-*indep* gadget $\mathcal{Z} = \mathcal{A} \wedge \mathcal{B}$ with $\mathcal{A} = (\mathtt{a}_0, \mathtt{a}_1)$ and $\mathcal{B} = (\mathtt{b}_0, \mathtt{b}_1)$ the input shares and $\mathcal{Z} = (\mathtt{z}_0, \mathtt{z}_1)$ the output sharing. In short, assuming that the input sharings are uniform, we want to find a secure way to compute $(\mathtt{z}_0 \oplus \mathtt{z}_1) = (\mathtt{a}_0 \oplus \mathtt{a}_1) \wedge (\mathtt{b}_0 \oplus \mathtt{b}_1)$. Hence, the process step computes the product terms $\mathtt{a}_0\mathtt{b}_0$, $\mathtt{a}_0\mathtt{b}_1$, $\mathtt{a}_1\mathtt{b}_0$, $\mathtt{a}_1\mathtt{b}_1$ and adds a fresh random share r to the cross-domain ones, that is, $\mathtt{a}_0\mathtt{b}_1$ and $\mathtt{a}_1\mathtt{b}_0$. Then, to assure non-completeness, registers ($\longrightarrow$) store the resulting shares $(\mathtt{x}_0, \mathtt{x}_1, \mathtt{x}_2, \mathtt{x}_3)$, as we can see in the Eq. (3).

$$
\begin{aligned}
f_0(\mathtt{a}_0, \mathtt{b}_0) &= \mathtt{a}_0\mathtt{b}_0 & \longrightarrow \mathtt{x}_0 \\
f_1(\mathtt{a}_0, \mathtt{b}_1) &= \mathtt{a}_0\mathtt{b}_1 \oplus r & \longrightarrow \mathtt{x}_1 \\
f_2(\mathtt{a}_1, \mathtt{b}_0) &= \mathtt{a}_1\mathtt{b}_0 \oplus r & \longrightarrow \mathtt{x}_2 \\
f_3(\mathtt{a}_1, \mathtt{b}_1) &= \mathtt{a}_1\mathtt{b}_1 & \longrightarrow \mathtt{x}_3
\end{aligned}
\tag{3}
$$

The processing step produces four shares. To reduce the number of output shares, there exists the compression step, as shown in the Eq. (4). Thanks to the register barrier between both steps and the fresh randomness, 1-glitch-extended security is satisfied.

$$z_0 = x_0 \oplus x_1$$
$$z_1 = x_2 \oplus x_3$$
(4)

Based on the domain-oriented scheme, Gross et al. proposed the first generic low-latency masking (GLM) in [12]. In their work, they skip the compression step after the non-linear operations, eliminating the registers after the shared processing. However, the number of shares grows quadratically after each masked multiplication. In consequence, the area and randomness costs increase substantially. In our work, we maintain the compression step and use a dual-rail synchronization element to obtain a generic low-latency masking.

2.3   The Dual-Rail Encoding

The dual-rail protocol encodes a bit using two signal wires: a wire x.t carries the logic value of a variable x while a second wire x.f transports its complement [8]. In this configuration, a valid token is obtained when one, and only one, signal wire is active (i.e. in a high logic state), although the null token is encoded when both wires are deactivated, that is, $\varnothing = (0,0)$. The encoding $(1,1)$ is never used, and the behavior of our design after the injection of this invalid token is out of scope of this work. Table 1 summarizes the dual-rail encoding.

Table 1. The dual-rail encoding.

Data	Token	
	x.t	x.f
null	0	0
$x = 0$	0	1
$x = 1$	1	0
not used	1	1

Moradi and Schneider presented the first work that borrowed asynchronous dual-rail techniques with the purpose of implementing low-latency masking [21]. Different from our choice of design, they designed a fully unrolled threshold implementation of two lightweight block-ciphers PRINCE and Midori using WDDL cells.

In contrast, we opted for an LAGS design, creating a single-cycle S-box within a synchronous AES architecture. The dual-rail encoding was also present in the implementation proposed by Sasdrich et al. [28] to create a first-order secure low-latency AES based on the LMDPL masking scheme [17]. Similar to the LMDPL, we employ the pre-charge/evaluation logic with monotonic functions to obtain a glitch-free circuit [25]. To eliminate the glitches, only regular AND and

OR gates are used to construct our dual-rail functions, due to their monotonic behavior [15].

We refer to Eqs. (5) and (6) for the AND and XOR functions, respectively, used in our work. We use the DPL_noEE AND gate [4], shown in Eq. (5), instead of the WDDL AND [30] to avoid the early propagation in the evaluation phase [16,19].

$$z = a \wedge b \Longleftrightarrow z.t = a.t \wedge b.t$$
$$z.f = (a.t \wedge b.f) \vee (a.f \wedge b.t) \vee (a.f \wedge b.f) \tag{5}$$

$$z = a \oplus b \Longleftrightarrow z.t = (a.t \wedge b.f) \vee (a.f \wedge b.t)$$
$$z.f = (a.f \wedge b.f) \vee (a.t \wedge b.t) \tag{6}$$

Encoded as dual-rail tokens, the information in the communication channel carries the data itself and the validity signal. For instance, the output token of a dual-rail logic gate is valid when $z.f \vee z.t = 1$, supposing $z.f \neq z.t$ to avoid the illegal case. Thus, the validity signal can be used to control the flow of tokens. Based on this idea, the data synchronization is managed by the tokens, eliminating the need of register layers, as we will explain in the next subsection.

2.4   Data Synchronization with the Muller C-Elements

Registers are important components in hardware masking due to their role in synchronizing the boundaries of different combinatorial blocks [9,27]. However, although limiting the combinatorial data path, registers increase the overall latency by requiring additional clock cycles to process the whole circuit.

In this work, we use an alternative state-holding element to obtain single-cycle S-box implementations. The state-holding module used in this work is built upon the Muller c-element [22], whose symbol is shown in Fig. 1 alongside with a gate-level implementation and a summary of its logical behavior. The Muller c-element operates as a Boolean function $f(a, b) = a \circ b$ that outputs 0 when all inputs have a low logic level, and when all inputs have a high logic level it outputs 1. Otherwise, the Muller c-element maintains its current steady state if at least one input is different from the others.

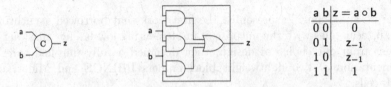

a b	$z = a \circ b$
0 0	0
0 1	z_{-1}
1 0	z_{-1}
1 1	1

Fig. 1. A Muller c-element symbol (left), a gate-level design (middle) and its truth table (right).

Recently, Nagpal et al. [23] presented a domain-oriented gadget built upon Muller c-elements and WDDL gates to create a generic single-cycle masking in hardware that is higher-order secure. We also use Muller cells, but with the purpose of designing self-timed latches to replace the register layers in any masking implementation, easing the application of our method in different scenarios. Moreover, we do not employ WDDL gates as they are vulnerable due to early propagation leakages [16].

Figures 2 and 3 show the 2-bit wide self-timed latches used in our designs. The dual-rail latches can be characterized as either strongly indicating or weakly indicating, depending on how their acknowledgement signal is computed. Figure 2, waits for all of its inputs to become valid, or null, before sending the respective acknowledgement. In contrast, a weakly indicating latch, Fig. 3, waits for only one specific input token to become valid or null before authenticating its current state [29].

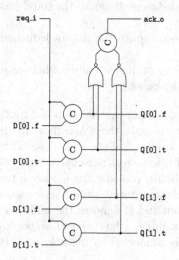
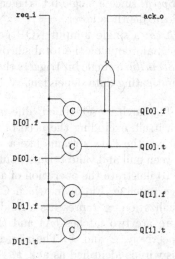

Fig. 2. A 2-bit wide strongly indicating asynchronous latch.

Fig. 3. A 2-bit wide weakly indicating asynchronous latch.

In both cases, weakly or strongly indicating, n pairs of Muller c-elements store a n-bit token $\overset{*}{x}$ and a regular 2-input NOR gate computes the validity signal for each pair — or a single pair for the weakly indicating version — to obtain the correspondent acknowledgement signal state.

The handshake logic contains two signals: a request input, denoted req_i, and an acknowledgement output, identified as ack_o. In fact, the acknowledgement signal indicates when the latch stores a valid (ack_o = 0) or a null (ack_o = 1) token. The request signal paces the data flow and is connected to one of the Muller c-element inputs. Thus, req_i = 0 requests a null token (i.e., the pre-charge), while req_i = 1 means that the combinatorial block following the latch is ready to evaluate a new valid token. Figure 4 shows the functioning of the handshake logic.

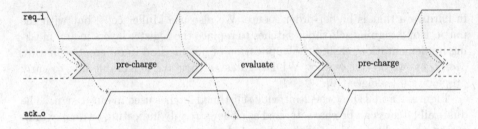

Fig. 4. Self-timed handshake in a pre-charge/evaluate logic.

In our designs, we favor the weakly indicating version based on three aspects.

1. *Speed*: since a single bit triggers the acknowledgement signal, the handshake logic depth is lower.
2. *Area*: a single 2-input NOR gate is used to compute the acknowledgement signal, reducing the total silicon area.
3. *Security*: a single bit triggers the acknowledgement, instead of the whole word, mitigating data dependent evaluation time leakages.

The latch is "self-timed" due to its handshake logic, that is managed by the data itself, excluding the need of a clock signal to pace the token flow. In this context, the data streams like a wave, with the intermediate states oscillating between null and valid tokens, configuring what is known as pre-charge logic [25].

To illustrate the operation of a self-timed circuit, consider the following two-stage pipeline, Fig. 5, in which C denotes a combinatorial circuit. For ease of visualization, ★ represents a random valid token and ∅ denotes the null token. There are two latches (A) and (B) in this example, whose initial states are, respectively, ∅ and ★. The req_i of the (A) is connected to the ack_o of (B). This wire is identified as ack_s.

Considering that C_A is pre-charged, the valid token is processed once it arrives at the pipeline input. The latch (A) keeps its logic state since its ack_s = 0. When req_i switches to 0, (B) absorbs the ∅ from (A) and sets ack_s = 1; The pre-charge phase of C_B is complete, which is signalized by (B) setting ack_s = 1. In consequence, (A) absorbs the valid token $C_A(★)$.

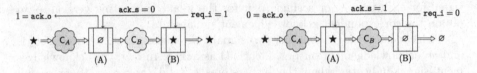

Fig. 5. Initial state.   **Fig. 6.** req_i switches to 0.

The external circuit issues a null token, pre-charging the combinatorial circuit C_A; the latches (A) and (B) stand by, since the request input remains constant at `req_i` = 0. Next, the `req_i` switches to 1, triggering the absorption of the output valid token $C_B(C_A(\bigstar))$, completing one self-timed processing cycle. This absorption sets the `ack_s` = 0.

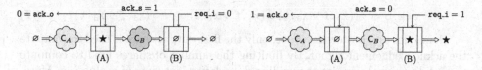

Fig. 7. The pre-charge of C_A.   **Fig. 8.** `req_i` switches to 1.

Although being a generic example, we aim at obtaining the same behavior for the AES S-box. Thus, in the following sections, we show how this can be implemented and discuss the implementation results for this solution.

3   Self-timed Masking Implementation

This section presents the design of a self-timed AES S-box based on the *simple* variant of the domain-oriented masking (DOM) proposed by Gross et al. in [13]. We start from a known implementation whose security performance has been already assured in the original work. Thus, we can study the resulting overheads when enabling self-timed features and evaluate first and higher-order security performance against side-channel analysis.

The dth order DOM gadget has $(d+1)^2$ registers, which will be replaced by dual-rail latches with the purpose of enabling the aforementioned self-timed features. To illustrate, consider the resulting 1st order DOM gadget shown in Fig. 9, in which all logic gate symbols represent their dual-rail variant.

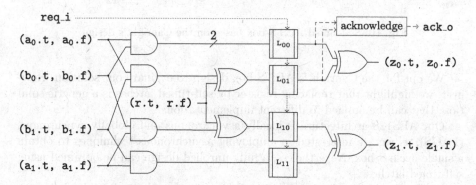

Fig. 9. The 1st order self-timed DOM gadget.

The request input is common to all latches and the Eq. 7 shows the complete detection logic for the general case, with the symbol ○ denoting the Muller c-element operation. In this manner, the acknowledgement signal indicates the token state stored in the current latch.

$$\texttt{ack\_o} = \neg((\texttt{x}_0.\texttt{f} \vee \texttt{x}_0.\texttt{t}) \circ (\texttt{x}_1.\texttt{f} \vee \texttt{x}_1.\texttt{t}) \circ \cdots \circ (\texttt{x}_d.\texttt{f} \vee \texttt{x}_d.\texttt{t})) \qquad (7)$$

Hence, for any protection order, only the first output share is used to compute the acknowledgement signal. By limiting the number of shares used to compute the validity signal, we aim at avoiding data dependent time of evaluation. Moreover, note that the more validity bits we use, the later the acknowledgement signal will be obtained, reducing the handshake speed.

Our S-box implementation, shown in the Fig. 10, is based on the Canright's design [6]. Basically, we implemented the *simple* variant shown in [13] using self-timed latches instead of registers. Each multiplier outputs a single acknowledgement signal and Muller c-elements are used to link the handshake of the two last multiplication stages. Indeed, the Muller c-elements are also employed in our design to join different acknowledgement signals.

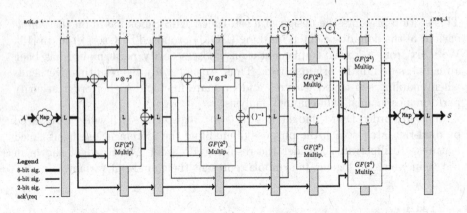

Fig. 10. Self-timed AES S-box based on the Canright's design.

We opt for the Canright's AES S-box design to evaluate our solution. However, we highlight that replacing register by self-timed latches is a generic solution, that can be applied in different implementations.

Our AES128 architecture is locally asynchronous and globally synchronous (LAGS), as we are interested in employing asynchronous techniques to obtain a single-cycle S-box. Nevertheless, a fully unrolled design can be obtained using self-timed latches.

Moreover, as the asynchronous pipeline can process several valid tokens, our AES architecture has a 32-bit single-rail data path with a single S-box. Thus, four substitution bytes are computed per clock cycle. In this context, a positive clock edge triggers the domino logic, allowing us to synchronize the computation of the correct tokens. If the clock period is adequate, the four SubBytes are ready before the next positive edge using only one S-box.

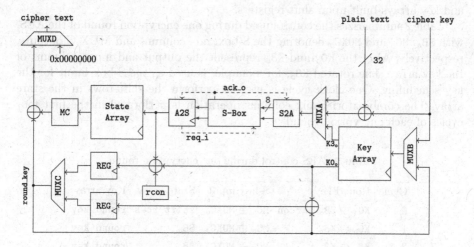

Fig. 11. The LAGS AES128 architecture with a 32-bit data path.

The interface between the synchronous and asynchronous worlds are managed by two modules: the synchronous to asynchronous (S2A) and the asynchronous to synchronous (A2S) blocks shown in the Fig. 11.

The S2A block converts the 32-bit single-rail data into four 8-bit dual-rail tokens. This block issues the tokens to the S-box input with the help of a multiplexer. The S2A is also able to identify whenever the circuit requests a pre-charge token. In other words, the acknowledgement output signal of the S-box block indicates when a token has been absorbed, triggering the pre-charge or the evaluation phase, depending on the acknowledgement signal state.

In parallel, the A2S module manages the S-box output — converting the valid token back to single-rail logic — and its request input signal. We count the number of occurrences of the positive acknowledgement edge in order to track the desired token progression in the pipeline. This block contains a 32-bit wide self-timed latch to store each output token. However, this latch has four acknowledgement signals, one for each output token, in order to identify the validity of each computation. The request signals are demultiplexed to obtain a single request output. At the end of the S-box computation, the four substitution bytes are available at the S-box output and the A2S module waits for the next positive edge of the clock signal to trigger the next computation. Since the A2S module stands by until the next positive edge of the clock signal, the request signal remains constant as well as the internal states of the S-box.

Different from previous low-latency AES128 implementations, such as [23, 28], we compute an AES128 round in five cycles. This is a choice of design. As we will show in the implementation results, this design allows us to obtain a relatively small AES architecture at the cost of higher encryption latency. Thus, to compute the AES128 encryption we need $10 \times 5 + 4 = 54$ clock cycles. Our AES architecture is based on the one by Moradi et al. [20], which relies on state and key arrays built upon shift registers.

Table 2 summarizes the control used during one encryption round of the AES, with SB_o, MC_o and $MUXC_o$ denoting the S-box, mix columns and MUX C outputs, respectively. Also, the KO_o and $K3_o$ represent the output and input columns of the key array. The rotated $K3_o$, for example, is used as the S-box input for the key scheduling. One clock cycle is used to perform the shift rows in the state array. The combinatorial mix columns operation is performed in the first four cycles of each encryption round.

Table 2. AES control during one encryption round.

Cycle	Round key	S-box input	State array	Key array
0	$KO_o \oplus SB_o \oplus$ rcon	$MC_o \oplus MUXC_o$	shift rows	round key
1	$KO_o \oplus K3_o$	$MC_o \oplus MUXC_o$	SB_o	round key
2	$KO_o \oplus K3_o$	$MC_o \oplus MUXC_o$	SB_o	round key
3	$KO_o \oplus K3_o$	$MC_o \oplus MUXC_o$	SB_o	round key
4	--	rotate[$K3_o$]	SB_o	stand by

Since we compute four S-boxes within one clock cycle, the system requires $144(d + d^2)$ refresh bits per clock cycle to protect the AES encryption, with d the masking protection order.

4   Implementation Results

We use Synopsys Design Compiler in order to synthesize our design using the CMOS 40-nm standard cell library with a target frequency of 100 MHz. The area results are normalized in terms of gate equivalents (GE) with a two-input NAND gate from the selected library as reference. No `compile_ultra` scripts were used in this work. We refer to Table 3, which reports the performance figures of our self-timed S-box implementations compared to the state-of-the-art.

Among the low-latency S-boxes, Sasdrich et al. [28] present the smallest first-order design. Nevertheless, our solution can be considered competitive in terms of gate counting and allows arbitrary protection order. Compared to the design proposed by Gross et al. [12], the area and randomness overheads of our solution show a better result, thanks to the presence of the compression step after the synchronization layers.

Nagpal et al. [23] propose two AES S-box designs: the Canright's [6] AES S-box based on composite fields and the Boyar and Peralta's [5] bit-slicing approach. Comparing the composite field implementations, our self-timed masking outperforms the area requirements of the solution proposed by Nagpal et al.,

Table 3. Performance figures of different masked S-box implementations.

Design	Masking order	Area [kGE]	Refresh [bits]	Latency [cycles]
Gross et al. [13]	1st	2.6	18	8
Arribas et al. [1]	1st	25.8	0	1
Gross et al. [12]	1st	60.7	2048	1
Sasdrich et al. [28]	1st	3.5	36	1
Nagpal et al. [23] {Canright's design}	1st	7.6	18	1
Nagpal et al. [23] {Boyar and Peralta's design}	1st	4.0	34	1
this work	1st	6.1	36	1
Gross et al. [12]	2nd	57.1	4446	1
Nagpal et al. [23] {Canright's design}	2nd	14.8	51	1
Nagpal et al. [23] {Boyar and Peralta's design}	2nd	9.3	102	1
this work	2nd	11.4	108	1

which is not the case when comparing our implementation to to their bit-slice AES S-box. However, the DOM gadget applied to the [5] multiplication may produce a flawed masking due to share's collision leakages [12].

Since we built our designs upon DOM gadgets, one S-box computation requires $36(d + d^2)$ fresh randomness per clock cycle. However, to securely compute the AES encryption, our AES design requires $144(d + d^2)$ refresh bits, since four SubBytes are computed within one clock cycle with a single S-box. However, despite being able to compute four SubBytes in a single-cycle, our design has shown to be slow compared to other solutions. Indeed, the throughput of our first-order S-box is approximately 5.3MB/s. Hence, our self-timed masking offers a trade-off between latency and throughput to designers. Although the latency is reduced, the clocked version of the same S-box implementation would perform better in terms of operating frequency, since the synchronous circuit has a smaller critical path.

One reason for this lack of performance is the number of S-box stages in our design. In fact, the *simple* variant of the DOM AES S-box has eight stages, as shown in the Fig. 10. Since the handshake signal needs to travel from the request input to the acknowledgement output to shift a token from a latch N to a latch $N + 1$, the higher is the number of stages, the slower is the token flow. Moreover, the latches have to be pre-charged after evaluating a valid token, which also compromises the speed. Based on this aspect, our solution may be more suitable for S-box designs with less combinatorial stages.

Thanks to our choice of design, our AES encryption can be computed in 54 clock cycles and the gate count results are significantly lower, compared to other low-latency solutions, as shown in Table 4. In this manner, low-latency masked S-boxes built within our 32-bit AES architecture would allow the hardware designer to obtain a smaller silicon area at the cost of increasing the number of clock cycles needed to perform an AES encryption.

Table 4. Performance figures of different low-latency masked AES implementations.

Design	Masking order	Area [kGE]	Refresh [bits]	Latency [cycles]	Area x Latency [kGE · cycles]
Sasdrich et al. [28]	1st	157.5	976	11	1,732.5
Nagpal et al. [23]	1st	104.9	680	11	1,153.9
this work	1st	14.2	144	54	766.8
Nagpal et al. [23]	2nd	203.9	2040	11	2,242.9
this work	2nd	23.4	432	54	1,263.6

5   Side-Channel Analysis

To verify the effectiveness of a countermeasure, it is common to simulate the power consumption of a device under attack. Moreover, implementing self-timed circuits on an FPGA is not straightforward, since the Muller c-elements and the handshake logic contains combinatorial loops. For these reasons, we use simulated traces in order to evaluate the side-channel vulnerability of our design.

To obtain a realistic acquisition, our simulations take into account the standard cell timing behavior so that the occurrence of glitches is possible. The logic simulation outputs value change dump (VCD) files which can be parsed to model the power consumption by processing the toggling activity of all wires in the device under test (DUT). Thus, we model the system's power consumption in a noiseless manner with a sampling frequency of 1 GHz. We refer to Fig. 12, which shows the block diagram illustrating the process to obtain the power consumption traces used in this work.

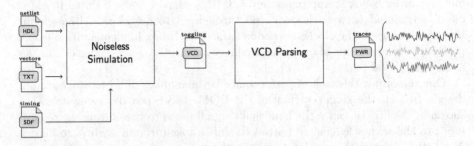

Fig. 12. Modeling power traces from VCD files generated after timing simulations.

We apply the *fixed vs random t-test* methodology[1] proposed by Goodwill et al. [10]. It uses the Welch's *t*-test to determine whether the difference of two dataset means provides sufficient evidence to reject the null hypothesis.

---

[1] We use SCALib for side-channel analysis: https://github.com/simple-crypto/scalib.

5.1   AES S-box Analysis

We start with a side-channel evaluation of our AES S-box. Figures 13 and 14 show the first and second-order univariate *t*-test results using one million simulated traces for the first-order implementation of our self-timed AES S-box. As expected, no exploitable side-channel leakages were identified in the first-order analysis. But second-order leakages were spotted for our first-order design.

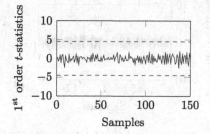 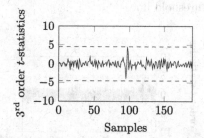

Fig. 13. 1st order TVLA results for the 1st order masked AES S-box based on one million simulated traces.

Fig. 14. 2nd order TVLA results for the 1st order masked AES S-box based on one million simulated traces.

Figures 15 and 16 show the second and third-order univariate *t*-test results using one million simulated traces for the second order implementation of our self-timed AES S-box.

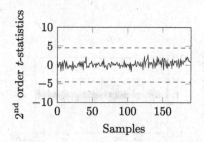 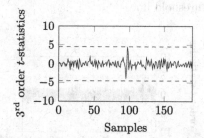

Fig. 15. 2nd order TVLA results for the 2nd order masked AES S-box based on one million simulated traces.

Fig. 16. 3rd order TVLA results for the 2nd order masked AES S-box based on one million simulated traces.

Again, the results confirm the robustness of our designs against side-channel analysis, even for the second-order implementation. These results are in adequacy with the DOM [13]. Thus, replacing registers by Muller c-element latches does not introduce weaknesses in the higher-order design.

5.2   Full Design Analysis

We evaluate the whole AES encryption in the same manner. Figures 17 and 18 show the first and second-order univariate t-test results using one hundred thousand simulated traces for the first-order implementation of our AES. Similar to the S-box, no first-order leakages were detected for the first-order design.

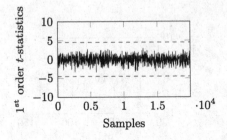
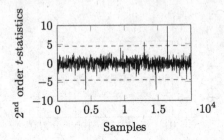

Fig. 17. 1st order TVLA results for the 1st order masked AES encryption based on one hundred thousand simulated traces.

Fig. 18. 2nd order TVLA results for the 1st order masked AES encryption based on one hundred thousand simulated traces.

Figures 19 and 20 show the second and third-order univariate t-test results using one hundred thousand simulated traces for the second-order implementation of our self-timed AES. Once again, the side-channel evaluation was performed on whole the encryption. As expected, the second-order evaluation shows no leakage, but third-order univariate analysis shows that some points cross the threshold line.

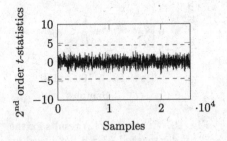
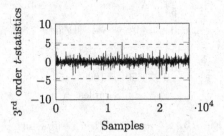

Fig. 19. 2nd order TVLA results for the 2nd order masked AES encryption based on one million simulated traces.

Fig. 20. 3rd order TVLA results for the 2nd order masked AES encryption based on one million simulated traces.

These results show that if time leakages are present in our design, they are difficult to exploit and to detect even in high resolution and low noise environment.

5.3 Bivariate Analysis

Figure 21 shows the bivariate analysis for the second-order AES S-box for one million traces, whereas Fig. 22 shows the t-test results for the second-order AES encryption using one hundred thousand traces. We stress that our side-channel evaluations were made in a noiseless high-resolution setting. This is due to measurement methodology used in this work, resulting in power consumption traces modeled from post-synthesis simulation taking into account the gate-level delays.

In both analysis, the upper triangle shows the results when random mask refresh is disabled, while the lower triangle illustrates the side-channel analysis when fresh randomness is employed. The result obtained from the unprotected setting uses only 10% of the amount of traces used in the protected scenario: one hundred thousand traces for the S-box and ten thousand traces for the AES encryption.

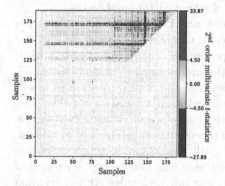

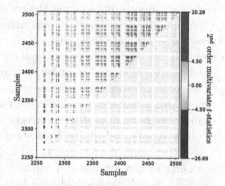

Fig. 21. Bivariate analysis of the 2nd order self-timed AES S-box implementation. The lower triangle shows the 2nd order t-test using one million traces with mask refresh enabled. The upper triangle shows the 2nd order t-test using one hundred thousand traces with mask refresh disabled. (Color figure online)

Fig. 22. Bivariate analysis of the 2nd order self-timed AES encryption implementation. The lower triangle shows the 2nd order t-test using one hundred thousand traces with mask refresh enabled. The upper triangle shows the 2nd order t-test using ten thousand traces with mask refresh disabled. (Color figure online)

The blue and red dots shown in the Figs. 21 and 22 represent sample points in which the multivariate t-statistics exceed the ± 4.5 threshold. For ease of visualization, only the first encryption round is shown in the AES analysis. As expected when the random number generator is shut off, second-order leakages are detected in our design, confirming the need of random refresh in DOM. It also confirms that our register replacement approach do not bring any weakness at higher-order in bivariate setting.

6   Conclusion

As previously stated, synchronizing the intermediate shares at the boundaries of combinatorial blocks is of high importance to obtain a secure masking implementation. Thus, this work presented a generic solution, that may be applied to different masked S-boxes, permitting the designer to obtain single-cycle implementations while assuring secure masking properties. Indeed, the main asset of our work is the reduction of the S-box latency to a single clock cycle, a feature achieved when replacing the register layers with self-timed latches. Nevertheless, the dual-rail logic adds a significant gate-count overhead to the final implementation, limiting its application in low area scenarios.

We observe that our solution has a low throughput compared to other low-latency solutions. This is due to the number of S-box stages in the pipeline, slowing the handshake logic propagation. For example, the throughput could be improved by reducing the number of S-box stages, enhancing the acknowledgement logic depth by reducing the number of intermediate handshakes.

Although being resistant against glitches, our DOM-based design uses the pre-charge/evaluate logic with monotonic cells, eliminating this hazard. Further research can be done to relax the security properties of the masked gadget when glitches are eliminated in order to reduce the overall implementation costs.

In order to evaluate our designs, we describe the implementation of a self-timed AES S-box and provide leakage assessment results based on noiseless side-channel analysis. One of the motivations behind a noiseless leakage assessment is to observe potential timing leakages due to the self-timed behavior of our S-box.

Furthermore, we present a 32-bit data path AES128 architecture to obtain a smaller silicon area design at the cost of computing one encryption round of the AES in five clock cycles. We highlight that our first-order AES encryption requires 54 clock cycles with a total area of 14 kGE approximately.

Finally, despite the area and throughput overheads, replacing the register by self-timed latches does not bring any weakness to the DOM implementation, even in second-order multivariate leakage analysis.

References

1. Arribas, V., Zhang, Z., Nikova, S.: LLTI: low-latency threshold implementations. IEEE Trans. Inf. Forensics Secur. **16**, 5108–5123 (2021). https://doi.org/10.1109/TIFS.2021.3123527
2. Barthe, G., et al.: Strong non-interference and type-directed higher-order masking. In: Weippl, E.R., Katzenbeisser, S., Kruegel, C., Myers, A.C., Halevi, S. (eds.) Proceedings of the 2016 ACM SIGSAC Conference on Computer and Communications Security, Vienna, Austria, 24–28 October 2016. pp. 116–129. ACM (2016). https://doi.org/10.1145/2976749.2978427
3. Batina, L., Robshaw, M. (eds.): CHES 2014. LNCS, vol. 8731. Springer, Heidelberg (2014). https://doi.org/10.1007/978-3-662-44709-3
4. Bhasin, S., Guilley, S., Flament, F., Selmane, N., Danger, J.: Countering early evaluation: an approach towards robust dual-rail precharge logic. In: Proceedings

of the 5th Workshop on Embedded Systems Security, WESS 2010, Scottsdale, AZ, USA, 24 October 2010, p. 6. ACM (2010). https://doi.org/10.1145/1873548. 1873554

5. Boyar, J., Peralta, R.: A small depth-16 circuit for the AES S-Box. In: Gritzalis, D., Furnell, S., Theoharidou, M. (eds.) SEC 2012. IAICT, vol. 376, pp. 287–298. Springer, Heidelberg (2012). https://doi.org/10.1007/978-3-642-30436-1_24

6. Canright, D.: A very compact S-Box for AES. In: Rao, J.R., Sunar, B. (eds.) CHES 2005. LNCS, vol. 3659, pp. 441–455. Springer, Heidelberg (2005). https://doi.org/10.1007/11545262_32

7. Chari, S., Jutla, C.S., Rao, J.R., Rohatgi, P.: Towards sound approaches to counteract power-analysis attacks. In: Wiener, M. (ed.) CRYPTO 1999. LNCS, vol. 1666, pp. 398–412. Springer, Heidelberg (1999). https://doi.org/10.1007/3-540-48405-1_26

8. Davis, A., Nowick, S.M.: Asynchronous circuit design: motivation, background, & methods. In: Birtwistle, G., Davis, A. (eds.) Asynchronous Digital Circuit Design, pp. 1–49. Springer, London (1995)

9. Faust, S., Grosso, V., Pozo, S.M.D., Paglialonga, C., Standaert, F.: Composable masking schemes in the presence of physical defaults & the robust probing model. IACR Trans. Cryptogr. Hardw. Embed. Syst. **2018**(3), 89–120 (2018). https://doi.org/10.13154/tches.v2018.i3.89-120

10. Goodwill, G., Jun, B., Jaffe, J., Rohatgi, P.: A testing methodology for side-channel resistance validation. In: NIST Non-invasive Attack Testing (NIAT) Workshop, vol. 7, pp. 115–136 (2011)

11. Goubin, L., Patarin, J.: DES and differential power analysis the duplication method. In: Koç, Ç.K., Paar, C. (eds.) CHES 1999. LNCS, vol. 1717, pp. 158–172. Springer, Heidelberg (1999). https://doi.org/10.1007/3-540-48059-5_15

12. Groß, H., Iusupov, R., Bloem, R.: Generic low-latency masking in hardware. IACR Trans. Cryptogr. Hardw. Embed. Syst. **2018**(2), 1–21 (2018). https://doi.org/10.13154/tches.v2018.i2.1-21

13. Groß, H., Mangard, S., Korak, T.: Domain-oriented masking: compact masked hardware implementations with arbitrary protection order. In: Bilgin, B., Nikova, S., Rijmen, V. (eds.) Proceedings of the ACM Workshop on Theory of Implementation Security, TIS@CCS 2016 Vienna, Austria, October 2016, p. 3. ACM (2016). https://doi.org/10.1145/2996366.2996426

14. Ishai, Y., Sahai, A., Wagner, D.: Private circuits: securing hardware against probing attacks. In: Boneh, D. (ed.) CRYPTO 2003. LNCS, vol. 2729, pp. 463–481. Springer, Heidelberg (2003). https://doi.org/10.1007/978-3-540-45146-4_27

15. Jukna, S.: Notes on hazard-free circuits. SIAM J. Discret. Math. **35**(2), 770–787 (2021). https://doi.org/10.1137/20M1355240

16. Kulikowski, K.J., Karpovsky, M.G., Taubin, A.: Power attacks on secure hardware based on early propagation of data. In: 12th IEEE International On-Line Testing Symposium (IOLTS 2006), 10–12 July 2006, Como, Italy, pp. 131–138. IEEE Computer Society (2006). https://doi.org/10.1109/IOLTS.2006.49

17. Leiserson, A.J., Marson, M.E., Wachs, M.A.: Gate-level masking under a path-based leakage metric. In: Batina, L., Robshaw, M. (eds.) CHES 2014. LNCS, vol. 8731, pp. 580–597. Springer, Heidelberg (2014). https://doi.org/10.1007/978-3-662-44709-3_32

18. Mangard, S., Popp, T., Gammel, B.M.: Side-channel leakage of masked CMOS gates. In: Menezes, A. (ed.) CT-RSA 2005. LNCS, vol. 3376, pp. 351–365. Springer, Heidelberg (2005). https://doi.org/10.1007/978-3-540-30574-3_24

19. Moradi, A., Immler, V.: early propagation and imbalanced routing, how to diminish in FPGAs. In: Batina, L., Robshaw, M. (eds.) CHES 2014. LNCS, vol. 8731, pp. 598–615. Springer, Heidelberg (2014). https://doi.org/10.1007/978-3-662-44709-3_33

20. Moradi, A., Poschmann, A., Ling, S., Paar, C., Wang, H.: Pushing the limits: a very compact and a threshold implementation of AES. In: Paterson, K.G. (ed.) EUROCRYPT 2011. LNCS, vol. 6632, pp. 69–88. Springer, Heidelberg (2011). https://doi.org/10.1007/978-3-642-20465-4_6

21. Moradi, A., Schneider, T.: Side-channel analysis protection and low-latency in action. In: Cheon, J.H., Takagi, T. (eds.) ASIACRYPT 2016. LNCS, vol. 10031, pp. 517–547. Springer, Heidelberg (2016). https://doi.org/10.1007/978-3-662-53887-6_19

22. Muller, D.E., Bartky, W.S.: A theory of asynchronous circuits. In: Proceedings of an International Symposium on the Theory of Switching, April 1957, Part I. the Annals of the Computation Laboratory of Harvard University, vol. XXIX, pp. 204–243. Cambridge University Press, Cambridge, MA, USA (1959). https://doi.org/10.2307/3611677

23. Nagpal, R., Gigerl, B., Primas, R., Mangard, S.: Riding the waves towards generic single-cycle masking in hardware. IACR Cryptol. ePrint Archive, p. 505 (2022), https://eprint.iacr.org/2022/505

24. Nikova, S., Rechberger, C., Rijmen, V.: Threshold implementations against side-channel attacks and glitches. In: Ning, P., Qing, S., Li, N. (eds.) ICICS 2006. LNCS, vol. 4307, pp. 529–545. Springer, Heidelberg (2006). https://doi.org/10.1007/11935308_38

25. Popp, T., Mangard, S.: Masked dual-rail pre-charge logic: DPA-resistance without routing constraints. In: Rao, J.R., Sunar, B. (eds.) CHES 2005. LNCS, vol. 3659, pp. 172–186. Springer, Heidelberg (2005). https://doi.org/10.1007/11545262_13

26. Rao, J.R., Sunar, B. (eds.): CHES 2005. LNCS, vol. 3659. Springer, Heidelberg (2005). https://doi.org/10.1007/11545262

27. Reparaz, O., Bilgin, B., Nikova, S., Gierlichs, B., Verbauwhede, I.: Consolidating masking schemes. In: Gennaro, R., Robshaw, M. (eds.) CRYPTO 2015. LNCS, vol. 9215, pp. 764–783. Springer, Heidelberg (2015). https://doi.org/10.1007/978-3-662-47989-6_37

28. Sasdrich, P., Bilgin, B., Hutter, M., Marson, M.E.: Low-latency hardware masking with application to AES. IACR Trans. Cryptogr. Hardw. Embed. Syst. **2020**(2), 300–326 (2020). https://doi.org/10.13154/tches.v2020.i2.300-326

29. Sparsø, J.: Introduction to asynchronous circuit design. DTU Compute, Technical University of Denmark (2020). https://www.amazon.com/dp/B08BF2PFLN

30. Tiri, K., Verbauwhede, I.: A logic level design methodology for a secure DPA resistant ASIC or FPGA implementation. In: 2004 Design, Automation and Test in Europe Conference and Exposition (DATE 2004), 16–20 February 2004, Paris, France, pp. 246–251. IEEE Computer Society (2004). https://doi.org/10.1109/DATE.2004.1268856

Evaluation Methodologies

An Evaluation Procedure for Comparing Clock Jitter Measurement Methods

Arturo Mollinedo Garay[1,2](✉)(iD), Florent Bernard[2](iD), Viktor Fischer[2,3](iD), Patrick Haddad[1](iD), and Ugo Mureddu[1](iD)

[1] STMicroelectronics, System Research Applications, 13790 Rousset, France
{arturo.mollinedogaray,patrick.haddad,ugo.mureddu}@st.com
[2] Hubert Curien Laboratory, Université Jean Monnet, Member of Université de Lyon, 42000 Saint-Etienne, France
{arturo.garay,florent.bernard,fischer}@univ-st-etienne.fr
[3] Faculty of Information Technologies, Czech Technical University in Prague, 160 41 Prague, Czech Republic

Abstract. According to recent security standards, the source of randomness of a true random number generator (TRNG) needs to be monitored online as well as inside the device to guarantee unpredictability and hence security. Monitoring is accomplished by measuring the physical parameters of the generator that determine the entropy rate per bit of the output bit-stream. The same parameters are preferably used as inputs for the stochastic model. Large majority of TRNGs implemented in logic devices use the clock jitter as a source of randomness. Consequently, the jitter is one of the main parameters to be characterized and observed. Several jitter measurement methods have been proposed in the last decade, but their precision and design constraints have not yet been objectively compared. We propose a simple yet useful methodology for the precise evaluation of jitter measurement methods including their design constraints. Our evaluation procedure relies on an analytical model of the jitter measurement method and simulations based on the model followed by stringent analysis of measurement errors. The new evaluation procedure is illustrated on four jitter measurement methods. The results clearly reveal differences in precision and in feasibility in hardware between the methods and confirm the usefulness of the new approach. In particular, the method presented in [7] proved to be the best performing, with an average measurement error of less than 10% and 93% implementable on an FPGA at that level of precision.

Keywords: Jitter measurement · True random number generator · Ring oscillators · Error analysis · Hardware implementation

1 Introduction

Random numbers play an essential role in cryptography: they are used as encryption keys, initialization vectors, nonces, padding values and recently also as random masks in side channel attack countermeasures. Since security relies on the statistical quality and unpredictability of the generated numbers, an evaluation of the random number generation mechanism is crucial in the security validation procedures required by security standards [1, 15, 17].

Cryptographic primitives and protocols are algorithmic functions that can be easily implemented in logic devices. However, True Random Number Generators (TRNGs) need some analog physical phenomena that can be exploited as a non-manipulable source of randomness. These analog sources are difficult to locate and exploit inside logic devices. Moreover, to prevent malfunction and attacks, the sources of randomness have to be continuously monitored, preferably inside the cryptographic system itself [1, 6]. Monitoring the input parameters of the TRNG stochastic model required in [1], is especially useful, since it helps to determine the TRNG output entropy rate in real time.

The most common source of randomness in logic devices is the instability in the timing of the clock signal generated in free-running oscillators, e.g. ring oscillators (ROs) – clock jitter [11, 16]. The clock jitter, i.e. fluctuations of the rising and falling edges of the clock signal over time can be seen as an analog quantity that needs to be converted into the digital domain to form random numbers. These fluctuations are caused by two different sources of noise: global and local.

Global noise sources (e.g. substrate and supply noise) should not be used as a source of randomness, since they are easy to manipulate by the attacker. Fortunately, their impact on generated numbers can be significantly reduced by using two or more identical oscillators and only exploiting the difference in their clock jitters.

Local jitter sources include uncorrelated noises (e.g. thermal noise) and correlated noises (e.g. low frequency random noises such as flicker noise or data dependent noises) [9]. To guarantee security, only uncorrelated and non-manipulable jitter sources should be considered to estimate entropy rates [2]. Although all jitter sources are somehow involved, it is essential to quantify the contribution of the uncorrelated thermal noise to the final clock jitter. The security evaluation of a TRNG using jitter as a source of randomness is linked to an accurate measurement of the jitter component caused by the thermal noise [3].

Several jitter measurement methods that can be embedded in cryptographic systems have been published over the last decade [7, 8, 12, 14, 18–20], but only a few are suitable for evaluation of jitter caused only by thermal noise [7, 19, 20]. Paradoxically, the authors of these papers do not estimate or even discuss precision on range of validity of their methods. No objective methodology for the analysis and comparison of their precision linked with their input parameters is currently available. Even though it would be useful when deciding whether a particular method deserves to be implemented in hardware and assessing its limits in terms of precision and exploration space, i.e. the space of possible input values. Although in [21], the authors analyzed the efficiency and precision of the method derived from [14], their approach cannot be applied to other jitter measurement methods.

In this paper, we propose a new objective procedure for the evaluation of jitter measurement methods based on the simulation of a given method using its own analytical model. This paper proves that the methods analysed here are only sufficiently accurate under certain conditions. In other words, this paper proves that it is necessary to delimit the parameters that make a jitter measurement method accurate. For example, the method presented in [20] should not be used unless a delay time of around 18 ps inside a chain of buffers can be assured. In practice, the simulation of the method is followed by the detailed analysis of its jitter estimation error and of the exploration space in which the estimation error is under the required threshold.

Of course, simulations using the model can help pinpoint sources of possible error, but if the evaluation shows that the measurement error is too high, the method being evaluated should definitely be rejected.

Our evaluation procedure is transferable to any jitter measurement method and can serve as a first assessment of the measurement principle. The methodology is illustrated and was confirmed using four published jitter measurement methods [7, 18–20] and clearly revealed their precision, advantages, weaknesses and exploration spaces. The simulation results can be used to target the implementation of a jitter measurement method in any hardware. In this work we targeted a Cyclone V FPGA as an example, but the same approach can be applied to other devices. All source-codes used have been made freely available at our Gitlab repository to ensure an easy reproduction of our results.

The paper is organized as follows. In Sect. 2, we present the proposed methodology for analytical modeling of jitter estimation methods for cryptography together with analytical models of four published methods. We illustrate the effectiveness of our methodology based on their detailed analysis in Sect. 3. The results are discussed and the selected methods are compared in Sect. 4. We conclude the paper in Sect. 5.

2   Modeling the Jitter Measurement Methods

2.1   Principle

In our approach, each jitter measurement method is represented by an analytical model given by a random function f_{σ_t}. This function has two kinds of inputs: 1) the standard deviation of the injected jitter (σ_t); 2) a limited number of parameters of the measurement method ($p_1, p_2, ..., p_n$). The model outputs the standard deviation of the measured jitter, i.e. $\sigma_{meas} = f_{\sigma_t}(p_1, p_2, ..., p_n)$. The objective of modeling is to evaluate the precision of the jitter measurement method depending on the model inputs. The precision of the method is evaluated using the relative measurement error:

$$err_\% = \frac{|\sigma_t - \sigma_{meas}|}{\sigma_t} 100\% \tag{1}$$

For each jitter measurement method, the evaluation procedure is performed in four steps:

1. First, we create an analytical model f_{σ_t} of the jitter measurement method.
2. Next, in a set of simulations, we apply the model while varying its input parameters throughout the exploration space.
3. Next, we evaluate the jitter estimation error as a function of the model input parameters and obtain the exploration space in which the method produces consistent results.
4. Finally, we analyze if this space is conceivable given the constraints implied by the TRNG stochastic model and with respect to feasibility of the jitter measurement method in the hardware concerned.

2.2   Model Assumptions

Since the proposed methodology requires construction of an analytical model of the method to be evaluated, a solid understanding of the method is needed. Model construction is based on some realistic assumptions, which are also made in [3,14,18–20]:

a) The jitter measurement methods need the average oscillator clock period μ to be known so as to be able to compute the jitter. This assumption is quite easy to meet, as discussed later.
b) We expect that a differential method of randomness extraction and of jitter measurement is applied. This method, which assumes that all oscillators feature identical topology, is very useful to eliminate the impact of operating temperature and power supply variations, i.e. global noise sources, on the jitter.
c) We assume that the clock jitters, which are used as the sources of randomness, namely jitter coming from the thermal noise, are independent and identically distributed (iid) in all clock signals generated. In this case each clock period follows a normal law $\mathcal{N}(\mu, \sigma^2)$, where σ represents the jitter coming from the thermal noise. In our simulations, we let $\sigma = 1\%_o\mu$, which is in the order of magnitude observed in hardware experiments [7,8,20].

The measurement of the absolute mean clock period can easily be embedded in logic systems: the number of clock periods is repeatedly counted during sufficiently long time periods and the mean counter value is then computed. Using the circuit presented in Fig. 1, the average period of RO_1 with respect to the period of RO_0 can be measured with a counter value obtained after a sufficiently long accumulation time. The absolute error of this measurement will converge to $\frac{1}{k}$ [13], where k is the number of the reference clock periods during the measurement interval. For now, we conclude that μ can be measured with sufficient accuracy if k is large enough and if the oscillators used have similar periods [18].

The third of the above-mentioned assumptions can significantly simplify the model. Since the jitters originating from the thermal noise are assumed to be independent, the jitter of the sampled clock characterized by its standard deviation σ_1 can be transferred to the reference clock. The sampled clock signal will consequently be jitter free, and the new total reference clock jitter σ_t, which will

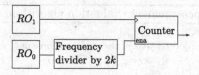

Fig. 1. Block diagram of the circuit allowing the measurement of the average period of RO_1 with respect to the period of RO_0. This diagram also represents the *counter method* as explained in Sect. 3.1.

appear in the reference clock, will include the contribution of both oscillators to the resulting jitter as follows [2,7]:

$$\sigma_t = \sqrt{\frac{\mu_0}{\mu_1}\sigma_1^2 + \sigma_0^2} \tag{2}$$

where μ_1 and μ_0 represent the average periods of the generated clocks, and σ_1 and σ_0 represent their jitters caused by thermal noise. To obtain consistent results, we used this technique in all the jitter measurement methods we evaluated.

2.3 Simulations Based on the Model and Their Precision

The jitter measurement methods are evaluated using Algorithm 1. Based on the analytical model of the jitter measurement method, this algorithm can be used to evaluate the precision of the method depending on its input parameters. Namely, it computes the jitter measurement errors depending on the injected jitter and on input parameters such as jitter accumulation time, the difference in mean frequencies, etc. Although the size of the injected jitter is one of the inputs of the analytical model, to simplify the simulations and to compare the methods as objectively as possible, we let the mean period of all oscillations be the same and we inject the same jitter in every simulation of the four jitter measurement methods. The obtained errors (mean, maximal and minimal error) are then depicted as a function of the input parameter in a graph, facilitating evaluation of the method and the choice of its exploration space.

When comparing the methods, we accept the space of input parameters in which the average measurement relative error is less than 10% and the maximal error is less than 25%. Of course, these limits can be modified at any time depending on the goals of the evaluation.

Here, we stress that the measurement error depends not only on the method itself, but also on the precision of its simulation. Since the jitter is a random variable, the measurement error also varies randomly. The precision of the simulation thus mainly depends on the size of the population (the number of experiments) used to calculate the variance of the jitter in one measurement (N_{exp}), and the number of measurement errors needed to compute the average error (N_{err}). On the other hand, as the simulation can be time consuming, we need to tune N_{err} and N_{exp} to find a good trade-off between precision and simulation time.

Algorithm 1. Jitter measurement evaluation algorithm

---

procedure $(\sigma_t, p_{min}, p_{max}, p_{step})$
        ▷ The injected jitter, the exploration space of an input parameter p
  $p \leftarrow p_{min}$ $err\%_{mean} \leftarrow []$ $err\%_{max} \leftarrow []$ $err\%_{min} \leftarrow []$
  while $p < p_{max}$ **do**
    $j \leftarrow 0$
    while $j < N_{err}$ **do**
      $err\% \leftarrow []$ $i \leftarrow 0$
      while $i < N_{exp}$ **do**
        $\sigma_{meas} \leftarrow f_{\sigma_t}(p)$ $i \leftarrow i + 1$
      end while
      $err\% \leftarrow [err\%, \frac{|\sigma_t - \sigma_{meas}|}{\sigma_t} 100\%]$ $j \leftarrow j + 1$
    end while
    $err\%_{max} \leftarrow [err\%_{max}, max(err\%)] err\%_{min} \leftarrow [err\%_{min}, min(err\%)]$
    $err\%_{mean} \leftarrow [err\%_{mean}, mean(err\%)]$ $p \leftarrow p + p_{step}$
  end while
  return $err\%_{mean}, err\%_{max}, err\%_{min}$
        ▷ Three lists containing the mean, maximal and minimal relative error
end procedure

We set the population size at $N_{exp} = 3\,000$, so that the measurement errors caused by simulation will converge to $1/\sqrt{N_{exp}}$ or 1.8%. At the same time, as we are interested in the average measurement error, a relatively small number of error calculations are enough to obtain it with sufficient precision. For this reason, we set the number of error calculations at $N_{err} = 50$.

As N_{exp} and N_{err} are set, we can deduce that an average error greater than 1.8% with a $1/50$ margin obtained in the simulations originates from the jitter measurement error and not from the imprecision of the simulation itself.

2.4   Hardware Constraints and Data Acquisition

The exploration space is determined by the measurement method and confirmed by simulation in Algorithm 1, but this space may also be limited by hardware constraints. Indeed, some input parameters, such as the oscillation frequency and the precision of its setup, can be limited by the hardware used. For this reason, we implemented the selected methods in an Intel Cyclone V FPGA. The objective of these experiments was to check if the hardware requirements of the method itself and those identified by our simulations, can be met. To simplify our experiments, we implemented ring oscillators and the necessary data acquisition circuitry in hardware, but the acquired data were processed in the computer.

The ring oscillators were instantiated and manually placed and routed in the FPGA using nine buffers and one NAND gate to enable the oscillations. The buffers were instantiated using LCELL structures [4]. This resulted in oscillators with a frequency of around 260 MHz, which corresponds to the frequency required by the experiment in Sect. 3.3 and that were also used in the simulations.

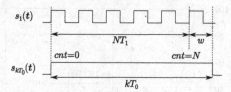

Fig. 2. Waveforms of the *counter method* signals.

To obtain comparable results, we made a particular effort to acquire measurement data in identical conditions for all the methods we evaluated. First, before acquisition, we waited for about 10 min for the hardware to warm up and for the operating temperature to stabilize. After this interval, counter values and clock samples (depending on the method tested) were acquired periodically in data streams in $t_m = 8\mu s$ intervals. The time period of the data acquisition intervals was determined by the most time consuming method. In this way, sets of $N_{mes} = 600\,000$ data samples were acquired and used to measure the input parameters of the method being tested.

As mentioned in Sect. 2.2, a jitter measurement method may require measurement of the average period of a ring oscillator. To this end, we used a quartz oscillator with a known frequency to define the time lapse t_m. Hence, it sufficed to instantiate a counter at the output of a RO to estimate its average period according to Eq. (3). In this equation, $\bar{c}$ represents the average of the N_{mes} acquired counter values after the measurement interval t_m and the measured average clock period μ_{meas} of a RO.

$$\mu_{meas} = \frac{t_m}{\bar{c}} \tag{3}$$

3   Application of the New Methodology

In this section, we apply the new evaluation procedure to four jitter measurement methods. For each method, we first provide the theoretical background including the principle of the method and the design constraints. We then present and discuss the simulation of the measurement method and its results.

3.1   Counter Method

Theoretical Background. The *counter method* presented in Fig. 1 consists in counting the clock periods of the RO_1 ring oscillator during a jitter accumulation time determined by number k of the reference clock periods generated by the second oscillator [18]. The obtained counter values vary depending on the size of the accumulated jitter, and the jitter size can be deduced from the variance of counter values. Similar jitter measurement methods are described in [14,21].

Assuming that RO_1 and RO_0 are perfectly aligned at $t = 0$ (e.g. as shown in Fig. 2), the precision of the method will be influenced by the remnant time

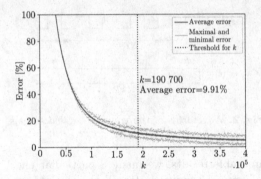

Fig. 3. Error as a function of k for the *counter Method*.

w between the last counted edge of RO_1 and the end of the measurement time. Consequently, the precision of the method can be increased by selecting an accumulation time that is long enough so the remnant time is negligible.

Simulation of the Method and Simulation Results. As discussed in the previous paragraph, the precision of the method depends on the jitter accumulation time kT_0. Therefore, we use parameter k as the input parameter of Algorithm 1:

$$\sigma_{meas} = f_{\sigma_t}(k) \tag{4}$$

According to the assumptions presented in Sect. 2, the time period kT_0 can be represented by the normal law in the following way:

$$kT_0 \sim \mathcal{N}\left(k\mu_0', k\sigma_t^2\right) \tag{5}$$

We established $k_{min} = 0$, $k_{max} = 400\,000$ and $k_{step} = 100$ for our simulations. Figure 3 shows the average error of the method as a function of k. It can be seen that it only falls below 10% at $k \approx 190\,700$.

Discussion. As discussed in Sect. 2, the entropy rate should be estimated depending on the contribution of thermal noise to the clock jitter, because the contribution of flicker noise is difficult to compute. Unfortunately, the results in [8] show that the effect of flicker noise on the total clock jitter can be neglected for short accumulation times, namely for $k < 300$. However, we have shown that the *counter method* needs $k > 190\,700$ to guarantee a measurement error of less than 10%.

The jitter measurement based on the counter method thus requires accumulation times that are too long to avoid the effect of flicker noise. The proportion of effects of the thermal and flicker noise on the total jitter can be estimated using the method introduced in [8] and applied in [21], but this method needs additional resources and can introduce additional errors in jitter estimation. We can conclude that the *counter method* is not immediately applicable for estimation of the clock jitter caused by the thermal noise.

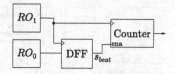

Fig. 4. Block diagram of the simulated *coherent sampling method*.

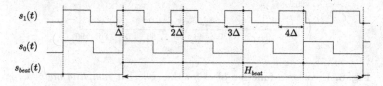

Fig. 5. Evolution of $s_1(t)$, $s_0(t)$ and $s_{beat}(t)$.

3.2 Coherent Sampling Method

Theoretical Background. The second method we considered in our study was the *coherent sampling method* [19]. In this method, the output of oscillator RO_1 is sampled on the output edges of oscillator RO_0 using a D flip-flop (DFF) as shown in Fig. 4. If the clock periods of the two oscillators are close enough, i.e. if their difference is comparable with the size of the jitter, the DFF output (signal s_{beat}) has a variable period whose variance depends on the jitter affecting the two clocks. This variation is then revealed by a counter that counts the periods of the output of RO_0 during half a s_{beat} period (H_{beat}).

Note that $\Delta := \mu_0 - \mu_1$ is the difference between the mean values of the two clock periods. The parameter Δ determines the length of H_{beat} in Fig. 5. Small Δ values result in long s_{beat} periods. Consequently, the influence of the flicker noise may become non-negligible. Big Δ values result in short s_{beat} periods. In this case, the variance of the counter output no longer reflects the jitter.

In the case of the jittered clocks, the counter value will vary if the jitter accumulated during the accumulation period is not negligible compared to Δ. It is therefore clear that the precision and hence the error of the *coherent sampling method* will strongly depend on Δ.

Simulation of the Method and Simulation Results. As before, the *coherent sampling method* was simulated using the analytical model according to Algorithm 1. The aim of the simulations was to study the dependence of the measurement error on Δ:

$$\sigma_{meas} = f_{\sigma_t}(\Delta) \tag{6}$$

In the simulations, recalling that $s_1(t)$ is supposed to be jitter free, the clock period T_0 and the period difference $T_\Delta := T_0 - T_1$ were represented by the normal law in the following way:

$$T_0 \sim \mathcal{N}\left(\mu_1 + \Delta, \sigma_t^2\right)$$
$$T_\Delta \sim \mathcal{N}\left(\Delta, \sigma_t^2\right) \tag{7}$$

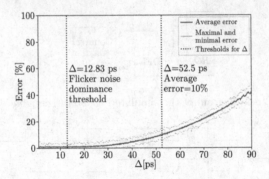

Fig. 6. Error as a function of Δ for the *coherent sampling method*.

We selected $\Delta_{min} = 1$ ps, $\Delta_{max} = 90$ ps and $\Delta_{step} = 0.5$ ps for our simulations. Figure 6 depicts the measurement error as a function of Δ, obtained using the model. The method reaches the targeted average error of 10% for $\Delta = 52.5$ ps. Additional simulations allowed us to confirm that the limits of Δ can be expressed as a function of μ_1. Therefore, the upper limit of Δ can be found for any value of μ_1. In other words, to guarantee that the average error is less than 10%, the following condition must hold: $\Delta < 0.014\mu_1$. The low boundary of Δ is determined by the impact of the flicker noise. To reduce the impact of the flicker noise, according to the results presented in [8], the accumulation time should be less than $300\mu_1$ and consequently $\Delta > 12.83$ ps, i.e. $0.003\mu_1$. We can conclude that, according to the simulation results, the difference between the two clock periods should respect the following condition: $0.003\mu_1 < \Delta < 0.014\mu_1$.

Discussion. Note that the boundaries of Δ are theoretically independent of the absolute clock period values. It is also clear that the method guarantees a measurement error of less than 10% only in a relatively small interval of Δ. This implies that the designer must be able to set the difference in clock periods very precisely.

To study the *feasibility of the method in hardware*, we implemented 16 identical ring oscillators. Each ring was followed by a 32-bit counter used to measure the average clock period at the ring's output according to Eq. (3). The measured average periods were used to calculate Δ for every possible pair of ROs according to Eq. (8). Each RO can be used to generate either the sampling or the sampled clock signal. Then a total of 240 different couples can then be identified among the 16 ROs. In Equation (8), j refers to the sampled RO and i refers to the sampling RO.

$$\Delta_{i,j} = \frac{|\mu_i - \mu_j|}{\mu_j}100\%, \quad i,j \in \{0,1,...,15\} \, \forall i \neq j \tag{8}$$

The results of this experiment are shown in Fig. 7. The figure shows that only 23.7 % of the Δ values were inside the boundaries determined by the simulations.

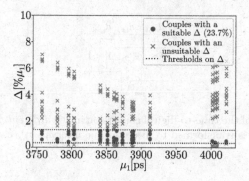

Fig. 7. The figure shows the measured Δ values as a percentage of μ_1 for the 240 instantiated couples of ROs. The circles appearing between the dotted lines correspond to the Δ values that satisfy the condition $0.33\%\mu_1 < \Delta < 1.36\%\mu_1$.

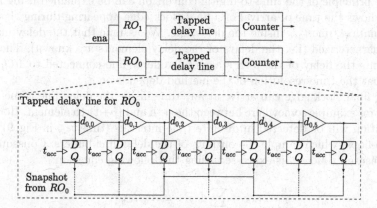

Fig. 8. Basic block diagram of the *differential delay line method* using two time-to-digital converters, each containing a counter and a tapped delay line.

In other words, only around 1 out of 4 pairs of oscillators implemented in the hardware could be used to perform the *coherent sampling method*. We conclude that the critical dependence of the method on the setup of the parameter Δ makes the method difficult to implement in hardware.

3.3   Differential Delay Line Method

Theoretical Background. The differential delay line method was first published in [20]. It is based on a time-to-digital conversion (TDC) using a counter and a series of chained buffers, as shown in Fig. 8. The measured time, which depends on the accumulated clock jitter, is composed of the integer number of clock periods counted by the counter at the delay line output and of the part of the clock signal period that remains in the delay line at the end of the jitter accumulation time.

Fig. 9. Illustration of the time-to-digital conversion in the *differential delay line method* the arrival time $t_{i,s}$ is transformed to discrete time intervals $\tau_{i,s}$ causing the measurement error w.

Since the time delays are measured using the delay lines, the delay elements (buffers) inside the delay lines have to be characterized – the delay corresponding to the delay element, which is a random variable, must be known.

The principle of the time-to-digital conversion can be explained using Fig. 9, which shows the time of arrival of a final clock edge appearing during the jitter accumulation time t_{acc} inside the delay line. We assume that the delay line has been characterized (i.e. the delays of the delay elements are known). Then $d_{i,p}$ represents the delay of the p-th stage of the delay line connected to RO_i. It is clear that the time resolution of the method depends on $d_{i,p}$.

The final clock edge can arrive at any time inside the delay line (time $t_{i,s}$ in Fig. 9), for example somewhere between the p-th and $(p+1)$-th element. However, its position will be detected in discrete time intervals (time $\tau_{i,s}$ in Fig. 9) using the flip-flops, which sample the outputs of the delay line buffers. Consequently, the measured time $\tau_{i,s}$ is expressed using the following equation:

$$\tau_{i,s} = \sum_{j=p}^{n} d_{i,j} + \mu_i Acnt_{i,s} \tag{9}$$

where n is the length of the delay line in terms of the number of buffers, μ_i is the mean clock period of RO_i and $Acnt_{i,s}$ is the registered counter value from Fig. 8.

Based on Eq. (9), we can see that the results depend on two measurements [20]: the measurement of the average period of oscillators (μ_i) and the characterization of the delay elements ($d_{i,j}$).

Simulation of the Method and Simulation Results. Note that our main objective is to assess the precision of the method depending on its implementation in hardware. From the analysis presented above, and since the mean clock period can be measured as precisely as required, the final precision of the method basically depends on the way the delay line is implemented and characterized. The model can thus be represented by Eq. (10):

$$\sigma_{meas} = f_{\sigma_t}(\mu_d, \sigma_d) \tag{10}$$

where μ_d is the mean delay of delay elements $d_{i,j}$ and σ_d is the standard deviation of the delay. In other words, we model the delays of the delay line elements with

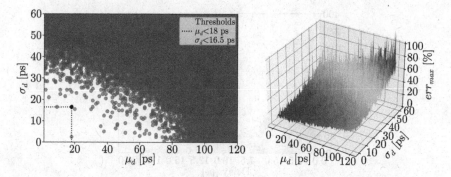

Fig. 10. Error analysis of the *differential delay line method*. The left panel presents the coordinates in the space (μ_d, σ_d), under which the method yields inaccurate measurements. The right panel shows the maximal error as a function of μ_d and σ_d.

a normal law that reflects the variations in the manufacturing of the underlying hardware. At each simulation, we generate a delay line as a succession of buffers with delays $d_{i,j}$ given by the following formula:

$$d_{i,j} \sim \mathcal{N}(\mu_d, \sigma_d^2) \quad ; \quad d_{i,j} > 0 \text{ ps} \quad ; \quad \forall i,j \tag{11}$$

It is important to note that only $d_{i,j}$ values bigger than 0 ps were used in our simulations. This means that the normal law characterizing the chain elements was truncated to exclude negative or null delays as they would be unrealistic.

As the authors explain in [20], to obtain correct results, the delay of the delay line must be longer than one and a half oscillation periods. This constraint is essential as it determines the number of delay elements depending on the frequency of oscillations. To be as close to reality as possible, we selected $\mu_{d_{min}} = 1$ ps, $\mu_{d_{max}} = 120$ ps, $\mu_{d_{step}} = 1$ ps, $\sigma_{d_{min}} = 0$ ps, $\sigma_{d_{max}} = 60$ ps and $\sigma_{d_{step}} = 0.5$ ps for our simulations.

Further, we set the accumulation time t_{acc} to 1 μs. Like in the previous jitter measurement methods, the frequency of oscillations was around 260 MHz, i.e. their period was about 3.85 ns. Consequently, the accumulation time lasted around 260 oscillation periods and the impact of the flicker noise was negligible, as required in [8].

We focused our attention on the repercussions of μ_d and σ_d on measurement precision. The results are shown in Fig. 10. The circles in the left panel show the coordinates in the (μ_d, σ_d) space, in which the precision requirements were not met, i.e. where the $err\%_{mean} > 10\%$ or $err\%_{max} > 25\%$. The right panel shows a surface representing the maximal error of the *differential delay line method* as a function of μ_d and σ_d over the exploration space.

The μ_d and σ_d thresholds were identified by looking for the space, in which no errors appear in the left panel of Fig. 10. Accordingly, we chose $\mu_d < 18$ ps and $\sigma_d < 16.5$ ps to ensure suitable jitter measurement precision based on our initial criteria.

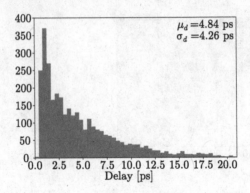

Fig. 11. The figure shows the histogram of the measured delays of the CARRY elements on a Cyclone V FPGA.

Discussion. Our analysis confirmed that the precision of the method significantly depends on its hardware implementation and specifically, that it is linked to the delays added by the buffers present in the chain. It is therefore crucial to measure these delays in the real hardware in which the method is intended to be implemented.

It is worth mentioning that the authors in [20] used just one look-up table (LUT) to implement ring oscillators and consequently they oscillated at a very high frequency (close to 1 GHz). The oscillation period was thus relatively small, nevertheless, 256 delay elements were needed to satisfy the minimal length of the delay line covering one and a half clock periods.

It was therefore clear that a trade-off is needed between the oscillation period and the number of elements in the delay line depending on the underlying hardware. For the above mentioned reasons, we let the oscillators oscillate at a lower frequency of about 260 MHz. Next, we implemented four ring oscillators followed by four tapped delay lines containing $n = 800$ buffers using carry chains in the selected Cyclone V FPGA family.

Our main objective was to precisely characterize the implemented delay lines using the dedicated carry chains available in the selected technology [4] and a set of n D flip-flops using the procedure described in [20]. To obtain accurate estimation of delays, it is crucial to place the D flip-flops equidistantly from their respective delay elements as illustrated in Fig. 8. We followed the placement and routing procedure described in [5,10] to fulfill this requirement.

The results are presented in Fig. 11. On average, a carry chain element in the targeted FPGA added a delay of 4.84 ps. The standard deviation of the delays was 4.2 ps. These values are well below the thresholds obtained using Fig. 10 ($\mu_d < 18$ ps and $\sigma_d < 16.5$ ps). The measured delays thus indicate that when implemented in the selected hardware, the method should give results with a suitable precision.

Although the mean delay of delay elements was shown to be small enough to guarantee the required precision, the number of required delay elements must also be taken into consideration. For example, the carry chain in the targeted

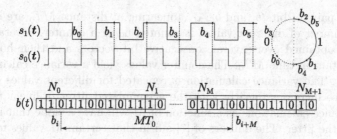

Fig. 12. Illustration of a sampling used as a basis for the *method testing the autocorrelation of distant samples*.

Cyclone V FPGA is limited to 1 000 elements. The mean delay is 4.84 ps, taking into account the standard deviation of the delay of 4.84 ps, the oscillation frequency should be higher than 400 MHz to save up at least one and a half in less than 1 000 elements. These high frequencies significantly increase power consumption. The only solution would be to increase the mean delay of delay elements at the expense of precision.

It appears that the hardware used in [20] was better suited for the implementation of the method. Indeed, bigger carry chain delays in the selected FPGA family reduced measurement precision, but at the same time the number of delay elements became reasonable. However, in our opinion, the required frequency of oscillations (close to 1GHz) was definitely beyond the space of practical values.

3.4 Method Testing the Autocorrelation of Distant Samples

Theoretical Background. As shown in Sect. 3.2 the *coherent sampling* method yields accurate measurements only if the sampled and sampling free running oscillators have very similar but not identical periods. This may be difficult to obtain in hardware. The *method testing the autocorrelation of distant samples* presented in [7] is not limited by the constraint linked to the frequencies of the two free running oscillators.

Like the coherent sampling method, the output of oscillator RO_1 is sampled at (rising or falling) edges of the clock signal generated by the second oscillator (RO_0). As depicted by the phase diagram in the right panel in Fig. 12 the samples b_i of signal $s_1(t)$ can be equal to one or zero, while creating a pattern that will depend on the frequencies of the oscillations and the relative phase. In order to avoid clusters of samples on the ring of the phase diagram, the ratio T_0/T_1 should not be a good approximation of a fraction of two small integers p and q. In this case, and if we suppose perfect jitter-free oscillations, a repetitive pattern would appear at the sampler output. The size and the shape of the pattern, which would be repeated for infinity, would depend on the frequencies of the oscillations and their mutual phase.

However, in the case of jittered oscillations, a bit vector N_0 containing N samples and a homologous bit vector N_M distant in time MT_0 would differ from each other with respect to the jitter accumulated over time MT_0. As shown in

Fig. 12, N pairs of bits b_i and b_{i+M}, appearing at distance MT_0 are compared and the number of different values is counted using a counter. Several counter values are obtained over several consecutive bit vectors and their homologous vectors distant in time MT_0. The counter values yield a variance calculation for a given M. The variance calculation is repeated for different values of M. As the variance increases linearly with the accumulated jitter, a linear regression of the variance calculations as a function of M will have a slope that is directly related to the jitter. The authors of [7] affirm that certain M values may result in counter values close to 0 or N. The variance of these counter values is not representative concerning the jitter and may bias the slope of the linear regression and thus influence the jitter measurement. Hence, we define cnt_{th} as a threshold that discards the counter values whose average is outside the defined limits $(cnt_{th}N; (1 - cnt_{th})N)$. The last steps of the method are crucial for the precision of the jitter measurement.

Simulation of the Method and Simulation Results. The linear regression that leads to the jitter measurement is crucial for this method. It seems logical to link the error of the method to the way the linear regression is drawn. Two extreme approaches can be used to reduce this error: either by calculating as many points as possible so that their individual errors are averaged out and hence less influential; or by using as few points as possible, i.e., only those that are very close to the path of the desired straight line. The linear regression is based on different M values. The limits of M were fixed at 150 and 300. As suggested by the authors, N was assumed to be 100. However, two other parameters of the method affect the number of points used. On one hand, the step of M defines the amount of calculated variances. On the other hand, the threshold on the average counter values, cnt_{th}, ensures that an obtained variance is a good reflection of the accumulated jitter over MT_0. A fair trade-off between these two parameters is thus required. To address this issue, we simulated the method using Algorithm 1 with M_{step} and cnt_{th} as input parameters (Eq. (12)).

$$\sigma_{meas} = f_{\sigma_t}(M_{step}, cnt_{th}) \tag{12}$$

Specifically for this method, the oscillators periods were fixed at $\mu_1 = 4\,003$ ps and $\mu_0 = 3\,704$ ps for the reasons explained in the following subsection. The results in the left panel of Fig. 13 use a fixed $M_{step} = 25$ to show the effect of the threshold cnt_{th} on the error. We chose $cnt_{th_{min}} = 0$, $cnt_{th_{max}} = 0.5$ and $cnt_{th_{step}} = 0.01$. The right panel uses $cnt_{th} = 0.25$ to show the effects of varying the M_{step}. Figure 13 shows that there is an $M_{step} = 39$ that always yields counter values close to 0 or N. However, low cnt_{th} thresholds include points that add to the error of the regression and high cnt_{th} thresholds use too few points to perform an accurate regression.

Discussion. Ultimately, the choice of the method parameters is based on the precise values of T_1 and T_0. The periods of the oscillators depend on their hardware implementation. It would thus be interesting to measure how many pairs

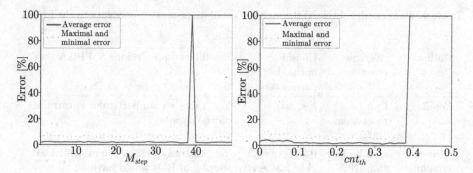

Fig. 13. Error of the *method testing the autocorrelation of distant samples* due to the linear regression. The left panel shows the effect of varying M_{step} while the right one shows the effects of varying cnt_{th}.

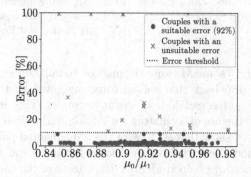

Fig. 14. Simulated average errors of the couples of oscillators implemented on the FPGA for *method testing the autocorrelation of distant samples.*

of oscillators on a given hardware enable a successful jitter measurement. To this end, we instantiated two groups of ring oscillators in our FPGA. Group A consisted of 16 ROs composed of 9 buffers and group B consisted of 16 ROs composed of 10 buffers. A total of 255 pairs of oscillators were formed by selecting an RO from group A as the sampling oscillator and an RO from group B as the sampled oscillator. Using the same approach described in Sect. 2.4 we approximated every RO period to their average periods. We used the measured average periods as inputs for our model. This time, we fixed the parameters $M_{step} = 10$ and $cnt_{th} = 0.25$ and studied the influence of μ_0/μ_1 on the measurement error.

The crosses in Fig. 14 show the pairs that would fail to yield an accurate measurement under these conditions. According to our simulations, 92% out of the 255 couples resulted in an average error $< 10\%$. Figure 14 allowed us to chose a pair of values μ_1 and μ_0 for the simulations in the previous subsection. A simulation where $\mu_1 = 4\,003$ ps and $\mu_0 = 3\,704$ ps yielded an average error of 1.8%. Therefore, these values correspond to a case where the influence of μ_0/μ_1 on the measurement error is not important and allow us to study the effects of the linear regression.

Table 1. Main results obtained through the simulations and experiments on the FPGA.

Method	Average error	Chosen optimal parameters	Feasibility in a Cyclone V FPGA
Counter	Can be more than 100%	$k < 300$	Yes, there are no particular constraints on the support
Coherent sampling	Can be less than 10%	$0.33\%\mu_1 < \Delta$ $\Delta < 1.36\%\mu_1$	Yes, but only 24% of the implemented couples of ROs would have an acceptable error
Differential delay line	Can be less than 10%	$\mu_d < 18$ ps $\sigma_d < 16.5$ ps	No, the measured μ_d demands for long delay lines that do not fit on the FPGA
Auto-correlation	Can be less than 10%	$cnt_{th} < 40\%$	Yes, 93% of the implemented couples of ROs would have a suitable error if $N = 100$; $cnt_{th} = 0.25$; $M_{step} = 10$

As shown in Fig. 13 some M cause the method to fail. Our simulations allowed us to confirm this drawback that was already suggested by the authors of the *method testing the autocorrelation of distant samples*. This method fulfills its purpose of allowing the use of oscillators with dissimilar periods during sampling. The results in Fig. 14 show that 92.97% of the implemented pairs of ROs can be used for this method. Nevertheless, the method still depend on the hardware. Our evaluation procedure helped identify the effects of the method parameters on the precision of its jitter measurement. Approximations of these parameters are given by the authors of [7] and inferred by our simulations.

4   General Discussion

Our methodology allowed us to draw conclusions concerning on the jitter measurement methods we evaluated. The *counter method* and the *coherent sampling method* proved to be ineffective for different reasons. While the limitations of the *counter method* are the theoretical assumptions it makes, the failure of the *coherent sampling method* was due to the high demands on its hardware. Nevertheless, both of their constraints could be overcome if the influence of flicker noise could be taken into account in their stochastic models. Although for now our methodology does not address this issue, we plan to do so in a future work. The simplicity of these methods renders this analysis compelling.

The *differential delay line method* and the *method testing the autocorrelation of distant samples* are a step ahead of the other two. However, the last method seems to be less dependent on its hardware. While its optimal parameters depend on the particular hardware implementation, it is possible to choose certain parameters that result in accurate jitter measurements.

Table 1 lists the most important results obtained with all the methods evaluated in this work. The first column lists their acceptable average errors obtained through simulations. The second column shows what we considered to be the optimal parameters of the methods both to obtain acceptable precision and to be certain not to risk flicker noise influencing real measurements (according to the results in [8]). The third column indicates if the method is implantable on the targeted Cyclone V FPGA and to what extent based on our experiments. The choice of a different hardware will result in different values that may not be contained within the ranges of the second column, thus affecting the feasibility of the method on that hardware.

5   Conclusions

In this paper, an evaluation procedure for jitter measurement methods has been proposed. This procedure is based on the analysis and simulation of the measurement principle. The simulation allows to control the injected jitter and to evaluate the accuracy of the measurement by comparing the result of the simulation with the injected value. Obtaining accurate jitter measurement results with the simulation is therefore a necessary condition to confirm validity of the method. Simulations can also help in finding the optimal ranges of parameters in the exploration space allowing to obtain the required precision. These ranges can guide the designer in the hardware implementation of this method.

Our evaluation procedure proved to be effective for the identification of the optimal conditions of accurate jitter measurement with each evaluated method. Furthermore, the method helps identify their limits and defects.

While the proposed methodology is unavoidable for validation of the measurement method, it is not sufficient: some hardware constraints, which depend on the selected technology, can still limit its implementation. Nevertheless, if the simulations show that the measurement is not precise enough or incorrect, there is no use to continue with tiring and time consuming hardware implementations - the method will not work better in hardware.

The procedure described in this article represents a simple low-time consuming way of evaluating jitter measurement methods that relies on current computational power. To further facilitate the design of measurement methods, our codes will be made open-source. This will allow designers to apply the proposed methodology using their own criteria.

We intend to encourage authors of a future jitter measurement method to use our procedure. In this way, they can push their proposal to its limits and identify its shortcomings reliably. Hopefully, our work will facilitate progress towards the ultimate goal of developing an on-board on-line jitter measurement method for the characterization of the entropy source of a TRNG to ensure its effectiveness.

References

1. A proposal for: functionality classes for random number generators, AIS 20 / AIS 31. Technical report, Bundesamt für Sicherheit in der Informationstechnik (BSI), Bonn (2011)

2. Allini, E.N., Skórski, M., Petura, O., Bernard, F., Laban, M., Fischer, V.: Evaluation and monitoring of free running oscillators serving as source of randomness. In: CHES'18, vol. 3, pp. 214–242 (2018)
3. Baudet, M., Lubicz, D., Micolod, J., Tassiaux, A.: On the security of oscillator-based random number generators. J. Cryptol. **24**(2), 398–425 (2011). https://doi.org/10.1007/s00145-010-9089-3
4. Corporation, I.: Cyclone V Device Handbook. Intel Corporation, Santa Clara (2020)
5. Dadouche, F., Turko, T., Uhring, W., Malass, I., Bartringer, J., Le, J.P.: Design methodology of TDC on low Cost FPGA targets case study: implementation of a 42 PS resolution TDC on a Cyclone IV FPGA Target. In: SENSORCOMM'15, pp. 29–34 (2015)
6. Fischer, V.: A closer look at security in random number generators design. In: Schindler, W., Huss, S.A. (eds.) COSADE 2012. LNCS, vol. 7275, pp. 167–182. Springer, Heidelberg (2012). https://doi.org/10.1007/978-3-642-29912-4_13
7. Fischer, V., Lubicz, D.: Embedded evaluation of randomness in oscillator based elementary TRNG. In: Batina, L., Robshaw, M. (eds.) CHES 2014. LNCS, vol. 8731, pp. 527–543. Springer, Heidelberg (2014). https://doi.org/10.1007/978-3-662-44709-3_29
8. Haddad, P., Teglia, Y., Bernard, F., Fischer, V.: On the assumption of mutual independence of jitter realizations in P-TRNG stochastic models. In: DATE 2014, pp. 1–6. IEEE (2014)
9. Hajimiri, A., Limotyrakis, S., Lee, T.H.: Jitter and phase noise in ring oscillators. IEEE J. Solid-State Circ. **34**(6), 790–804 (1999)
10. Khaddour, W., Dadouche, F., Uhring, W., Frick, V., Madec, M.: Design methodology and timing considerations for implementing a TDC on a Cyclone V FPGA target. In: NEWCAS 2020, pp. 126–129 (2020)
11. Koç, Ç.K. (ed.): Cryptographic Engineering. Springer, Boston (2009)
12. Lu, Y., Liang, H., Yao, L., Wang, X., Qi, H., Yi, M., Jiang, C., Huang, Z.: Jitter-quantizing-based TRNG robust against PVT variations. IEEE Access **8**, 108482–108490 (2020)
13. Lubicz, D., Bochard, N.: Towards an oscillator nased TRNG with a certified entropy rate. IEEE Trans. Comput. **64**(4), 1191–1200 (2015)
14. Ma, Y., Lin, J., Chen, T., Xu, C., Liu, Z., Jing, J.: Entropy evaluation for oscillator-based true random number generators. In: Batina, L., Robshaw, M. (eds.) CHES 2014. LNCS, vol. 8731, pp. 544–561. Springer, Heidelberg (2014). https://doi.org/10.1007/978-3-662-44709-3_30
15. FIPS 140-3 Derived Test Requirements (DTR): CMVP Validation Authority Updates to ISO/IEC 24759. Special publication (NIST SP) SP 800–140, National Institute of Standards and Technology, Gaithersburg, MD (2020)
16. Petura, O., Mureddu, U., Bochard, N., Fischer, V., Bossuet, L.: A survey of AIS-20/31 compliant TRNG cores suitable for FPGA devices. In: FPL 2016. IEEE (2016)
17. Turan, M.S., Barker, E., Kelsey, J., McKay, K.A., Baish, M.L., Boyle, M.: Recommendation for the entropy sources used for random bit generation. Technical report NIST SP 800–90B, National Institute of Standards and Technology, Gaithersburg, MD (2018)
18. Valtchanov, B., Aubert, A., Bernard, F., Fischer, V.: Modeling and observing the jitter in ring oscillators implemented in FPGAs. In: DDECS 2008, pp. 158–163 (2008)

19. Valtchanov, B., Fischer, V., Aubert, A.: A coherent sampling based method for estimating the jitter used as entropy source for true random number generators. In: SAMPTA'09 (2009)
20. Yang, B., Rozic, V., Grujic, M., Mentens, N., Verbauwhede, I.: On-chip jitter measurement for true random number generators. In: AsianHOST'17, pp. 91–96 (2017)
21. Zhu, S., Chen, H., Fan, L., Chen, M., Xi, W., Feng, D.: Jitter estimation with high accuracy for oscillator-based TRNGs. In: Bilgin, B., Fischer, J.-B. (eds.) CARDIS 2018. LNCS, vol. 11389, pp. 125–139. Springer, Cham (2019). https://doi.org/10. 1007/978-3-030-15462-2_9

Comparing Key Rank Estimation Methods

Rebecca Young[1], Luke Mather[2], and Elisabeth Oswald[1,3(✉)]

[1] Department of Computer Science, University of Bristol, Bristol, UK
elisabeth.oswald@aau.at
[2] PQShield, Bristol, UK
[3] Digital Age Research Center (D!ARC), University of Klagenfurt,
Klagenfurt, Austria

Abstract. Recent works on key rank estimation methods claim that algorithmic key rank estimation is too slow, and suggest two new ideas: replacing repeat attacks with simulated attacks (PS-TH-GE rank estimation), and a shortcut rank estimation method that works directly on distinguishing vector distributions (GEEA). We take these ideas and provide a comprehensive comparison between them and a performant implementation of a classical, algorithmic ranking approach, as well as some earlier work on estimating distinguisher distributions. Our results show, in contrast to the recent work, that the algorithmic ranking approach outperforms GEEA, and that simulation based ranks are unreliable.

Keywords: Key rank · Estimation

1 Introduction

From a real-world adversary's point of view, a side channel attack is successful if it reveals enough information about the unknown secret key such that this key can be found via a biased brute-force key search. Such a biased brute-force key search works by ranking all keys according to their distinguishing scores. This pragmatic viewpoint is also taken in the context of formal evaluations where the actual demonstration of an attack may include the argument of how much effort remains for a biased brute-force search.

The computational effort for an adversary to perform a biased brute-force key search, which can be estimated by an evaluator by calculating the position of a known secret key within the (ranked) key space, were first addressed by the academic community in [18,19]. Then, in quick succession, a number of better (faster and tighter) key rank and key enumeration algorithms appeared, e.g. [1,8,9,11,14] to name some approaches. The algorithms by [8] and [14] enable both key enumeration as well as key ranking, and it was later shown that they are mathematically equivalent [12]. After an early comparison [15] there appeared several improvements to existing algorithms; the algorithm by [14] in particular was improved in [11] but also in [13], and there exists a fast implementation

I. Buhan and T. Schneider (Eds.): CARDIS 2022, LNCS 13820, pp. 188–204, 2023.
https://doi.org/10.1007/978-3-031-25319-5_10

[8] by [16]. There is also [4] (with further work [3,5]), which is of comparable speed as [8], but enables better bounds. A short cut estimator for the expected key rank (aka "Guessing Entropy", short GE) called GM (for "Massey Guessing Entropy") was proposed in [2] to deal with long keys. This estimator turns out to be highly problematic in the experiments by [20], who proposed two new ideas for GE estimation.

1.1 The Challenge in Practice

A *single run* of any existing ranking algorithm is unproblematic: for a key space associated with a typical symmetric encryption algorithm, even a few minutes of computation time are of no concern in an evaluation. However memory and time requirements for the mentioned ranking algorithms don't scale linearly, and a single side channel experiment is insufficient to judge an implementation (unless it is trivially vulnerable). Consequently, the challenge is to have an approach that enables to estimate key ranks over many (repeat) experiments for potentially long keys very quickly.

Many Ranks: a single experiment can only give "circumstantial evidence", thus repeating experiments, and computing some (summary) statistics is necessary for a sound interpretation. Beside the question of *which statistic to use to report outcomes*, the challenge in practice is to produce *multiple experiments* within the time and financial constraints of a product evaluation cycle.

Long Keys: whilst a typical symmetric algorithm takes 128 bits of key material, long term security requires the use of 256 bits of key material in the symmetric setting. The idea of biased brute-force attacks also applies to asymmetric cryptography, which then leads to the requirement of dealing with keys that are considerably longer than 128 bits. Thus the challenge in a practical evaluation setting is to also *deal with long cryptographic keys*.

1.2 Recent Conceptual Advances

The latest work [20] proposes two ideas to make average key rank computations faster[1]. Their first new idea is called PS-TH-GE (pseudo-theoretic GE), and the principle behind it can be applied to both average key ranks and GE computation. Their second idea exclusively applies to GE computation and it is called GEEA (GE estimation algorithm).

The idea behind the PS-TH-GE goes back to [6,7,10,17], which all observed that it is possible to statistically characterise the distribution of distinguishing vectors resulting from correlation, distance of means, and template attacks. Whilst the previous work [17] derived the distributions for specific distinguishers,

---

[1] It is important to distinguish between the average over multiple key rank experiments, and the expected key rank aka Guessing Entropy. Both metrics capture the average behaviour, but the latter requires information about the entire key rank distribution.

[20] suggest to simply use the plug in estimator (empirical mean and covariance) based on repeat attack samples from the actual device. Thus, [20] can deal with any "additive" distinguisher. This is a potential solution for the problem of needing many experiments: instead of (re)sampling distinguishing vectors from attacks on real device data, the suggestion is to sample them from an initial characterisation that is based on the real device data (we call this synthetic data).

The additional idea behind the GEEA is that to derive the full key rank distribution, it suffices to estimate the distribution related to "pair-wise" success rates. With this, the key rank distribution over arbitrarily many distinguishing vectors (representing a full key rather than a single subkey) can be easily derived (i.e. long keys are easy to deal with), and by sampling from this distribution, a GE estimate can be produced.

1.3   Assumptions, Gaps in Knowledge, and Our Contribution

The premise behind [20], but also other recent works like [2] is that even the fastest ranking algorithms [14] and [8] (despite the various optimisations) are still too slow to be of practical use in the context of long keys and also repeat experiments. In [20] the cost of running a histogram based ranking algorithm is given as roughly 17s for a 16-byte key. Many of the reported experiments in [20] are for either single byte keys or short keys (where calculating the key rank is trivial)[2]. Only a few experiments are for realistic key sizes, where notably a comparison with ranks derived (algorithmically via e.g. [8] or [14]) from actual attacks for long keys is missing. Furthermore, a direct comparison with the previous work of [17] is missing as well: the previous work [17] derived explicit expressions for distinguishers like correlation and templates, whereas [20] base most experiments on deep learning distinguishers where we do not have explicit formulas for the distinguishing vectors. It is thus unclear how the use of the plug-in estimator (as suggested by [20]) compares to the carefully derived estimators from Rivain.

The lack of substantial large key experiments, and experiments with known distinguisher distributions creates multiple gaps and makes it impossible to understand how well the PS-TH-GE and GEEA perform in comparison to a well implemented classical ranking algorithm, and/or using estimators for known distinguisher distributions. It is also unclear how good synthetically produced ranks are in contrast to fresh-attack-based ranks.

We believe that this creates an overall gap to the needs of real world evaluations where the rank of the full key matters, and guarantees are required about the behaviour of average rank estimation metrics. Our contribution aims to narrow this gap. First we compare our customised implementation of an open-source algorithmic ranking algorithm (based on [11]) with an implementation of GEEA in Sect. 2 to understand the difference in computational performance. We then

---

[2] The style of experiments in [20] is in line with the single byte experiments in previous work such as [17].

compare ranks based on Rivain's method as well as the methods by Zhang et al. (PS-TH-GE and GEEA) with each other as well as with ranks based on fresh data in Sect. 3 (we use simulations for this purpose). For completeness we also work with some real device data in Sect. 4.

Our results show that the (optimised) algorithmic ranking based on [11] is significantly faster than GEEA, but more importantly that it can cope with keys of up to 4096 bits. There is in fact a severe performance penalty "hidden" in the GEEA algorithm, which is how many keys need to be sampled (the parameter M), see Sect. 2. We then show that in the context of classical distinguishers (correlation and templates) both PS-TH-GE and GEEA offer considerably less accurate (average) ranking results than when using a performant ranking algorithm implementation. We also observe that Rivain's formulas for correlation and template based distinguishing vectors are delivering by far better results than just using the plug-in estimator i.e. using the PS-TH-GE. We derive these conclusions from experiments based on simulated data, in Sect. 3. We demonstrate the same behaviour based on data from a real device in Sect. 4, and provide a more in-depth conclusion in Sect. 5. Our comparisons are deliberately shown via the use of box plots: this technique visualises significant distribution characteristics, and with it we provide evidence that any metric that only captures an average behaviour (like the GE) misses significant information about the "worst case" or "lucky" adversary.

2 Classical Key Rank Estimation vs. GEEA

We want to challenge the premise of [20] and [2] that algorithmic key rank implementations are not performant enough to be used "at scale" during evaluations. To challenge this premise, we took the open source library https://github.com/sca-research/labynkyr that was released alongside [11]. In their work [11] examine a large number of key rank implementation options (for the purpose of enumeration) and provide optimised C++ libraries to support large scale, parallel key enumeration experiments. We extracted the key ranking part from it, and wrote our own interface for the library so that we could use it conveniently with both simulated and real data. In order to cater for very long keys, we use Boost's multi-precision library to efficiently accumulate the large rank values. Finally we took advantage of the concept of compile time evaluation, where one can supply the compiler with the values of certain variables and have it "pre-compute" the output of relevant functions, with the aim of achieving some (slight) performance advantage during run time (e.g. because we know the dimension of any specific key rank experiment a priori, we can compile for these specific dimensions).

All our experiments were conducted on standard computing equipment, we use laptops (1.6 GHz and 2.3 GHz), and the implementations are all single threaded.

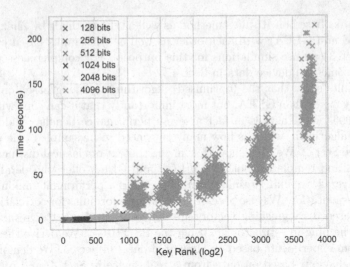

(a) Distribution of execution time across increasing key space size.

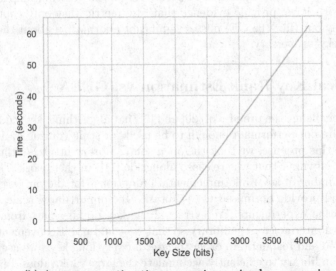

(b) Average execution time across increasing key space size.

Fig. 1. Key rank performance evaluation.

2.1 Evaluating a Modern Algorithmic Ranking Algorithm

We then ran a set of simulated attacks in order to quickly generate distinguishing scores to profile the performance of our ranking implementation. In the simulated attacks, all distinguishing scores are generated via standard DPA style attacks using correlation as a distinguisher. We fixed the simulated device leakage model as well as the adversarial prediction model to Hamming weight, used the usual

AES SubBytes as a target function, and varied the additive Gaussian noise to "shift" keys (when fixing the number of attack traces). In this way we were able to generate a very large number of repeat AES attack simulations (for a single subkey), which we used to generate data for attacks with keys of increasing length (by simply grouping single key experiments). We went from the standard AES key size of 128 bits, up to 256 bits, and then further up using typical asymmetric key sizes 512, 1024, 2048 and 4096, by simply increasing the number of columns. For each of these key sizes we took three random keys, and went through simulations with added noise creating SNR values ranging from 0.1 up to 0.5. In total we performed 9000 key rank experiments on our simulated data.

Figures 1a and b show the results. In both plots, the x-axis represents the rank of a returned key, and the y-axis the time that it took to compute its rank. Each x represents an outcome, and we indicated groupings based on the size of the key space with different colours. The plot in Fig. 1a shows the time that it takes for one key rank experiment to finish. Across all key space sizes, keys which have a low rank (thus are easy to find) are returned faster, and there is a small gradual increase for deeper keys. Obviously, for very long keys (4096 bits), there is a more marked time difference for key rank experiments returning likely vs unlikely keys. The slowest experiments are, obviously, the experiments that return unlikely keys from the 4096 bit case, where it took up to 220 s to complete a single run of key rank. Short keys, such as 128 bit keys, take in the worst case 0.01 s to complete. The plot in Fig. 1b shows a simple arithmetic average over the execution time for each key space size: the time does not scale linearly, but key lengths of interest are all within "easy reach".

Our experiments dispel the myth that algorithmic key ranking is too slow for practice. For short and medium keys (128–2024 bits) the time to return a single key rank (including deep keys) is extremely short. For very long keys (4096), the time to return a single (deep key) experiment is such that repetitions are practically possible.

2.2 Evaluating GEEA Performance Characteristics

The trick behind GEEA is that one can compute the GE based on working with comparison vectors (i.e. the difference between the score of the correct key and the incorrect key candidates), which enables writing the GE as a sum of "pairwise" comparison scores. This trick helps to reduce the number of parameters that need to be estimated (co-variances between key candidates are now ignored), and enables us to easily move from the distribution associated with a single subkey to the full key distribution. The resulting GE estimation algorithm is given as Alg. 1 in [20].

A crucial parameter of the GEEA algorithm is the factor M, which is the number of samples that one takes from the full key rank distribution. Each sample of the distribution gives an estimated value for the key rank. As in all key rank estimation approaches, a single sample only gives circumstantial evidence. Thus more than one sample is required, which leads to the immediate question of how many samples M are necessary to get an accurate key rank estimation?

In their paper [20] say that choosing M in the order of 10^7 leads to good results, but in their experiments they set $M = N$ (N would be the number of repeat attacks in a conventional setting). In the context of their concrete experiments, this results in N being in the order of 10,000, which is evidently much smaller than 10^7.

Our GEEA implementation runs in Python, and we ensured to match the speed of [20]: they report that GEEA for $M = 1$ takes about 10^{-4} s). The most costly part of computing GEEA is the CDF computation: for each sample, this look-up is required, thus the cost of GEEA is directly proportional to M.

Figures 2a and b show two representative results from running GEEA with an increasing value of M for a key with 16 bytes. Figure 2a shows the linear increase of the execution time as a function of M. Figure 2b side shows the distribution of the ranks for increasing values of M. We used a box plot to visualise the distribution. A box plot gives a "five number summary" and contains a box and "whiskers". The box represents where most of the ranks sit, with the middle line being the median (the boundaries are given by the first and third quartile). The whiskers represent the first/third quartile plus 1.5 of the inter-quartile range, and any additional points are the outliers. Figure 2b shows very clearly how the rank estimation quality of GEEA changes as M increases.

We argue that at the very least M needs to be chosen equal or larger than 10,000 for any reasonable estimation quality, which would indicate that a single run of GEEA is considerably slower than an equivalent run of our good implementation of an algorithmic ranking algorithm. Figure 3 provides evidence for this claim: we compared the execution time for ranking keys (16 byte full key) at different depths of the key space. GEEA was always slower than algorithmic key rank. Perhaps with a more optimised implementation of GEEA (e.g. switching to C or C++, and using an optimised implementation for the CDF computation, which we do via a standard library call) one could bring a GEEA with a reasonable choice for M to the performance of key rank for a 16 byte key. Evidently though, we find no supporting evidence that GEEA would outperform algorithmic ranking.

3   Challenging Synthetic Data Based Key Ranks

The idea of using synthetic distinguishing scores to speed up success rate estimations dates back to [17] who derived the distribution of the distinguishing vectors of correlation and template based DPA style attacks. Using the distribution of such distinguishing vectors [17] demonstrate that drawing from these distributions, rather than conducting fresh experiments (aka experiments that require the measurement of new traces from a device), enables a very accurate prediction of success rates for the corresponding attacks. Three distributions were characterised: two modified correlation coefficients $\dot{\rho}$ and $\ddot{\rho}$, as well as a least squares estimate L_k.

The key insight by [20] was to notice that the distribution of any "additive distinguisher" [17] can be characterised by a multivariate normal distribution

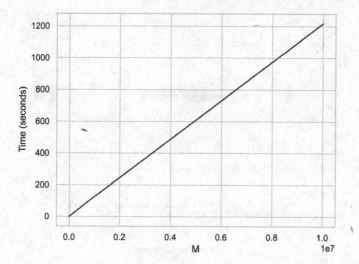

(a) GEEA performance cost with increasing M.

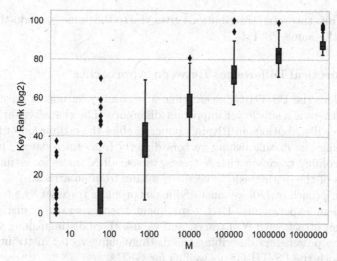

(b) GEEA rank outcomes with increasing M.

Fig. 2. GEEA performance evaluation as M increases.

(MVN) and that the parameters of the MVN can be estimated by the plug-in estimators (i.e. by computing the average and the empirical co-variance matrix describing a distinguishing vector directly by observing multiple such vectors arising from attacks on real traces). Once the statistical characterisation has been obtained, further attack outcomes can be simulated without having to take new measurements from a device (the idea of simulating attacks from the

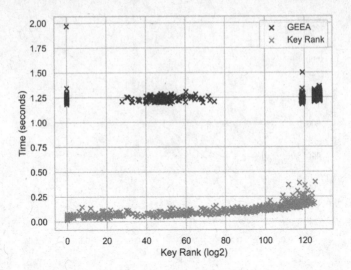

Fig. 3. GEEA vs key rank.

estimated (aka theoretic) distributions gives rise to the name "pseudo-theoretic", leading to the name PS-TH-GE).

3.1   Theoretical Differences Between Approaches

At face value, the PS-TH-GE seems like a natural extension of Rivain's approach, but there is a subtle yet important difference. The characterisation of the distinguisher distributions in Rivain's paper enables the estimation of the MVN characterising the distinguishing vectors directly from *trace* data. This means that in a profiling trace set with N traces, we use all N traces for estimating the parameters of the distinguishing vector resulting from an attack.

In the approach by [20] we must divide our profiling trace set by q to generate q independent experiments. Each experiment then generates a distinguishing vector for an attack with N/q traces, and we use these distinguishing vectors to estimate the parameters describing the distinguishing vector distribution. This holds for both the PS-TH-GE as well as for GEEA.

Thus the estimator of Rivain uses a single profiling set of size N, but the empirical estimator (i.e. in the context of the PS-TH-GE and GEEA) requires the use of multiple profiling sets of size N/q.

All of these discussed estimations are thus based on a set, i.e. a finite amount of available data for profiling. From a statistical perspective, one can thus use them to make predictions about attack outcomes using up to some number of traces: N for Rivain's approach and N/q for the PS-TH-GE and GEEA. It is not clear that it is possible to use the estimations for attacks with more than N, resp. N/q traces, and we will demonstrate in a practical experiment that the quality of the estimations gets poorer when trying to predict outcomes of attacks with more than N, and resp. N/q traces.

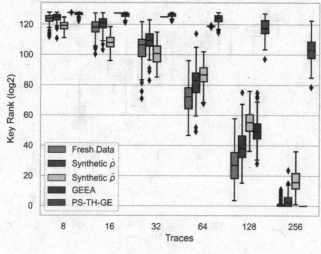

(a) Full summary statistic

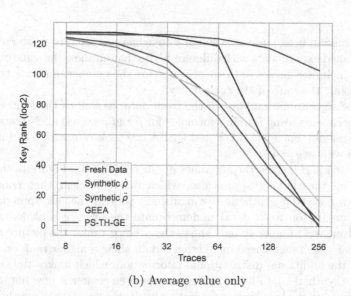

(b) Average value only

Fig. 4. Comparison between key rank estimation algorithms for correlation attacks.

3.2   Practical Differences Between Approaches: Correlation

We first focus on a comparison between the approaches in the context of a typical correlation based attack. We based our experiments on simulations (16 key bytes), which enable us not only to fully control the attacks (and thereby key depths) but also to do many attacks extremely quickly for all rank estimation methods. We use the same type of simulations as described before (HW leakage

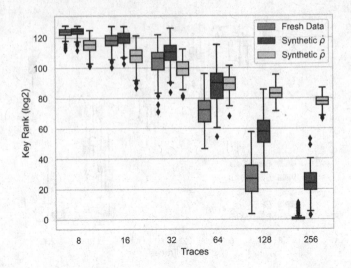

Fig. 5. Average key rank estimations using correlation based models, initial trace set of 32 traces.

model, Gaussian noise) and set the noise to achieve a signal to noise ratio of 0.1 (we performed experiments with different SNRs that all show the same outcomes, thus only include a subset of them). By varying the number of attack traces, we can then shift the rank of the correct key.

For the algorithmic ranking on the fresh data, as well as the synthetic data based experiments using Rivain's formulas for $\dot{\rho}$ and $\ddot{\rho}$ we use all N traces (given along the x-axis). For PS-TH-GE and GEEA, we took N/q with $q = 8$ if $N > 8$; for $N = 8$ we use $q = 4$.

Figures 4a and b show the outcomes of our comparison. Figure 4a shows box plots for all the ranking approaches, which makes it clear that none of the approaches based on synthetic distinguishing vectors (i.e. all approaches bar classical ranking on fresh data) underestimate the average rank and also the "lucky adversary". Figure 4b only shows the average ranks: the picture here also makes clear that ranks based on synthetic data show a higher rank on average. However, the additional distributional information, which makes the shortcomings of the synthetic data based ranks perhaps even clearer is now not available.

We argue that these outcomes not only demonstrate that synthetic data based ranks underestimate the average case adversary, but they also demonstrate the need to report more statistical information than just the average, because lucky adversaries are important for practical security.

Next we want to know how much the number of initial profiling traces N impacts on the outcomes of synthetic data based key ranks. We set up an experiments where we estimated $\dot{\rho}$ and $\ddot{\rho}$ based on $N = 32$ traces, and the performed key rank estimations using Rivain's formulas (N is a parameters in these formulas) for varying N. Figure 5 shows that synthetic data based key ranking experiments with $N \leq 32$ well approximate the actual key rank (as per freshly

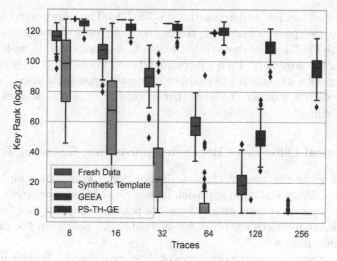

(a) Full summary statistic

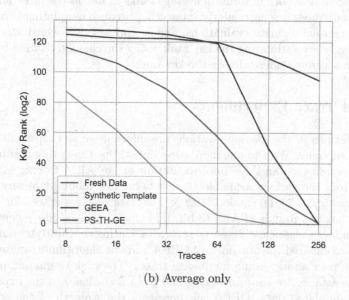

(b) Average only

Fig. 6. Comparison between key rank estimation algorithms for Gaussian template attacks.

sampled attack data), whereas key ranking experiments with $N > 32$ significantly diverge from the true key rank (they underestimate the vulnerability).

These experiments show clearly that if synthetic data based ranking would be considered in practical security evaluations, then it is clearly preferable to use the precisely characterised distributions from Rivain over a simple plug-in estimator based approach, which underpins the PS-TH-GE. They also demonstrate the need for ensuring plenty of profiling traces (or examples of attack outcomes when using the PS-TH-GE), because synthetic data based ranks tend to underestimate the power of the adversary. Thus a recommendation would be to "overdimension" the profiling data set, which will enable to generate reasonably good simulations for attacks with a number of traces that is smaller than the number of traces used for profiling.

3.3   Practical Differences Between Approaches: Gaussian Templates

Rivain also gives the precise characterisation for a Gaussian template based distinguisher. Using the same approach as described before, we ran experiments to compare several key rank estimations. We note at this point that because full rank covariance matrices are often not-invertible, we only keep the diagonal of the covariance matrix.

Figures 6a and b show the outcomes of these experiments. It is clear from the outcomes that all synthetic data based key ranks are severely misestimating the key rank outcomes: both in terms of average ranks as well as the lucky adversary.

These outcomes are particularly interesting because templating is of importance in practical security evaluations, and it appears that whilst the PS-TH-GE is severely overestimating the key rank, using the characterisation by Rivain appears to severely underestimate the key rank.

4   Real Trace Experiments

The simulations in the previous sections are informative and representative for key rank outcomes. We add one more experiment, this time using a public trace set corresponding to an AES implementation on an ARM Cortex M3, which we took from the various available trace sets on the public repository Zenodo https://doi.org/10.5281/zenodo.5710205. The data features traces for multiple keys, and has two rounds of AES (8-bit, table look-up based implementation).

We conduct a classical differential style attack using the SubBytes output as a target. We compared the outcomes of GEEA and our algorithmic ranking implementation, over an increasing number of traces. The rank estimation outcomes from these attacks are show in Fig. 7a and b. In accordance to the experiments based on simulated data, GEEA misrepresents the strength of the adversary, which is problematic in an evaluation context. We see once more that reporting the distribution of the key rank estimations gives a more nuanced picture of the strength of attacks at certain levels of number of attack traces.

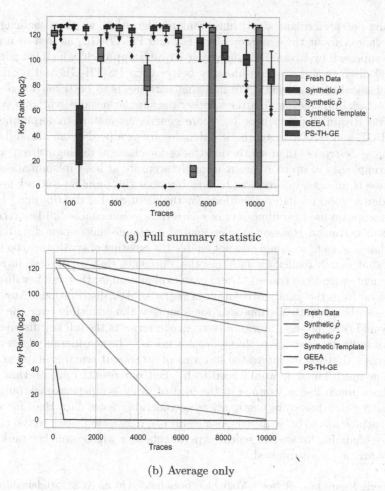

(a) Full summary statistic

(b) Average only

Fig. 7. Comparison between key rank estimation algorithms based on real data (8-bit AES implementation).

5 Conclusion

Our contribution analyses and challenges in some ways the premise of some of the most recent works on key rank estimation. Two assumptions were made in previous work: firstly, that algorithmic key ranking algorithms are too slow to deal with long keys or many experiments and secondly, that it is possible to create synthetic experiments based on characterising a single profiling dataset. Our own algorithmic ranking implementation, which is a more easily useable and customisable version of an open source library for parallel key enumeration, outperforms the supposedly best competitor GEEA, at all key lengths and depths. We also show that it is possible to rank keys with up to 4096 bits.

Using our performant algorithmic ranking algorithm, we then challenge the recent claims about the accuracy of GEEA and PS-TH-GE, as well as a much earlier approach by Rivain. We show that Rivain's approach, which is limited to classical distinguishers, is considerably better than PS-TH-GE and GEEA for these classical distinguishers. We find that care needs to be taken when simulating attack outcomes, no matter whether the simulations use synthetic vectors based on Rivain's approach or the more generic estimation underpinning the PS-TH-GE and GEEA. If synthetic outcomes are produced by an initial dataset containing N traces, then synthetic attacks for classical distinguishers, with a trace complexity of up to N are reasonably accurate at least in simulations.

There is an open question as to how big does the profiling dataset need to be to derive good enough estimations for the distinguisher distribution? In our experiments we used profiling sets of a size that would enable full key recovery with near certainty. However, the estimated distinguishing vector distributions from these datasets seemingly do not enable the creating of synthetic attack outcomes that match reality for *full keys* (i.e. outcomes that we observe in repeat experiments with fresh traces). There is an explanation for this: full key distributions arise from the joint distribution of *many* subkey distributions. Any small divergence between the estimations for a subkey thus multiply over the many subkey and turn into a significant divergence in terms of the full key distribution.

This indicates that the profiling dataset needs to be considerably larger than the dataset that we are interested in terms of attacks. It remains an open question how much larger it would need to be. But our results indicate that there is perhaps much less advantage in the idea of using synthetic attack outcomes because for the (preceding) estimation, considerably more data than for a successful attack has to be acquired. As a result one may as well stick to the current practice of aiming for several repeat experiments over which some key rank summary statistics are computed.

Acknowledgements. Rebecca Young has been funded by an NCSC studentship. Elisabeth Oswald was supported in part by the ERC via the grant SEAL (project reference 725042).

References

1. Bernstein, D.J., Lange, T., van Vredendaal, C.: Tighter, faster, simpler side-channel security evaluations beyond computing power. IACR Cryptology ePrint Archive 2015:221 (2015)
2. Choudary, M.O., Popescu, P.G.: Back to Massey: impressively fast, scalable and tight security evaluation tools. In: Fischer, W., Homma, N. (eds.) CHES 2017. LNCS, vol. 10529, pp. 367–386. Springer, Cham (2017). https://doi.org/10.1007/978-3-319-66787-4_18
3. David, L., Wool, A.: Fast analytical rank estimation. In: Polian, I., Stöttinger, M. (eds.) COSADE 2019. LNCS, vol. 11421, pp. 168–190. Springer, Cham (2019). https://doi.org/10.1007/978-3-030-16350-1_10

4. David, L., Wool, A.: Poly-logarithmic side channel rank estimation via exponential sampling. In: Matsui, M. (ed.) CT-RSA 2019. LNCS, vol. 11405, pp. 330–349. Springer, Cham (2019). https://doi.org/10.1007/978-3-030-12612-4_17

5. David, L., Wool, A.: Rank estimation with bounded error via exponential sampling. J. Cryptogr. Eng. **12**(2), 151–168 (2022)

6. Fei, Y., Ding, A.A., Lao, J., Zhang, L.: A statistics-based success rate model for DPA and CPA. J. Cryptogr. Eng. **5**(4), 227–243 (2015)

7. Fei, Y., Luo, Q., Ding, A.A.: A statistical model for DPA with novel algorithmic confusion analysis. In: Prouff, E., Schaumont, P. (eds.) CHES 2012. LNCS, vol. 7428, pp. 233–250. Springer, Heidelberg (2012). https://doi.org/10.1007/978-3-642-33027-8_14

8. Glowacz, C., Grosso, V., Poussier, R., Schüth, J., Standaert, F.-X.: Simpler and more efficient rank estimation for side-channel security assessment. In: Leander, G. (ed.) FSE 2015. LNCS, vol. 9054, pp. 117–129. Springer, Heidelberg (2015). https://doi.org/10.1007/978-3-662-48116-5_6

9. Grosso, V.: Scalable key rank estimation (and key enumeration) algorithm for large keys. In: Bilgin, B., Fischer, J.-B. (eds.) CARDIS 2018. LNCS, vol. 11389, pp. 80–94. Springer, Cham (2019). https://doi.org/10.1007/978-3-030-15462-2_6

10. Lomné, V., Prouff, E., Rivain, M., Roche, T., Thillard, A.: How to estimate the success rate of higher-order side-channel attacks. In: Batina, L., Robshaw, M. (eds.) CHES 2014. LNCS, vol. 8731, pp. 35–54. Springer, Heidelberg (2014). https://doi.org/10.1007/978-3-662-44709-3_3

11. Longo, J., Martin, D.P., Mather, L., Oswald, E., Sach, B., Stam, M.: How low can you go? Using side-channel data to enhance brute-force key recovery. IACR Cryptology ePrint Archive 2016:609 (2016)

12. Martin, D.P., Mather, L., Oswald, E.: Two sides of the same coin: counting and enumerating keys post side-channel attacks revisited. In: Smart, N.P. (ed.) CT-RSA 2018. LNCS, vol. 10808, pp. 394–412. Springer, Cham (2018). https://doi.org/10.1007/978-3-319-76953-0_21

13. Martin, D.P., Mather, L., Oswald, E., Stam, M.: Characterisation and estimation of the key rank distribution in the context of side channel evaluations. In: Cheon, J.H., Takagi, T. (eds.) ASIACRYPT 2016. LNCS, vol. 10031, pp. 548–572. Springer, Heidelberg (2016). https://doi.org/10.1007/978-3-662-53887-6_20

14. Martin, D.P., O'Connell, J.F., Oswald, E., Stam, M.: Counting keys in parallel after a side channel attack. In: Iwata, T., Cheon, J.H. (eds.) ASIACRYPT 2015. LNCS, vol. 9453, pp. 313–337. Springer, Heidelberg (2015). https://doi.org/10.1007/978-3-662-48800-3_13

15. Poussier, R., Grosso, V., Standaert, F.-X.: Comparing approaches to rank estimation for side-channel security evaluations. In: Homma, N., Medwed, M. (eds.) CARDIS 2015. LNCS, vol. 9514, pp. 125–142. Springer, Cham (2016). https://doi.org/10.1007/978-3-319-31271-2_8

16. Poussier, R., Standaert, F.-X., Grosso, V.: Simple key enumeration (and rank estimation) using histograms: an integrated approach. In: Gierlichs, B., Poschmann, A.Y. (eds.) CHES 2016. LNCS, vol. 9813, pp. 61–81. Springer, Heidelberg (2016). https://doi.org/10.1007/978-3-662-53140-2_4

17. Rivain, M.: On the exact success rate of side channel analysis in the Gaussian model. In: Avanzi, R.M., Keliher, L., Sica, F. (eds.) SAC 2008. LNCS, vol. 5381, pp. 165–183. Springer, Heidelberg (2009). https://doi.org/10.1007/978-3-642-04159-4_11

18. Veyrat-Charvillon, N., Gérard, B., Renauld, M., Standaert, F.-X.: An optimal key enumeration algorithm and its application to side-channel attacks. In: Knudsen, L.R., Wu, H. (eds.) SAC 2012. LNCS, vol. 7707, pp. 390–406. Springer, Heidelberg (2013). https://doi.org/10.1007/978-3-642-35999-6_25

19. Veyrat-Charvillon, N., Gérard, B., Standaert, F.-X.: Security evaluations beyond computing power. In: Johansson, T., Nguyen, P.Q. (eds.) EUROCRYPT 2013. LNCS, vol. 7881, pp. 126–141. Springer, Heidelberg (2013). https://doi.org/10.1007/978-3-642-38348-9_8

20. Zhang, Z., Ding, A.A., Fei, Y.: A fast and accurate guessing entropy estimation algorithm for full-key recovery. IACR Trans. Cryptogr. Hardw. Embedded Syst. 26–48 (2020)

Cycle-Accurate Power Side-Channel Analysis Using the ChipWhisperer: A Case Study on Gaussian Sampling

Nils Wisiol[1] , Patrick Gersch[1(✉)] , and Jean-Pierre Seifert[1,2]

[1] Security in Telecommunications, Department of Electrical Engineering and Computer Science, TU Berlin, 10587 Berlin, Germany
{nils.wisiol,patrick.gersch,jean-pierre.seifert}@tu-berlin.de
[2] FhG SIT, 64295 Darmstadt, Germany

Abstract. This paper presents an approach to uncover and analyze power side-channel leakages on a processor cycle level precision. By carefully designing and evaluating the measurement setup, accurate trace timing is enabled, which is used to overlay the trace with the corresponding assembly code. This methodology allows to expose the sources of leakage on a processor cycle scale, which allows for evaluating new implementations. It also exposes that the default ChipWhisperer configuration for STM32F4 targets used in prior work includes wait cycles that are rarely used in real-world applications, but affect power side-channel leakage.

As an application for our setup, we target the widely used Sign-Flip function of Gaussian sampling code used in multiple Post-Quantum Key-Exchange Mechanisms and Signature schemes. We propose new implementations for the Sign-Flip function based on our analysis on the original implementation and further evaluate their leakage.

Our findings allow the conclusion that unmasked cryptographic implementations of schemes based on Gaussian random numbers for STM32F4 cannot be secure against power side-channel, and that masking just the Gaussian sampler is not a viable option.

Keywords: Power side-channel analysis · ChipWhisperer · Processor cycle level analysis · Gaussian sampler · Sign-Flip · GALACTICS · FALCON · FrodoKEM

1 Introduction

The US National Institute of Standards and Technology (NIST) started the Post-Quantum Cryptography (PQC, [22]) process in November 2017 to select quantum-resistant key-exchange mechanisms and signature schemes as a preemptive response to the emergence of large-scale quantum computers. As part of this process, NIST also considers the resistance to side-channel attacks an important criterion for algorithm selection; submitters are encouraged to provide

© The Author(s), under exclusive license to Springer Nature Switzerland AG 2023
I. Buhan and T. Schneider (Eds.): CARDIS 2022, LNCS 13820, pp. 205–224, 2023.
https://doi.org/10.1007/978-3-031-25319-5_11

implementations of their schemes optimized for microcontrollers and FPGAs. The ARM Cortex-M4 turned out as a baseline for many performance comparisons, because of projects like [14] and works based on it [15,30]. Therefore, this paper will also focus on this specific microcontroller.

The Round 3 finalist FALCON [10], Round 3 alternative candidate FrodoKEM [2] and BLISS [8] (the constant-time GALACTICS implementation [8]), one of the earliest post-quantum schemes, all make use of Gaussian sampling. In FALCON and BLISS, Gaussian random numbers are required for signature generation; in FrodoKEM, they are required for key encapsulation and decapsulation. All of their Gaussian sampler implementations are based on a Sign-Flip function. Multiple attacks on BLISS implementations are based on side-channel leakage that revealed information about the signs of the Gaussian samples used [20,32]. In this paper, we turn our attention to leakage observable via the power side-channel, which is a typical attack scenario against embedded and IoT devices. The ChipWhisperer [28] provides a complete side-channel lab setup to enable rapid analysis.

Contributions

- We show that the Gaussian sampling routines of the GALACTICS implementation of BLISS as well as FALCON and FrodoKEM on the STM32F4 all leak information about the generated samples. We propose several new variants of these implementations, and show that the leakage can be reduced. However, based on our findings in this case study, we argue that the number representation format of the STM32F4 results in side-channel leakage that cannot be mitigated by modifications to the sampling routine.
- Using the ChipWhisperer, we present a novel approach for in-depth power analysis on the processor cycle level. We demonstrate a detailed breakdown of the required setup and condition to execute cycle level analysis on the STM32F4. The core of this cycle level analysis is to accurately overlay the traces with the corresponding assembly instructions.
- We demonstrate a flaw in the latest release version of the ChipWhisperer, potentially affecting previous power analyses based on the ChipWhisperer STM32F4 setup.

Organization of This Paper. In Sect. 3, we describe our measurement and analysis setup in detail and argue for its validity. Furthermore, we argue that the current default configuration of the ChipWhisperer STM32F4 target board does not reflect real-world scenarios, which may influence the validity of previous power side-channel analyses based on the ChipWhisperer setup (Sect. 4). Afterwards, in Sect. 5, we apply our analysis setup to the Sign-Flip subroutines used by GALACTICS, FALCON and FrodoKEM and demonstrate that information about the sampled sign is leaked. This section also covers advancements of the Sign-Flip implementation that are able to reduce the amount of leakage, but cannot fully mitigate it. Finally, we argue that the number representation format of the STM32F4 makes it impossible to implement a side-channel resistant Sign-Flip function. We discuss our findings in Sect. 6.

Related Work. Many works use the ChipWhisperer platform to evaluate implementation for their power side-channel resistance [13,20,29].

The implementation of the PQC candidate SIKE for the ARM Cortex-M4 was successfully attacked by Genêt and Kaluterović using the ChipWhisperer through single-trace power analysis [12]. A masked implementation of Saber for the Cortex-M4 was also successfully attacked using the ChipWhisperer. The leakage is thought to be related to the Hamming distance between the old and new values of pipeline registers [27]. Askeland and Rønjom attack the NTRU implementation with a ChipWhisperer. They identify the Hamming weight of secrets as a major reason for power side-channel leakage in implementations [5].

Leakage simulation aims to construct the power trace for a given set of assembly instruction. ELMO is a prominent leakage simulator for the Cortex-M0/M4 processor. Even though such simulators can give a good estimation of the power leakage, only experiments on real hardware can confirm their accuracy [21].

Tibouchi and Wallet present a timing attack on the Sign-Flip in BLISS, resulting in a full exposure of the signing key [32]. A power side-channel leakage in the Sign-Flip is used for two attacks against the Gaussian sampler resulting in full key recovery in Marzougui et al. [20].

2   Preliminaries

Notation. Binary strings are represented as a string with a bit symbol prepended, e.g. b'1000 1010' and hexadecimal numbers with a '0x' pre-pended, e.g. 0x1234. Numbering is in LSB_0 scheme, meaning it starts at zero for the least significant bit (LSB). The LSB is the right-most bit. The most significant bit (MSB) is thus the highest-order bit, which is the left-most bit.

Negative Number Representation. In writing, negative numbers are usually represented by a minus sign. However, in digital circuits, different ways to represent negative numbers in binary exist. First, the *sign-magnitude* represents negative numbers using a single sign bit, often the most-significant bit. Second, the *ones' complement* representation performs a bitwise NOT operation to get from positive to negative numbers. Third, the *two's complement* representation transforms from positive to negative numbers by applying a bitwise NOT operation and adding one afterwards. The preferred choice for modern processors is the *two's complement*, as it allows for efficient arithmetic operations and has only one representation of zero.

In Table 1, we give examples of 16-bit two's complement number representation along with their Hamming weight. The Hamming weight is a metric for the number of 1's in a binary string. Especially the difference in Hamming weight between positive and negative numbers should be noted.

Table 1. *Two's complement* numbers from -4 to 4 in 16-bit representation

Number	16-bit two's complement	Hamming weight
-4	b'11111111 11111100'	14
-3	b'11111111 11111101'	15
-2	b'11111111 11111110'	15
-1	b'11111111 11111111'	16
0	b'00000000 00000000'	0
1	b'00000000 00000001'	1
2	b'00000000 00000010'	1
3	b'00000000 00000011'	2
4	b'00000000 00000100'	1

Sign-Flip. Lattice-based cryptography such as FALCON [10], FrodoKEM [2], and BLISS [8] requires random numbers X drawn from discrete Gaussian distribution with mean $\mu = 0$ and variance σ^2, i.e.

$$X \sim \lceil \mathcal{N}(\mu, \sigma) \rceil.$$

$$\rho_{\mu,\sigma}(x) = e^{-\frac{|x-\mu|^2}{2\sigma^2}} \tag{1}$$

$$D_{\mathbb{Z},\mu,\sigma}(x) = \frac{\rho_{\mu,\sigma}(x)}{\sum_{z \in \mathbb{Z}} \rho_{\mu,\sigma}(z)} = \frac{1}{\sqrt{2\pi}\sigma} e^{-\frac{|x-\mu|^2}{2\sigma^2}} \tag{2}$$

(Note that the definitions for the distributions vary slightly across the different schemes.)

Several implementations for generation of random Gaussian numbers exist. Implementations based on the cumulative distribution table (CDT) have been shown to achieve the highest throughput and the smallest resource utilization [16]. Such implementations contain a hard-coded, precomputed table of the cumulative distribution table. Taking advantage of the symmetry of the distribution about the mean, the table size can be reduce by half to reduce resource consumption by only storing values for positive x. In this case, a sample is first drawn from distribution $|\lceil \mathcal{N}(\mu, \sigma) \rceil|$. In a second step, the Sign-Flip function

$$\text{Sign-Flip}(x, c) = \begin{cases} -x & c = 0 \bmod 2, \\ x & \text{otherwise,} \end{cases} \tag{3}$$

is applied to determine the sign, resulting in a sample from $\lceil \mathcal{N}(\mu, \sigma) \rceil$. In Eq. 3 the sample from the distribution $|\lceil \mathcal{N}(\mu, \sigma) \rceil|$ is x and c is a sample from a uniformly random distribution. The side-channel security of CDT samples has previously been studied [17].

The Sign-Flip function is used by GALACTICS [8], FrodoKEM [2] and FAL-CON [10] as part of their discrete CDT Gaussian sampling procedure. Table 2 shows the output for a chosen set of example inputs, represented in two's complement.

Table 2. Sign-Flip function value table. Output is shown in decimal and two's complement representation.

Input x	If $c = 0 \pmod 2$	If $c = 1 \pmod 2$
0	b'00000000 00000000' (0)	b'00000000 00000000' (0)
1	b'11111111 11111111' (-1)	b'00000000 00000001' (1)
2	b'11111111 11111110' (-2)	b'00000000 00000010' (2)
3	b'11111111 11111101' (-3)	b'00000000 00000011' (3)

Side-Channel Analysis. Kocher et al. [18] have shown that *power side-channel attack* can reveal information about the internal state of hardware that processes cryptographic algorithms, including the secret key, by measuring and analyzing the power consumed by the hardware over time (*power trace*). Subsequently, Chari et al. [7] proposed *template attacks*, where the attacker has full control to a device identical to the device under attack. This device can be used to build a model of power-consumption behavior in dependence of the internal state, which afterwards is deployed to deduce the internal state from the measured power consumption. Lerman et al. [19] proposed to use machine learning techniques for this modeling. This work also uses a machine learning based template attack; for creating the model, we use a multilayer perceptron.

3 Measurement Setup

This section details how we collected power traces using the ChipWhisperer for this work. It focuses on high temporal precision to enable matching of measured power consumption of the hardware to the instructions executed at the given time. Our complete analysis was conducted using the ChipWhisperer-Lite and an UFO target board with an STM32F4 target mounted. In the standard Chip-Whisperer configuration that we use, the power measurement is started using a dedicated GPIO pin of the target board.

3.1 Target: STM32F4

The STM32F4 target (CW308T-STM32F, [26]) was mounted on a ChipWhisperer CW308 UFO target board (NAE-CW308-04, [25]). The UFO board includes an oscillator, power supply and on-board LC low-pass filter for the connected target board. A 7.3728 MHz crystal was used for all experiments and set as default clock source. Therefore, each processor clock cycle takes approximately 135 ns. Based on the STM32F4, the STM32F405RGT6 [31] target board was designed to conform with the UFO target board requirements. The power consumption of the STM32F4 target is measured via a voltage measurement on a shunt resistor.

The STM32F4 is an ARM Cortex-M4 [3] architecture, implementing the 32-bit *Armv7E-M* instruction set architecture (ISA). The *Armv7E-M* has 16 32-bit registers named *r0–r15*. The Cortex-M4 has a 3-stage pipeline (fetch, decode, execute).

GPIO Toggle Speed. A GPIO pin from the STM32F4 is used to signal the start and end of the power trace by inserting toggle commands using software. Therefore, the toggle speed of the STM32F4 is of utmost importance as the accuracy of the start and end of the power trace relies on it. If the toggle of a GPIO pin takes longer than one clock cycle of the processor, this delay needs to be considered when matching the power trace against the executed program.

The GPIO toggle speed of the STM32F4 depends on the OSPEEDRy hardware register (b'10'), the capacitive load and the operation voltage (3.3 V) [31, datasheet p. 117]. Depending on the capacitative load, the rise/fall time lies between 4 ns and 6 ns and thus does certainly not exceed the cycle time of 135 ns. The GPIO toggle speed can thus be ignored for the purpose of matching power consumption against executed instructions in our setup.

Memory Caching & ART Accelerator. Memory caches are important to speed up memory access of the processor, but highly stateful. Memory access times can vary dramatically depending on the current state of the cache, with cache hits being processed much faster than cache misses. To match the executed instructions against the recorded power trace, we turned off all caching on our target board.

We disable the cache as enabling the cache raises the question how to initialize the cache state. As in practical attacks, the cache state depends on prior usage of the device, there is no generally valid answer to this question. For real-time applications, the cache may be turned off even in real-world scenarios.

The used STM32F405RGT does not include a flexible static memory controller (FSMC) [31, datasheet p. 14]. There is also no cache in between the ARM Cortex and the SRAM. All code will be placed on the Flash memory. The Flash memory is accessed from the processor by the adaptive real-time memory (ART) accelerator with a build in cache. Section 4 will discuss the ART in greater details.

Compiler Instruction Re-ordering. For all experiments, the *arm-none-eabi-gcc* compiler on version 9.2.1 has been used. It supports different compiler optimization flags to improve various performances like memory usage, code size and program speed. Optimization level 3 (*-O3*) is the one which is most often used for NIST PQC candidates [14], resulting in fast performance of the compiled binaries. However, due to instruction re-ordering by the compiler, this may falsify side-channel leakages.

This is a particular problem when using the GPIO-triggering used by the default ChipWhisperer [24] configuration, as instructions for triggering the start and end of the trace could be moved to earlier or later locations in the program.

Using the ChipWhisperer, tracing a function might look like *"trigger_high(); func_under_test(); trigger_low();"*. As *trigger_high()* and *trigger_low()* operate on the same hardware registers, the compiler is aware that they depend on each other. The same is not true for the function we are trying to trace. Therefore, the compiler might re-order the *func_under_test()* before or after the triggers. In some situations, the compiler might even inline a *func_under_test()* if the function is relatively short (depending on compiler flags and limits). After inlining the function, the compiler can re-order the inlined instructions with the trigger functions. As a result, the tracing might not capture some instructions of the *func_under_test()*, not trace the function at all, or include leakage caused by code originally intended to run before the start of the trigger or after the intended end of the trigger.

If the compiler supports the function attributes *no_reorder* or *noinline* then these can be used to mitigate the issue. Unfortunately the used *arm-none-eabi-gcc* compiler does not and therefore simply ignores these attributes. The only way to reliably prevent the re-ordering is the *optimize(2)* function attribute to set the optimization level lower for the code section of our *func_under_test*. While instruction reordering is one of the fundamental concepts to enable most optimizations, compilers do not have a model for time, but only for the result of instructions. Thus, reordering only guarantees that the outcome of the program is according to the code, but not intermediate steps [6].

3.2 Trace: ChipWhisperer

The ChipWhisperer-Lite (NPCB-CWLITECAP-01 CW1173 [23]) in our experiments collects traces using a high-gain Low Noise Amplifier (LNA) with a 10-bit Analog Digital Converter (ADC). In our setup, the tracing is triggered when the ChipWhisperer detects a rising edge on it's trigger input. I.e., the tracing starts when signal goes from *low* to *high* and stops when switching back from *high* to *low*.

Trigger Speed. After examining the toggle speed of the GPIO in Sect. 3.1, we will have a look at how accurate the ChipWhisperer-Lite notices the trigger and therefore starts/stops the tracing. Any delay by one or more clock cycle needs to be known to overlay the trace with the assembly code. To evaluate the delay at which the ChipWhisperer starts tracing after the rising edge on the trigger signal, we ran the code shown in Listing 1.1 on the target board.

After preparation, it uses a single instruction (line 8) to output a rising edge on the GPIO pin that we connected to the ChipWhisperer trigger input. In the instruction immediately after, the GPIO is set to low again.

Listing 1.1. ARM assembly code to toggle the trigger GPIO

```
1   # r11 stores the MMIO address of the trigger GPIO
2   # r10 stores the value to set the trigger GPIO to high
3   # r12 stores the value to set the trigger GPIO to low
4   ldr r11, =0x40020000
5   ldr r10, =0x1000
6   ldr r12, =0x10000000
7   .align 4
8
9   # Using a single 'STR' instruction
10  # to set the trigger GPIO to high
11  str r10, [r11, #24]
12
13  # Pre-calculated address and value in r11 and r12
14  # to set the trigger GPIO to low in one instruction
15  str r12, [r11, #24]
```

A single 'STR' instruction on an STM32F4 takes 2 cycles [4] to execute, but every consecutive 'STR' instruction takes only 1 cycle. We thus expect that the trigger GPIO is high for exactly 1 cycle. We call this assembly version of toggling the trigger GPIO using fixed constants in specific registers *fast_trigger*, as opposed to the ChipWhisperer's standard implementation, which is wrapped in C functions and hence requires multiple cycles to execute.

Our experimental results show that the ChipWhisperer collects a power trace of length 4. The default configuration of the ChipWhisperer collects 4 samples each processor cycle. Therefore, the trigger GPIO is high for exactly 1 cycle according to the ChipWhisperer-Lite trigger speed. In experiments that traced the power consumption of a single instruction, we confirmed that the triggering of the ChipWhisperer does not incur significant delay by comparing the power traces of an instruction with high and low Hamming weight. Also the findings presented in Sect. 5.2 show high correlation between Hamming distance of register updates and power trace, providing evidence that there is no delay introduced by the triggering. We therefore conclude that no extra offset or precaution needs to be taken when working with the ChipWhisperer-Lite trigger mechanism.

3.3 Analyze: Overlaying the Trace with Corresponding Assembly Code

To remove a side-channel vulnerability from an implementation, it is required to determine the code location that caused the leakage of secret information. In this work, we locate leaking code locations by matching the collected power traces against the assembly instructions executed at the time. After leakage locations in the trace and program have been identified, this information can be used to improve attack performance and accuracy (by restricting attention to affected areas) or to harden the implementation against side-channel attacks (by avoid using leaking instructions in the code).

After making sure that the start and end of the power trace is accurate (see Sect. 3.1 and 3.2), no caches are activated (Sect. 3.1) and the instruction order is preserved (Sect. 3.1) a simple mapping between clock cycle in the power trace and instructions in the assembly code can be applied. Each instruction takes a predefined amount of clock cycles according to the microcontroller's documentation [4].

For simplicity, we use this methodology only on branch free code. For code containing loops, we collect the power-trace of a single iteration instead of the full loop or unroll the loop. For code containing function calls, we trace the function separately or inline it.

4   ChipWhisperer Firmware

The ChipWhisperer comes with an open-source software package [24] to assist in side-channel analysis. At the time of writing, the latest release is version 5.5.2 from May 5, 2021. The ChipWhisperer firmware for the STM32F4 target prepares the device for side-channel analysis, which includes configuring the ART (adaptive real-time memory) accelerator, which handles memory accesses from the processor to the flash memory. The memory is accessed upon execution of load instructions (D-Code) and to read the next instructions of the program in execution (I-Code).

To access data from flash memory, the address is sent to the ART accelerator, which reads 128-bit of data from memory[1] and caches it. For subsequent memory access, if the requested data is present in the cache, it can be read out without any delays. However, if not, a fixed number of cycles are needed for the ART accelerator to read the Flash memory and provide the data.

The ChipWhisperer 5.2.2 firmware configures the ART with disabled data cache, disabled instruction cache, disabled prefetching, and five wait states latency. This means that after 128-bit of instructions, 5 cycles of waiting are required to read the next 128-bit from memory. 128-bit of instructions can be comprised of 4 instructions of 32-bit (in ARM mode), 8 instructions of 16-bit (in thumb mode), or a mixture of both. We observed the presence of wait states in the length of the power traces that we collected during our experiments.

However, the microcontroller can also be operated without wait states at the voltage and frequency used by the ChipWhisperer platform [31, reference manual p. 80]. Given the significant performance loss caused by the wait states, it is unlikely that real-world applications would use a configuration with wait states.

---

[1] To the best of our knowledge, the alignment of this process is not documented. However, our experiments provide some evidence that a 4-word alignment is used on our target device.

Accordingly, on 21 September 2021, the ChipWhisperer development branch was updated (commit 5863217) to configure STM32F4 target boards to not apply wait states (0 WS).

However, at the time of writing, this change was not yet contained in any released ChipWhisperer software package; the ART configuration with wait states is default since the introduction of STM32F4 support in ChipWhisperer in March 2017. Consequently, power traces of the STM32F4 target board collected using the ChipWhisperer contain wait states unless the target board ART configuration was changed manually or, after 21 September 2021, the ChipWhisperer development code was used instead of installing the software as described in the documentation [24].

The presence of wait cycles in the execution of the program influences the internal state of the processor, which may influence the leakage of implementations in side-channel analysis. This casts some doubt on the validity of previous side-channel analysis of the STM32F4 on the ChipWhisperer platform.

As a case study on the impact of wait states on side-channel leakage, we compare template attacks on the Sign-Flip function running on the STM32F4 using five wait states (5 WS) and no wait states (0 WS). Marzougui et al. [20] demonstrated that a generic MLP classifier could recover the sampled sign in 99.9% and 100.0% of cases, respectively, when using 5 WS. For direct comparison, we re-ran this experiment with an identical setup (Sect. 3) and machine learning attack (see Sect. 5.1 for details on the machine learning attack) for both the 5 WS and 0 WS configuration. The firmware version with 0 WS results in 98.0% (Sect. 5.1) while the version with 5 WS reaches 99.9%. A detailed breakdown of the classification performance depending on training set size can be seen in Fig. 1.

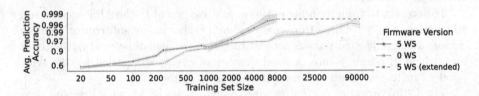

Fig. 1. Classifier comparison for the firmware with 5 WS and 0 WS for different training sizes. Validation set size is fixed at 1000. Each data point shown is based on multiple runs of the attack. The darker colored lines show the mean value, while the lighter areas show the range of accuracies. The accuracy for 5 WS with a training set size of 8000 to 98000 is constant as it already reaches 0.999% at a size of 8000. (Color figure online)

While this reduced prediction accuracy, it is still very high and continues to pose a security threat to GALACTICS on the STM32F4. Of the two attacks presented by Marzougui et al. [20, Sec. 6.1, 6.3] that use leakage of the Sign-Flip function, one will need slightly more samples of the target under attack, and one will remain largely unaffected as predictions are the result of majority voting across a very large population.

5   Sign-Flip Analysis

In this section, we will first present three different implementations of the Sign-Flip function as found in three different post-quantum cryptography schemes. Next, we will show that all of them are vulnerable to profiling power analysis attacks using machine learning (see Sect. 5.1).

Subsequently, in Sect. 5.2, we present advanced and modified implementations of the Sign-Flip function and attack using the same profiling attack. We find that, even though some implementations leak less under this attack, none of them is immune to our attack.

Additionally, we present an implementation of a two-sided CDT Gaussian sampler, which avoids the need for a Sign-Flip function at the cost of increasing the CDT table size (Sect. 5.3). Our findings show that also this approach cannot remove significant leakage on the sign of the sampled values.

5.1   Power Analysis of Different Implementations

We analyze the Sign-Flip implementations of GALACTICS [9] and of the reference implementations of FrodoKEM [1] and FALCON [11] by subjecting them to identical attacks.

First, for each scheme, we located and isolated the code implementing the Sign-Flip function. While similar, the schemes all use slightly different implementations of the same functionality, as shown below. (Variable names have been changed to match the Sign-Flip(x, c) definition of Eq. 3.)

In GALACTICS, given `uint32` inputs x and c, Sign-Flip returns `int32` defined by

$$(x \ \& \ -(c \ \& \ 1)) \ \text{\textasciicircum} \ (-x \ \& \ \text{\textasciitilde}(-(c \ \& \ 1))).$$

In FrodoKEM, given `uint32` inputs x and c, Sign-Flip returns `int32` defined by

$$(x \ \text{\textasciicircum} \ -c) \ + \ c.$$

In FALCON, given `uint16` inputs x and c, Sign-Flip returns `int16` defined by

$$(-c \ \text{\textasciicircum} \ x) \ + \ c.$$

Note that the implementations of FrodoKEM and FALCON differ in the used data types. Note that the - and ˜ have higher operator precedence than &.

Second, to be able to run the identified code snippet independently from the rest of the scheme, we determined the random distribution of values for x and c in each scheme.

Third, using the identified code and input random distributions, we ran each snippet 10,000 times and recorded input, output, and power traces using the setup described in Sect. 3.

Finally, to run an attack, the recorded data is randomly partitioned in a training set of 9,000 examples and a test set of 1,000 samples. (Fig. 2 displays a subset of collected traces for each scheme.) Using the training set, the sections of the power trace showing large difference across the recorded output sign are identified. Then, an neural network is trained given the identified sections and output sign values. Our neural network uses the *adam* optimizer, a batch size of 200 and one hidden layer with 20 neurons. Using the test set, the prediction accuracy of the trained network is evaluated. While other parameters for the neural network may increase performance or prediction accuracy, we found that these parameters already reach high accuracy and thus reveal sensitive information.

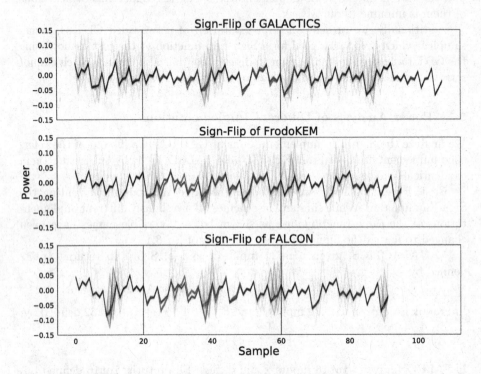

Fig. 2. 1,000 power traces of Sign-Flip implementations. Red and blue lines indicate negative and positive output, respectively. The section between the vertical black lines is used for training the neural network. (Color figure online)

We find that for all three schemes, our attack yields near-perfect prediction accuracy of the sampled sign. The Sign-Flip implementation of GALACTICS was previously under attack using a profiling power analysis attack [20]. For GALACTICS, our experiment resulted in a prediction accuracy of 98.0%.

In the case of FALCON and FrodoKEM we observed similar-looking power traces, which is expected due to the similar implementations. We obtained prediction accuracy of 99.7% and 100.0% for FALCON and FrodoKEM, respectively.

The Sign-Flip implementations in FALCON and FrodoKEM are originally not encapsulated in a subroutine. To confirm that extracting the code snippet and running it individually did not influence the leakage significantly, we re-ran our experiment on the full FALCON and FrodoKEM Gaussian sampler. These experiments confirm our findings, with prediction accuracy reaching 97.3% and 100.0%, respectively. We note that the Hamming distance for register updates in FALCON is smaller than in FrodoKEM, as 16 bit values are used, which could explain the difference in prediction accuracy [27].

5.2   Analyzing Various New Versions

To harden cryptographic schemes based on Gaussian random numbers against power side-channel analysis, we explore candidate implementations of the Sign-Flip function in this section. To understand the issue in the existing implementation, we matched the power trace against the instruction executed at the given time as described in Sect. 3.3.

As the implementation used by GALACTICS showed the least leakage in above analysis, we chose it as a starting point. It's assembly code is shown in Table 3, together with example register values for the cases $c = 0$ and $c = 1$ (mod 2). The distribution for input value x is a half-normal distribution (positive half) with mean zero and variance $\sigma = \frac{205}{256}$ [8, p. 14], input value c is uniformly distributed in $\{0, \dots, 255\}$.

Table 3. Assembly instruction of the GALACTICS Sign-Flip implementation, displayed along with assigned register values for the cases of $c = 0$ and $c = 1$ (mod 2). The code expects x in r3 and c in r4 and stores the result in r5.

Instruction	Pseudo-code	Computed function	Assigned register values	
			If $c = 0 \bmod 2$	Otherwise
1. and.w r1, r4, #1	r1 = r4 & 1	$= c$ & 1	0x00000000	0x00000001
2. subs r4, r1, #1	r4 = r1 - 1	$= (c$ & 1) - 1	0xFFFFFFFF	0x00000000
3. negs r5, r3	r5 = -r3	$= -x$	$-x$	$-x$
4. negs r1, r1	r1 = -r1	$= -(c$ & 1)	0x00000000	0xFFFFFFFF
5. and.w r0, r1, r3	r0 = r1 & r3	$= [-(c$ & 1)] & x	0x00000000	x
6. ands r5, r4	r5 = r5 & r3	$= -x$ & [-(c & 1)]	$-x$	0x00000000
7. eors r5, r0	r5 = r5 & r0	$= - \oplus ([-(c$ & 1) & x)	$-x$	x

In Fig. 3, we display a number of power traces of the Sign-Flip implementation, colored by the sign of the return value. The power trace is labeled with the instruction currently in the execution stage of the processor's pipeline. As this stage is responsible for executing the calculation on the arithmetic logic unit (ALU), we assume that the leakage comes from there. However, we note that the fetch and decode stage of the pipeline might also leak information [21]. The plot is shorter than the original trace (Fig. 2) as the *fast_trigger* from Sect. 3.2 was used. This results in 0 cycles following the GPIO *high* instruction and only 1 cycle for the GPIO *low* instruction (denoted as "Trigger").

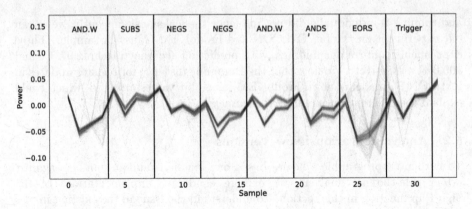

Fig. 3. GALACTICS Sign-Flip power trace annotated with the corresponding assembly instructions in the execute stage of the pipeline. Traces shown in red and blue had negative and positive return value, respectively. (Color figure online)

Our power traces show that the instructions 1, 2, 4, 5, and 6 have visible leakage. From Table 3, we know that instructions 1, 2, and 4 differ by 15 and 16 in Hamming weight, and instructions 5 and 6 are expected to differ by 8, when comparing $c = 0$ and $c = 1 \pmod 2$ cases. Applying the same methodology, we found similar patterns of leakage in the Sign-Flip implementations of FALCON and FrodoKEM.

To reduce leakage, we propose four ARM assembly implementations of the Sign-Flip function aimed at avoiding large difference in Hamming weight when comparing $c = 0$ and $c = 1 \pmod 2$ cases. An implementation based on the MUL instruction is shown in Table 4, an implementation based on shifting in Table 5. Table 6 shows an implementation built on the XOR operation, and finally, Table 7 shows an implementation based on the SXTH instruction.

We base one alternative Sign-Flip implementation on the MUL instruction, shown in Table 4. First, depending on c, the value 2 or 3 is generated. (Note the similar Hamming weight.) Then, the value $-2x$ is computed by negating x and summation. Finally, x is multiplied with 1 or 3 (depending on c), and the result is added to $-2x$.

While this reduces the precision of x by 2 bits, even in the case of FALCON, only 5-bit precision are needed, as x only is in the range of $[0, 18]$. FrodoKEM and GALACTICS use even less precision. While this avoids some of the large Hamming weight differences seen in the GALACTICS implementation, instruction 7 still shows Hamming weight significantly differing between the cases of $c = 0$ and $c = 1 \pmod 2$.

Table 4. Sign-Flip implementation based on the MUL instruction. The code expects x in r3 and c in r4 and stores the result in r5.

Instruction	Pseudo-code	Computed function	Assigned register values	
			If $c = 0 \bmod 2$	Otherwise
1. lsl r4, r4, #1	r4 = r4 «1	= c «1	c «1	c «1
2. and.w r1, r4, #3	r1 = r4 & 0x3	= c & 0x3	0x00000000	0x00000002
3. orr r1, r1, #1	r1 = r1 | 0x1	= (c & 0x3) | 0x1	0x00000001	0x00000003
4. negs r5, r3	r5 = -r3	= $-x$	$-x$	$-x$
5. add r5, r5, r5	r5 = r5 + r5	= $(-x) + (-x)$	$-2x$	$-2x$
6. mul r6, r1, r3	r6 = r1 * r3	= $(($c$ & 0x3) | 0x1) * x$	x	$3x$
7. add r5, r5, r6	r5 = r5 + r6		$-x$	x

Table 5. Sign-Flip implementation based on the SHIFT instruction. The code expects x in r3 and c in r4 and stores the result in r5.

Instruction	Pseudo-code	Computed function	Assigned register values	
			If $c = 0 \bmod 2$	Otherwise
1. negs r5, r3	r5 = -r3	= -x	-x	-x
2. add r6, r5, r5	r6 = r5 + r5	= $(-x) + (-x)$	-2x	-2x
3. add r5, r6, r5	r5 = r6 + r5	= $((-x) + (-x)) + (-x)$	-3x	-3x
4. and.w r1, r4, #1	r1 = r4 & 1	= c & 1	0x00000000	0x00000001
5. add r1, r1, 1	r1 = r1 + 1	= (c & 1) + 1	0x00000001	0x00000002
6. lsl r6, r3, r1	r6 = r3 «r1	= x «((c & 1) + 1)	2x	4x
7. add r5, r5, r6	r5 = r5 + r6		-x	x

Second, we propose an implementation of Sign-Flip based on the SHIFT instruction, shown in Table 5. Like the implementation based on MUL above, it avoids the computation of (c & 1) - 1 and hence does not show large Hamming weight difference between the two cases. Also it requires a reduction in precision of x by 3 bit as we need to compute $4x$. After computing computing $-3x$ in instructions 1–3, the values $2x$ or $4x$ are generated in instructions 4–6 using a left shift operation. Then, the result is calculated by adding these results in instruction 7, which also results in the majority of leakage which we observed. We remark that we also observed leakage of the left shift operation, even though there is no difference in Hamming distance between $c = 0$ and $c = 1$ (mod 2).

Third, in Table 6, we propose an implementation based on the XOR-Operation. In this implementation, the constants 0xAAAAAAAA and 0x55555555 are used to "mask" the values x and $-x$. As both constants have Hamming weight 16, this equalizes the Hamming distance for assigned register values across the cases $c = 0$ and $c = 1$ (mod 2). On the downside, it uses twice the number of instructions that the MUL and SHIFT implementations require. Also, we found significant leakage of the instructions 10 and 15. Instruction 10

Table 6. Sign-Flip implementation based on the XOR instruction. The code expects x in r3 and c in r4 and stores the result in r5.

Instruction	Pseudo-code	Assigned register values	
		If $c = 0 \bmod 2$	Otherwise
1. mov r5, 0xAAAAAAAA	r5 = 0xAAAAAAAA	0xAAAAAAAA	0xAAAAAAAA
2. and.w r1, r4, #1	r1 = r4 & 1	0x00000000	0x00000001
3. negs r4, r3	r4 = -r3	r4 = -x	r4 = -x
4. lsr r6, r5, r1	r6 = r5 «r1	0xAAAAAAAA	0x55555555
5. and.w r7, r5, r3	r7 = r5 & r3	b'x[31]0 ... x[1]0'	b'x[31]0 ... x[1]0'
6. and.w r8, r5, r4	r8 = r5 & r4	b'(-x)[31]0 ... (-x)[1]0'	b'(-x)[31]0 ... (-x)[1]0'
7. mov r5, #0x55555555	r5 = 0x55555555	0x55555555	0x55555555
8. and.w r9, r5, r4	r9 = r5 & r4	b'0x[30] ...0x[0]'	b'0x[30] ... 0x[0]'
9. and.w r10, r5, r3	r10 = r5 & r3	b'0(-x)[30] ... 0(-x)[0]'	b'0(-x)[30] ... 0(-x)[0]'
10. lsl r5, r5, r1	r5 = r5 «r1	0x55555555	0xAAAAAAAA
11. eor r7, r7, r9	r7 = r7 ⊕ r9	b'x[31](-x)[30] ... x[1](-x)[0]'	b'x[31](-x)[30] ... x[1](-x)[0]'
12. eor r8, r8, r10	r8 = r8 ⊕ r10	b'(-x)[31]x[30] ... (-x)[1]x[0]'	b'(-x)[31]x[30] ... (-x)[1]x[0]'
13. and.w r5, r7, r5	r5 = r7 & r5	b'0(-x)[30] ... 0(-x)[0]'	b'x[31]0 ... x[1]0'
14. and.w r6, r8, r6	r6 = r8 & r6	b'(-x)[31]0 ... (-x)[1]0'	b'0x[30] ...0x[0]'
15. eor r5, r5, r6	r5 = r5 ⊕ r6	-x	x

is leaking even though there is no difference in assigned Hamming weight, as it shifts depending on the value of c. Instruction 15 is leaking as it has unavoidable Hamming weight difference required by the Sign-Flip function's definition.

Fourth and last, in Table 7, we propose an implementation based on the SXTH instruction. While this implementation avoids the SHIFT instruction and thus the leakage it caused in the SHIFT and MUL versions above, we can still observe leakage of the ROR instruction. Again, we are also observing leakage of the final SXTH instruction, as it extends the computed 16 bit value of x and $-x$, respectively, having large Hamming weight difference across the cases $c = 0$ and $c = 1 \pmod 2$.

Table 7. Sign-Flip implementation based on the SXTH instruction. The code expects x in r3 and c in r4 and stores the result in r5.

Instruction	Pseudo-code	Assigned register values	
		If $c = 0 \bmod 2$	Otherwise
1. negs r7, r3	r7 = -r3	-x	-x
2. and.w r8, r4, #1	r8 = r4 & 1	0x00000000	0x00000001
3. lsl r8, r8, #4	r8 = r8 «4	0x00000000	0x00000008
4. pkhbt r6, r7, r3, LSL #16	r6 = (r3<31:16>,r7<15:0>)	(x<31:16>,-x<15:0>)	(x<31:16>,-x<15:0>)
5. ror r5, r6, r8	r5 = r6 ROR r8	(x<31:16>,-x<15:0>)	(-x<31:16>,x<15:0>)
6. sxth r5, r5	r5 = r5<15:0> → r5 <31:0>	r5 = -x	r5 = x

We subjected all four implementation of the Sign-Flip function to the power side-channel attack presented in Sect. 5.1. We find that while the four novel versions can reduce attack success to some extend, all proposed implementations are still vulnerable to attacks with significant prediction accuracy. In our tests, the SXTH implementation achieved the best results by reducing the prediction accuracy of the attack to 91.6%. A detailed comparison of all presented Sign-Flip implementations with respect to the predictive power of our attack can be found in Table 8; for completeness, we also include the result of an implementation with branching instructions. It is possible that different hyperparameters of the neural network, more measurements per clock cycle, and/or more examples in the training set can further increase the predictive power of attacks on our proposed implementations, but the presented finding are enough to argue that none of the implementation is sufficiently secure.

Table 8. Comparison of all different Sign-Flip versions using an MLP classifier. Trivial accuracy would be 50.0% as it is a coin flip.

Algorithm	Branching	Post-quantum crypto			This work			
		GALACTICS	FALCON	FrodoKEM	MUL	SHIFT	XOR	SXTH
Accuracy	99.8%	98.0%	99.7%	100.0%	95.2%	94.4%	93.2%	91.6%

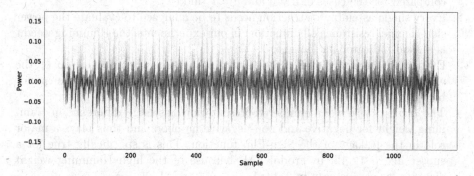

Fig. 4. 1,000 power traces of the FrodoKEM Gaussian sampler, modified to act as a two-sided sampler. Runs resulting in a negative sample are drawn in red, others in blue. The section marked by vertical lines shows the section used for training the classifier. (Color figure online)

5.3   Two-Sided CDT Gaussian Sampler

The Sign-Flip function used by the CDT Gaussian sampler in GALACTICS, FALCON and FrodoKEM allows reducing the CDT size by half and thus increases performance and decreases required randomness (Sect. 2).

To avoid leakage of the function, the Sign-Flip function can be avoided at the cost of doubling the CDT table size by adding negative numbers. We modified the FrodoKEM implementation of the Gaussian sampler [1] to operate as a two-sided CDT sampler without a Sign-Flip function.[2]

---

[2] While confirming that the random distribution did not change and is still according to the specification of $\sigma = 2.8$, we found that, due to the 16-bit constraint on the table entries, the original one-sided implementation has $\sigma \approx 2.8146$ and our two-sided adaptation has $\sigma \approx 2.8138$.

Again using the attack shown in Sect. 5.1, we found that the resulting prediction accuracy on the two-sided CDT sampler is 100.0%. The recorded power traces are displayed in Fig. 4.

6   Conclusion

In this paper, we demonstrated that the implementations of the Sign-Flip function in GALACTICS, FALCON and FrodoKEM all leak information on the sign of the produced Gaussian sample when attacked using power analysis with a simple, generic neural network classifier. Novel implementations that avoid large differences in Hamming weight during the execution of the Sign-Flip code could only reduce the leakage under our attack to a small extent.

We draw the following conclusions:

1. The identified weakness in the original implementations is caused by the Hamming weight distance between the register assignments in each instruction, compared for the cases of $c = 0$ and $c = 1 \pmod 2$.
2. Every single assembly instruction needs to be analyzed to evaluate the power side-channel resistance of a function; in our experiments, the Hamming weight is a good indicator for leakage.
3. Even when carefully crafting an assembly version of the Sign-Flip or the whole Gaussian sampler, leakage on the sign of the Gaussian sample cannot be significantly reduced.
4. The *two's complement* number representation has large difference in Hamming weight for negative and non-negative numbers and thus plays a major role in the leakage of the Sign-Flip function. This is specifically true for a subset like $[-12, 12]$ in FrodoKEM, because of the high Hamming weight distance, as can be seen in Table 1.

From these findings, we conclude that, on the STM32F4, no secure implementation of the Sign-Flip function exists as long as the return value is the unmasked sample value. However, as the sample value typically is added to other internal values, this requires either unmasking the sample (which leaks information) or masking the entire scheme.

References

1. Alkim, E., et al.: Frodokem implementation (2021). https://github.com/Microsoft/PQCrypto-LWEKE
2. Alkim, E., et al.: Frodokem learning with errors key encapsulation (2021). https://frodokem.org/files/FrodoKEM-specification-20210604.pdf
3. ARM: ARM Cortex-M4. https://developer.arm.com/Processors/Cortex-M4
4. ARM: ARM Cortex-M4 instruction cycle count. https://developer.arm.com/documentation/ddi0439/b/CHDDIGAC
5. Askeland, A., Rønjom, S.: A side-channel assisted attack on NTRU. Cryptology ePrint Archive, Paper 2021/790 (2021). https://eprint.iacr.org/2021/790

6. Carruth, C.: Why statement order can not be enforced. Stackoverflow (2016). https://stackoverflow.com/a/38025837

7. Chari, S., Rao, J.R., Rohatgi, P.: Template attacks. In: Kaliski, B.S., Koç, K., Paar, C. (eds.) CHES 2002. LNCS, vol. 2523, pp. 13–28. Springer, Heidelberg (2003). https://doi.org/10.1007/3-540-36400-5_3

8. Ducas, L., Durmus, A., Lepoint, T., Lyubashevsky, V.: Lattice signatures and bimodal Gaussians. Cryptology ePrint Archive, Paper 2013/383 (2013). https://eprint.iacr.org/2013/383

9. Ducas, L., Durmus, A., Lepoint, T., Lyubashevsky, V.: Galactics implementation (2019). https://github.com/espitau/GALACTICS

10. Fouque, P.A., et al.: Falcon: fast-Fourier lattice-based compact signatures over NTRU (2020). https://falcon-sign.info/falcon.pdf

11. Fouque, P.A., et al.: Falcon implementation (2020). https://falcon-sign.info/

12. Genêt, A., Kalulterović, N.: Single-trace clustering power analysis of the point-swapping procedure in the three point ladder of Cortex-M4 SIKE. Cryptology ePrint Archive, Paper 2022/364 (2022). https://eprint.iacr.org/2022/364

13. Kamucheka, T., Fahr, M., Teague, T., Nelson, A., Andrews, D., Huang, M.: Power-based side channel attack analysis on PQC algorithms. Cryptology ePrint Archive, Paper 2021/1021 (2021). https://eprint.iacr.org/2021/1021

14. Kannwischer, M.J., Rijneveld, J., Schwabe, P., Stoffelen, K.: PQM4: post-quantum crypto library for the ARM Cortex-M4. https://github.com/mupq/pqm4

15. Kannwischer, M.J., Rijneveld, J., Schwabe, P., Stoffelen, K.: pqm4: Testing and benchmarking NIST PQC on ARM Cortex-M4. Cryptology ePrint Archive, Paper 2019/844 (2019). https://eprint.iacr.org/2019/844

16. Khalid, A., Howe, J., Rafferty, C., Regazzoni, F., O'Neill, M.: Compact, scalable, and efficient discrete Gaussian samplers for lattice-based cryptography. In: 2018 IEEE International Symposium on Circuits and Systems (ISCAS) (2018). https://doi.org/10.1109/ISCAS.2018.8351009

17. Kim, S., Hong, S.: Single trace analysis on constant time CDT sampler and its countermeasure. Appl. Sci. 8(10) (2018). https://doi.org/10.3390/app8101809. https://www.mdpi.com/2076-3417/8/10/1809

18. Kocher, P., Jaffe, J., Jun, B.: Differential power analysis. In: Wiener, M. (ed.) CRYPTO 1999. LNCS, vol. 1666, pp. 388–397. Springer, Heidelberg (1999). https://doi.org/10.1007/3-540-48405-1_25

19. Lerman, L., Bontempi, G., Markowitch, O., et al.: Power analysis attack: an approach based on machine learning. Int. J. Appl. Cryptogr. 3(2), 97–115 (2014)

20. Marzougui, S., Wisiol, N., Gersch, P., Krämer, J., Seifert, J.: Machine-learning side-channel attacks on the GALACTICS constant-time implementation of BLISS. CoRR abs/2109.09461 (2021). https://arxiv.org/abs/2109.09461

21. McCann, D., Oswald, E., Whitnall, C.: Towards practical tools for side channel aware software engineering: 'grey box' modelling for instruction leakages. Cryptology ePrint Archive, Paper 2016/517 (2016). https://eprint.iacr.org/2016/517

22. National Institute of Standards and Technology (NIST): Post-Quantum Cryptography Standardization. https://csrc.nist.gov/Projects/Post-Quantum-Cryptography

23. NewAE Technology Inc.: ChipWhisperer-Lite 32-Bit. https://www.newae.com/products/NAE-CWLITE-ARM

24. NewAE Technology Inc.: ChipWhisperer software. https://github.com/newaetech/chipwhisperer

25. NewAE Technology Inc.: CW308 UFO Target Board. https://www.newae.com/products/NAE-CW308

26. NewAE Technology Inc.: STM32F4 Target for CW308. https://www.newae.com/ufo-target-pages/NAE-CW308T-STM32F4

27. Ngo, K., Dubrova, E., Johansson, T.: Breaking masked and shuffled CCA secure Saber KEM by power analysis. Cryptology ePrint Archive, Paper 2021/902 (2021). https://eprint.iacr.org/2021/902

28. O'Flynn, C., Chen, Z.D.: ChipWhisperer: an open-source platform for hardware embedded security research. Cryptology ePrint Archive, Paper 2014/204 (2014). https://eprint.iacr.org/2014/204

29. Park, J., et al.: PQC-SEP: power side-channel evaluation platform for post-quantum cryptography algorithms. Cryptology ePrint Archive, Paper 2022/527 (2022). https://eprint.iacr.org/2022/527

30. Ravi, P., Roy, D.B., Bhasin, S., Chattopadhyay, A., Mukhopadhyay, D.: Number "not used" once - practical fault attack on pqm4 implementations of NIST candidates. Cryptology ePrint Archive, Paper 2018/211 (2018). https://eprint.iacr.org/2018/211

31. STMicroelectronics: STM32F405/415. https://www.st.com/en/microcontrollers-microprocessors/stm32f405-415.html#overview

32. Tibouchi, M., Wallet, A.: One bit is all it takes: a devastating timing attack on bliss's non-constant time sign flips. Cryptology ePrint Archive, Paper 2019/898 (2019). https://eprint.iacr.org/2019/898

Attacking NTRU

Reveal the Invisible Secret: Chosen-Ciphertext Side-Channel Attacks on NTRU

Zhuang Xu[1]([✉])[ID], Owen Pemberton[2][ID], David Oswald[2][ID], and Zhiming Zheng[3,4,5]

[1] School of Mathematical Sciences and Shenyuan Honors College,
Beihang University, Beijing, China
xu_zhuang@buaa.edu.cn
[2] School of Computer Science, University of Birmingham, Birmingham, UK
o.m.pemberton@pgr.bham.ac.uk, d.f.oswald@bham.ac.uk
[3] Institute of Artificial Intelligence, LMIB, NLSDE and Beijing Advanced Innovation
Center for Future Blockchain and Privacy Computing, Beihang University,
Beijing, China
zzheng@pku.edu.cn
[4] Peng Cheng Laboratory, Shenzhen, China
[5] Institute of Medical Artificial Intelligence, Binzhou Medical University,
Yantai, China

Abstract. NTRU is a well-known lattice-based cryptosystem that has been selected as one of the four key encapsulation mechanism finalists in Round 3 of NIST's post-quantum cryptography standardization. This paper presents two succinct and efficient chosen-ciphertext side-channel attacks on the latest variants of NTRU, i.e., NTRU-HPS and NTRU-HRSS as in Round 3 submissions. Both methods utilize the leakage from the polynomial modular reduction to recover the long-term secret key. For the first attack, although the side-channel leakage does not directly reveal the secret polynomial $\mathbf{f}$, we recover differences between adjacent coefficients using appropriately chosen ciphertexts, and finally reconstruct $\mathbf{f}$ through linear algebra. The second attack is based on the inherent relation between the secret key and the public key in NTRU-HPS: we first reveal the "invisible" secret polynomial $\mathbf{g}$ with chosen ciphertexts and then use $\mathbf{g}$ and the public polynomial $\mathbf{h}$ to compute $\mathbf{f}$. In theory, these attacks only need 4 and 2 ciphertexts, respectively. We then practically apply those attacks on all reference implementations of four instances in the PQClean library and show that the accuracy of secret-key recovery can reach 100% with only few traces (4 to 24 and 2 to 6, respectively). We also observe similar leakage in optimized implementations in the pqm4 library and propose an according analysis scheme.

Keywords: Lattice-based cryptography · NTRU · Chosen-ciphertext attack · Side-channel analysis

I. Buhan and T. Schneider (Eds.): CARDIS 2022, LNCS 13820, pp. 227–247, 2023.
https://doi.org/10.1007/978-3-031-25319-5_12

1 Introduction

When will a practical quantum computer that can break current Public-Key Cryptography(PKC) based on RSA or elliptic curve cryptography be available? Quantum computing experts anticipate that it might take 10 to 15 years. But one thing is for sure: we need cryptographic algorithms that are secure even in the context of quantum computing to replace current PKC schemes. This kind of cryptography is also known as Post-Quantum Cryptography (PQC).

To select sound successors, the National Institute of Standards and Technology (NIST) held a public competition-like PQC standardization project in 2016 [2]. After two rounds of competition, four Key Encapsulation Mechanisms (KEMs) and three signature schemes have been chosen as finalists in Round 3 [1]. Out of the four KEMs, three are lattice-based schemes: CRYSTALS–Kyber [7], Saber [9] and NTRU [8]. The security of the first two are based on the hardness of Learning with Errors (LWE) problem variants, i.e., the Module-LWE problem and the Module Learning with Rounding problem; while that of NTRU is based on the NTRU problem, a different assumption. As the one with the longest history among them, NTRU has stood the test of time, i.e., NTRU-based schemes had been widely studied in theoretical and implementation security. Therefore, it is a very competitive finalist. Recently, NIST published the report on the third round. Although currently NTRU is not considered for future standardization as per the evaluation and selection results, NIST stated that NTRU may replace Kyber as a standard if the patent issue on Kyber is not resolved by the end of 2022 [3]. So, it is still significant to study the latest variants of NTRU in depth.

Theoretical security, performance and characteristics (in terms of algorithm and implementation) are core evaluation criteria in the first two rounds [1]. The implementation security of finalists is also a key criterion in Round 3. A deeper understanding of the threats to the implementation can help designers create more effective countermeasures to mitigate such attacks. In this paper, we focus on Side-Channel Analysis (SCA), a classic threat exposing the relation between secret intermediate values and side-channel leakage when the cryptographic algorithm runs on a specific hardware/software platform. More specifically, we focus on Simple Power Analysis (SPA) [18] using chosen ciphertexts for the NTRU KEM. To our knowledge, this is the first attempt to apply this kind of analysis on NTRU instances in Round 3.

1.1 Related Work

SCA on NTRU. There are many side-channel attacks [4,6,19,29] targeting the early version of NTRU. Most attacks exploit the leakage from the polynomial multiplication which involves the secret polynomial and a controllable/-known polynomial. But these attacks cannot be directly applied to the latest NTRU instances that use Toom-Cook and Number Theoretic Transform (NTT) techniques to optimize the multiplication process in the PQClean and pqm4 [16] library respectively, instead of "multiply and accumulate" approach in early NTRU implementations. Recently, Mujdei et al. [20] exploited the leakage from polynomial multiplication through Toom-Cook and NTT strategies and mounted successful correlation power analysis.

In addition to multiplication, other operations have also been targeted to mount SCA. In [17], Karabulut *et al.* showed the vulnerability of the sorting algorithm in the key generation step of NTRU. Askeland *et al.* [5] performed a partial key recovery by utilizing the leakage from the unpacking operation and the $\mathbb{Z}_3$ to $\mathbb{Z}_q$ mapping in a single-trace setting. They then applied the Block Korkin-Zolotarev (BKZ) lattice reduction algorithm to find the remainder of the secret key. Ueno *et al.* [26] presented a deep-learning-based SCA on the Fujisaki-Okamoto transformation and its variants and demonstrated that the scheme can be applied to most candidates in Round 3 (including NTRU). In [22], Ravi *et al.* proposed several chosen-ciphertext collision attacks on the NTRU family of KEMs through side-channel-based oracles. Their method is inspired by a Chosen-Ciphertext Attack (CCA) [15] on early versions of NTRU.

There are also attacks which target the session key instead of the long-term secret key. For example, Sim *et al.* [24] utilized the leakage from message encoding in encapsulation to recover the secret shared message which can be used to calculate the shared session key.

Chosen-Ciphertext SPA on Other Lattice-Based Schemes. As lattice-based KEMs share a common characteristic—the coefficients of the secret polynomial are sampled from a small value range (e.g., $\{-1, 0, 1\}$ in NTRU)—they are vulnerable under chosen-ciphertext-assisted SPA which can take advantage of this feature to recover the secret key using a small number of traces. This kind of attack becomes more practical when an attacker cannot collect enough traces to mount Differential Power Analysis (DPA) [18] or template attacks.

Park *et al.* [21] proposed the first chosen-ciphertext SPA targeting Ring-LWE encryption on an 8-bit microcontroller. In [13], several power analysis techniques were applied to NTRU Prime by Huang *et al.* One of them is a chosen-input SPA targeting the polynomial multiplication. With the assistance of chosen ciphertexts, an adversary can recover the key according to special patterns in the trace. Xu *et al.* [27] devised an efficient SPA on Montgomery reduction in the reference implementation of Kyber. With four traces from different crafted ciphertexts, they achieved a high accuracy of secret-key recovery.

1.2   Contributions

In this paper, we explore the feasibility of chosen-ciphertext SCA on all NTRU instances in Round 3. We also determine a lower bound of the required number of traces, exploiting a unique property in three NTRU-HPS instances.

Concretely, we extend the chosen-ciphertext SPA method from [27] to Round-3 NTRU. Interestingly, the relation between the public key and the secret key in NTRU-HPS simplifies this kind of attack. Our main contributions are:

1) We identify a leaky function, the $\mathcal{R}_q$ to S_3 polynomial modular reduction in decryption in the decapsulation phase, which leaks the Hamming Weight (HW) of a secret-dependent intermediate value.

2) We mount chosen-ciphertext Electro-Magnetic (EM) SCA on the reference implementations of all instances (i.e., `ntruhps2048509`, `ntruhps2048677`, `ntruhps4096821` and `ntruhrss701`) in PQClean, and recover the core secret polynomial **f** successfully with 4 ciphertexts supported by 4 to 24 traces.

3) We propose a more efficient attack utilizing an inherent property of NTRU-HPS—for that, we first recover the "invisible" secret polynomial **g** with 2 ciphertexts (2 to 6 traces) and then compute **f** from **h** and **g**.

4) We discuss the extension of our analysis to implementations from the optimized `pqm4` library.

Compared to several previous SCA schemes on NTRU [20,22,26], our schemes need less queries from the victim device. A stronger leakage point (compared to [5]) is exploited, so no additional lattice reduction is required.

1.3   Organization

The remainder of this paper is structured as follows: in Sect. 2, we introduce the necessary notations, background, and adversary model. In Sect. 3, we present our two types of chosen-ciphertext SPA on reference implementations of NTRU. Next, we propose an analysis strategy on optimized implementations in Sect. 4. Finally, we conclude in Sect. 5.

2   Preliminaries

2.1   Notation

Let q be a positive integer. Centered modular reduction is represented as $r' = r \mod q$ where $r' \in [-q/2, q/2 - 1]$ when q is even or $r' \in [-(q-1)/2, (q-1)/2]$ when q is odd. The non-negative modular reduction $r' = r \mod^+ q$ means $r' \in [0, q-1]$. Here, $\mathbb{Z}_q$ is the ring of integers modulo q with centered modular reduction. A polynomial is represented by a bold-font lowercase letter, e.g., **a**, its i-th coefficient is represented as a_i, and then $\mathbf{a} = \sum_i a_i x^i$. Polynomial multiplication is denoted using the $\cdot$ operator, whereas for scalar multiplication, the operator is omitted. We define $\boldsymbol{\Phi}_1 = x - 1$, $\boldsymbol{\Phi}_n = (x^n - 1)/(x - 1)$, for some integer n. The quotient rings of polynomials are $\mathcal{R}_q := \mathbb{Z}_q[x]/(x^n - 1) = \mathbb{Z}[x]/(q, \boldsymbol{\Phi}_1 \cdot \boldsymbol{\Phi}_n)$, $S_q := \mathbb{Z}_q[x]/(\boldsymbol{\Phi}_n)$. The polynomial multiplication of **a** and **b** in the above rings is represented by $\mathbf{a} \cdot \mathbf{b} \mod (q, \boldsymbol{\Phi}_1 \cdot \boldsymbol{\Phi}_n)$ and $\mathbf{a} \cdot \mathbf{b} \mod (q, \boldsymbol{\Phi}_n)$, respectively. $\boldsymbol{\Phi}_1 \cdot \boldsymbol{\Phi}_n = 0$ (or $\boldsymbol{\Phi}_n = 0$) is applied to terms which have a degree greater than or equal to n (or $n - 1$). We refer to the side-channel measurement (i.e., trace) as $p(t)$, where t is discretized time in sample points.

2.2   NTRU in NIST PQC Round 3

NTRU is one of the most-studied lattice-based schemes. Several variants have been proposed since it was first proposed by Hoffstein et al. in 1996 [11]. As our targets are the latest variants that have advanced to Round 3 in the NIST PQC standardization, we refer the reader to [23] for more information about the comparisons among different variants.

The submission of NTRU KEM in Round 3 is a merge of NTRU-HPS [12] and NTRU-HRSS [14]. Here, we briefly describe the core algorithms; for a full description, please refer to [8].

Algorithm 1. NTRU.PKE.KeyGen (*seed*) [8]

1: $(\mathbf{f}, \mathbf{g}) = \text{Sample\_fg}(seed)$
2: $\mathbf{f}_p = (1/\mathbf{f}) \bmod (p, \mathbf{\Phi}_n)$ /* $p = 3$ */
3: $\mathbf{f}_q = (1/\mathbf{f}) \bmod (q, \mathbf{\Phi}_n)$
4: $\mathbf{h} = (3\mathbf{g} \cdot \mathbf{f}_q) \bmod (q, \mathbf{\Phi}_1 \cdot \mathbf{\Phi}_n)$/*HPS*/ or $(3\mathbf{g} \cdot \mathbf{\Phi}_1 \cdot \mathbf{f}_q) \bmod (q, \mathbf{\Phi}_1 \cdot \mathbf{\Phi}_n)$/*HRSS*/
5: $\mathbf{h}_q = (1/\mathbf{h}) \bmod (q, \mathbf{\Phi}_n)$
6: $sk = \text{Pack\_S3}(\mathbf{f}) \parallel \text{Pack\_S3}(\mathbf{f}_p) \parallel \text{Pack\_Sq}(\mathbf{h}_q)$
7: $pk = \text{Pack\_Rq0}(\mathbf{h})$
8: **return** (sk, pk)

Algorithm 2. NTRU.PKE.Enc (*pk, packed_rm*) [8]

1: $(\mathbf{r}, \mathbf{m}) = \text{Unpack\_S3}(packed\_rm)$
2: $\mathbf{m}' = \text{Lift}(\mathbf{m})$
3: $\mathbf{h} = \text{Unpack\_Rq0}(pk)$
4: $\mathbf{c} = (\mathbf{r} \cdot \mathbf{h} + \mathbf{m}') \bmod (q, \mathbf{\Phi}_1 \cdot \mathbf{\Phi}_n)$
5: **return** $ct = \text{Pack\_Rq0}(\mathbf{c})$

Algorithm 3. NTRU.PKE.Dec (*sk, ct*) [8]

1: $\mathbf{c} = \text{Unpack\_Rq0}(ct)$
2: $(\mathbf{f}, \mathbf{f}_p, \mathbf{h}_q) = \text{Unpack}(sk)$
3: $\mathbf{v} = (\mathbf{c} \cdot \mathbf{f}) \bmod (q, \mathbf{\Phi}_1 \cdot \mathbf{\Phi}_n)$ /* Target step in this paper */
4: $\mathbf{m} = (\mathbf{v} \cdot \mathbf{f}_p) \bmod (p, \mathbf{\Phi}_n)$ /* $p = 3$ */
5: $\mathbf{m}' = \text{Lift}(\mathbf{m})$
6: $\mathbf{r} = ((\mathbf{c} - \mathbf{m}') \cdot \mathbf{h}_q) \bmod (q, \mathbf{\Phi}_n)$
7: $packed\_rm = \text{Pack\_S3}(\mathbf{r}) \parallel \text{Pack\_S3}(\mathbf{m})$
8: **if** $(\mathbf{r}, \mathbf{m}) \in \mathcal{L}_r \times \mathcal{L}_m$, $fail = 0$
9: **else** $fail = 1$
10: **return** $(packed\_rm, fail)$

Table 1. Parameters for different NTRU instances

NTRU	NTRU-HPS			NTRU-HRSS
	ntruhps2048509	ntruhps2048677	ntruhps4096821	ntruhrss701
n	509	677	821	701
q	2048	2048	4096	8192

Algorithm 4. NTRU.KEM.KeyGen $(seed)$ [8]

1: $(sk, pk) = $ NTRU.PKE.KeyGen $(seed)$
2: $s \leftarrow_\$ \{0, 1\}^{256}$
3: **return** $((sk, s), pk)$

Algorithm 5. NTRU.KEM.Encaps (pk) [8]

1: $coins \leftarrow_\$ \{0, 1\}^{256}$
2: $(\mathbf{r}, \mathbf{m}) = $ Sample_rm$(coins)$
3: $packed_rm = $ Pack_S3$(\mathbf{r}) \parallel$ Pack_S3$(\mathbf{m})$
4: $ct = $ NTRU.PKE.Enc $(pk, packed\_rm)$
5: $K = H_1 (packed\_rm)$ /* Shared Key */
6: **return** (ct, K)

Algorithm 6. NTRU.KEM.Decaps $((sk, s), ct)$ [8]

1: $(packed_rm, fail) = $ NTRU.PKE.Dec (sk, ct)
2: $K_1 = H_1 (packed\_rm), K_2 = H_2 (s, ct)$
3: **if** $fail = 0$ **return** K_1 /* Shared Key */
4: **else return** K_2 /* Random Key */

The NTRU KEM with Indistinguishability under Chosen-Ciphertext Attack (IND-CCA) is built by a generic transformation from a deterministic NTRU Public-Key Encryption (PKE) scheme. The deterministic PKE consists of Algorithms 1 to 3. All elements and corresponding computations are in three rings: $\mathcal{R}_q$, S_q and S_p, where q varies for different instances but $p = 3$ for all schemes. The specific parameters are shown in Table 1. $\mathbf{f}$, $\mathbf{g}$, $\mathbf{r}$ and $\mathbf{m}$ are polynomials in S_3, i.e., ternary polynomials of degree at most $n - 2$. The specific sample spaces are different: in NTRU-HPS, $\mathbf{g}$ and $\mathbf{m}$ have exactly $(q/16 - 1)$ coefficients equal to 1 and $(q/16 - 1)$ coefficients equal to -1; in NTRU-HRSS, $\mathbf{f}$ and $\mathbf{g}$ satisfy the non-negative correlation property [14]. As the polynomial $\mathbf{g}$ only appears in the key generation step and is not a component of the secret key sk, we call it the "invisible" secret polynomial in this paper. The transformations between polynomial and bytes are implemented by the Pack and Unpack functions. Lift$(\mathbf{m}) = \mathbf{m}$ for NTRU-HPS, but equal to $\Phi_1 \cdot (\mathbf{m}/\Phi_1 \mod (3, \Phi_n))$ in NTRU-HRSS.

The main steps of the IND-CCA NTRU KEM are shown in Algorithms 4 to 6. A successful key exchange will let two communication parties share a same session key, otherwise a random key will be generated. s and $coins$ are strings of uniform random bits. H_1 and H_2 are hash functions.

2.3 Threat Model

Target. The adversary's target is to obtain the secret key sk. Although it is recommended that applications do not reuse the secret key, in many scenarios (e.g., resource-constrained embedded devices), sk will be reused or stored locally.

Once the long-term secret key is leaked, the attacker can intercept KEM session involving the victim and other party and then recover the shared session key. Moreover, the attacker can impersonate the victim. The secret key of NTRU consists of three parts: $\mathbf{f}$, $\mathbf{f}_p$ and $\mathbf{h}_q$, where $\mathbf{h}_q$ can be computed using the public key. If one can recover $\mathbf{f}$, the other components can be computed accordingly. Therefore, the ternary polynomial $\mathbf{f}$ in the decryption phase during decapsulation becomes the core target of our analysis.

Attacker Capabilities. We assume a standard side-channel attacker who can initiate a KEM process with the victim, and further:

1) Encapsulate maliciously chosen ciphertexts and send them to the victim;
2) Has physical access to EM traces collected from the victim device while it executes the decapsulation algorithm.

Experimental Setup. For all experiments, we compile target implementations using ARM GCC version 10.3-2021.10 with the default (maximum) optimization -O3 and run on an STM32F407G discovery development board. The STM32 chip runs at a clock frequency of 168 MHz. We made minimal modifications to the source code to perform our experiments: we raise a General Purpose I/O (GPIO) pin to trigger our oscilloscope, output the randomly generated secrets $\mathbf{f}$ and $\mathbf{g}$ over serial to confirm our results, and add functions which can change the ciphertext as required in encryption step. A Langer RF-U 5-2 H-field probe is placed at a fixed position over microcontroller as shown in Fig. 1. The probe is connected to a Langer PA 306 preamplifier to amplify the signal by 30 dB. We use a PicoScope 6404C digital oscilloscope to sample traces at 1.25 GHz, which are then downsampled to 625 MHz before analysis. In some cases, we average across multiple traces of the same ciphertext to reduce the impact of measurement noise, as detailed in our attack discussion.

All experiments for the reference implementation and ARM-optimized implementation of NTRU use the `PQClean` library commits `964469d` and `pqm4` library commits `0b50e72`, respectively. We provide our data sets and code under the following link: https://mega.nz/folder/DFBTlb4A#oMeh9Z8DgNSxUHyzgYYjCA.

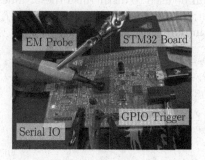

Fig. 1. Hardware setup

3   SPA on NTRU Reference Implementation

In this section, we discuss an SPA on implementations of NTRU in the PQClean library. We start with a preliminary idea and then propose two types of practical chosen-ciphertext attacks, exploiting the leakage from polynomial modular reduction. The first attack is applicable to all the four NTRU instances. The second attack utilizes an inherent property of NTRU-HPS, so it can only be applied to three NTRU-HPS instances.

3.1   A Preliminary Idea

Our first intuition was to recover $\mathbf{f}$ from the leakage of the output in line 3 in Algorithm 3 following the method from [27]—when the ciphertext is generated in the encryption of the encapsulation phase, if we choose the polynomial $\mathbf{c}$ so that it consists only of a nonzero constant term c_0, then the coefficient of $\mathbf{c} \cdot \mathbf{f}$ only has three possible values: $\{-c_0, 0, c_0\}$. If we can find a difference in power consumption between these values during computation, then we can recover the coefficients of $\mathbf{f}$. However, there is an implicit restriction:

$$\mathbf{c} \equiv 0 \quad (\mathrm{mod}\ (q, \mathbf{\Phi}_1)), \tag{1}$$

which means $\mathbf{c}(1) = \sum_{i=0}^{n-1} c_i \equiv 0 \pmod{q}$. We thus cannot simply set $\mathbf{c} = c_0$.

3.2   Attack 1: Recovering Differences Between Adjacent Coefficients

In order to overcome the above restriction, we can choose the polynomial $\mathbf{c}$ with the following Type-I structure:

$$\text{Type-I} : \mathbf{c} = c_0 + (q - c_0)x, \ c_0 \in \{1, 2, \ldots, q - 1\}. \tag{2}$$

In the decryption step, the coefficient of $\mathbf{v}$ (before $\mathrm{mod}q$) is:

$$v_i = c_0(f_i - f_{(i-1)\ \mathrm{mod}\ +n}) \ \mathrm{mod}^+ q \in \{-2c_0, -c_0, 0, c_0, 2c_0\} \ \mathrm{mod}^+ q. \tag{3}$$

As Eq. (3) shows, there are at most five possible values for each coefficient. If side-channel leakage can be used to distinguish these values, then we can recover all differences between adjacent coefficients, i.e., $\delta_i = f_i - f_{(i-1)\ \mathrm{mod}\ +n}$. The equations are written in matrix form as follows:

$$\begin{pmatrix} \delta_{n-1} \\ \delta_{n-2} \\ \vdots \\ \delta_1 \\ \delta_0 \end{pmatrix} = \begin{pmatrix} 1 & -1 & & & \\ & 1 & -1 & & \\ & & \ddots & \ddots & \\ & & & 1 & -1 \\ -1 & & & & 1 \end{pmatrix} \begin{pmatrix} f_{n-1} \\ f_{n-2} \\ \vdots \\ f_1 \\ f_0 \end{pmatrix}. \tag{4}$$

Because $\mathbf{f} \in S_3$, $f_{n-1} = 0$, we have:

$$
\begin{pmatrix} \delta_{n-2} \\ \vdots \\ \delta_1 \\ \delta_0 \end{pmatrix} = \begin{pmatrix} 1 & -1 & & \\ & \ddots & \ddots & \\ & & 1 & -1 \\ & & & 1 \end{pmatrix} \begin{pmatrix} f_{n-2} \\ \vdots \\ f_1 \\ f_0 \end{pmatrix},
\tag{5}
$$

$$
\begin{pmatrix} f_{n-2} \\ \vdots \\ f_1 \\ f_0 \end{pmatrix} = \begin{pmatrix} 1 & 1 & \cdots & 1 \\ & \ddots & \ddots & \vdots \\ & & 1 & 1 \\ & & & 1 \end{pmatrix} \begin{pmatrix} \delta_{n-2} \\ \vdots \\ \delta_1 \\ \delta_0 \end{pmatrix}.
\tag{6}
$$

We can then recover each coefficient by computing $f_i = \sum_{j=0}^{i} \delta_j$.

Target After the First Polynomial Multiplication. A question remains of whether we can find leakage suitable as a classifier. In the reference implementation from the PQClean library, we find a function[1] immediately after the multiplication between $\mathbf{c}$ and $\mathbf{f}$ that has significant leakage related to v_i. Our target is the iteration over coefficients (lines 4–8) in Listing 1.1.

```
1 void PQCLEAN_NTRUHPS2048509_CLEAN_poly_Rq_to_S3(poly *r, const poly *a) {
2     int i; uint16_t flag;
3     //1. Translation: non-negative integer ==> representative in [-q/2,q/2)
4     for (i = 0; i < NTRU_N; i++) {
5         r->coeffs[i] = MODQ(a->coeffs[i]);
6         flag = r->coeffs[i] >> (NTRU_LOGQ - 1);
7         r->coeffs[i] += flag << (1 - (NTRU_LOGQ & 1));
8     }
9     //2. Reduce mod (3, Phi)
10    PQCLEAN_NTRUHPS2048509_CLEAN_poly_mod_3_Phi_n(r);
11 }
```

Listing 1.1. Polynomial reduction from $\mathcal{R}_q$ to S_3 in reference C implementation

After the call to modular reduction MODQ() in line 5, each coefficient is in $\{0, 1, \ldots, q-1\}$. The code in lines 6–7 then implements the centered modular reduction (i.e., $\mathrm{mod}q$): if the output of line 5 is less than $q/2$, then the consequent two computations do not change its value, or else $(-q) \bmod{}^+ 3$ (i.e., 1 for $q = 2048, 8192$; 2 for $q = 4096$) will be added. As the coefficient of a polynomial is non-negative in the implementation setting, $-q$ is taken modulo 3 in advance to avoid the output becoming a negative number and also make sure that the result after modulo 3 is correct. From experimental observation, one can find a periodic pattern in the trace where the target code runs. We choose the local downward peak in each small interval as Point of Interest (PoI) (in total n PoI corresponding to n intermediate values), as it is related to the HW of the output in line 7 (see Fig. 2).

From the result in Fig. 2, we conclude that:

---

[1] https://github.com/PQClean/PQClean/blob/964469d/crypto_kem/
ntruhps2048509/clean/owcpa.c#L139.

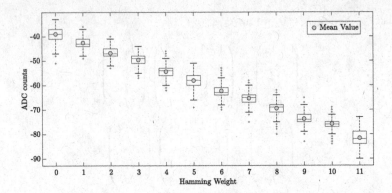

Fig. 2. $p\,(\text{PoI})$ vs. HW of intermediate value in `PQClean` (1000 PoI for each HW value, red line indicates median, red "+" indicates outliers). (Color figure online)

$$|p\,(\text{PoI}_i)| \propto h(v_i) = \begin{cases} \text{HW}\,(v_i)\,, & \text{if } v_i < q/2; \\ \text{HW}\,(v_i + (-q)\bmod{^+}3)\,, & \text{otherwise.} \end{cases} \qquad (7)$$

This means that a significant difference in the HWs of intermediate values can be observed in the corresponding sub-traces. Next, we choose a special c_0 to distinguish different possible values of δ via the HW leakage.

Precomputation. For our chosen ciphertext $\mathbf{c} = c_0 + (q - c_0)x$, the corresponding v_i is in $\{(-2c_0)\bmod{^+}q, (-c_0)\bmod{^+}q, 0, c_0\bmod{^+}q, 2c_0\bmod{^+}q\}$. By going through all the possible values of c_0, we establish a data set (h vs. (c_0, δ)) by evaluating the function $h\,()$ for the above five values (denoted by h_δ when c_0 is fixed). A partial data set is shown below. From Fig. 3, we know that when $c_0 = 1$, we can distinguish -2 from $\{-1, 0, 1, 2\}$, because the corresponding h is substantially larger.

As mentioned in [27], the task to find appropriate c_0 can be considered as a constrained clustering problem [25]. To some degree, more clusters reduces the

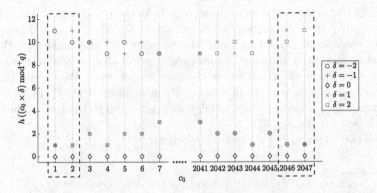

Fig. 3. Selected $h\,((c_0 \times \delta)\bmod{^+}q)$ vs. c_0 (dotted box indicates our chosen c_0).

distance between different clusters in our scenario. In order to achieve better separation in real traces with measurement noise, we only use c_0 which can divide five possible values of δ into two clusters according to high and low h_δ values (represented by "H" and "L"). This ensures that points within any one cluster are similar and the difference between clusters is big. Specifically, for $i, j \in \{-2, -1, 0, 1, 2\}$, we define the similarity within any one cluster B as $S_B = \max\limits_{i,j \in B}(|h_i - h_j|)$ and the distance between two clusters H and L as $D_{\text{H-L}} = \min\limits_{i \in H, j \in L}(|h_i - h_j|)$. We first run a hierarchical clustering on the data set for different c_0 individually to separate all the five possible values into two clusters. We then refine the process of selecting the best c_0: for the same partition, i.e., each cluster has exactly the same elements, we choose the c_0 which has the highest value of $D_{\text{H-L}} - 0.5(S_H + S_L)$.

Actual Analysis. In our actual analysis, we only use the best c_0 for good separability. For example, in the analysis of `ntruhps2048509`, we choose four `Type-I` ciphertexts where $c_0 = 1, 2, 2046, 2047$. Following the notation of "partition with HW feature tags" in [27], we can acquire four different partitions:

$$\{\text{H} : \{-2\}, \text{L} : \{-1, 0, 1, 2\}\}, \text{when } c_0 = 1;$$
$$\{\text{H} : \{-2, -1\}, \text{L} : \{0, 1, 2\}\}, \text{when } c_0 = 2;$$
$$\{\text{H} : \{2, 1\}, \text{L} : \{-2, -1, 0\}\}, \text{when } c_0 = 2046; \quad (8)$$
$$\{\text{H} : \{2\}, \text{L} : \{-2, -1, 0, 1\}\}, \text{when } c_0 = 2047.$$

Based on these partitions, we can build a decision tree (see Fig. 4). Next, we send the chosen ciphertexts to the victim and collect traces while it runs the decapsulation. After locating the section of each trace related to our target function, we find the n PoI and use k-means clustering to separate them into two clusters (H and L) according to the amplitude. With the knowledge of the clustering results under four ciphertexts and the decision tree, we can recover δ. This process is visually illustrated in Fig. 5.

However when analyzing the traces, we found it hard to correctly classify the first PoI (corresponding to δ_0). Consequently, instead of focusing on the recovery

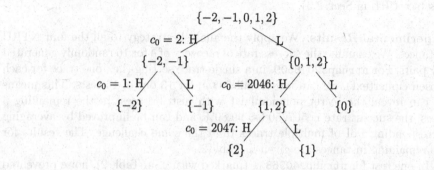

Fig. 4. Decision tree for recovery of δ_i.

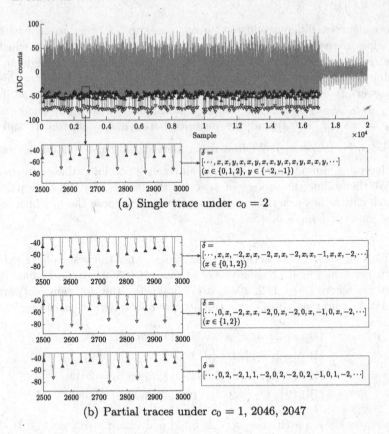

(a) Single trace under $c_0 = 2$

(b) Partial traces under $c_0 = 1, 2046, 2047$

Fig. 5. Traces used to recover δ_i; yellow (or purple) triangle indicates the corresponding δ_i is clustered to "H" (or "L"). (Color figure online)

of the top $(n-1)$ δ_i, we turned to aim at the recovery of all δ_i except δ_0, i.e., $\delta_1, \delta_2, \ldots, \delta_{n-1}$. From Eq. (4), δ_0 can be calculated using the remaining δ_i, i.e., $\delta_0 = -\sum_{i=1}^{n-1} \delta_i$. We can then recover the secret polynomial $\mathbf{f}$ from Eq. (6). As all instances of NTRU call this reduction function, our analysis scheme is universal. The chosen ciphertexts for the other three instances are provided in our data sets (see URL in Sect. 2.3).

Experimental Results. We apply the analysis strategy to all the four NTRU instances. We examine the success rate of recovery of δ for 16 randomly generated key pairs. For `ntruhps2048509`, in a single-trace setting (i.e., one trace for each chosen ciphertext), we can recover all 508 δ_i for 13 out of 16 tests. This means we can reveal the secret polynomial $\mathbf{f}$ with just 4 traces. In the remaining 3 cases, the success rate is $507/508 \approx 99.80\%$, and can be improved by averaging corresponding PoI of multiple traces from the same challenge. The results for the remaining instances are given in Table 2.

In one test for `ntruhps4096821` (marked with $\perp$ in Table 2), noise prevented recovery of both δ_0 and δ_1 even after averaging 10 repeated traces. Since more

Table 2. Keys with 100% success rate and lower success rate for all NTRU instances, the number after percentage indicates the needed number of traces in averaging for each ciphertext in order to improve the SR to 100%

Instance	$\#\delta_i$	#SR=100%	#SR<100%	SR in <100% cases
ntruhps2048509	508	13	3	99.80% (2), 99.80% (4), 99.80% (5)
ntruhps2048677	676	16	0	–
ntruhps4096821	820	13	3	99.88% (4), 99.88% (6), 99.88% ($\perp$)
ntruhrss701	700	15	1	99.86% (2)

than one δ_i was incorrect we could not use Eq. (4) to reconstruct $\mathbf{f}$. To solve this issue, we altered ciphertext Type-I to rotate the PoI as shown below:

$$\text{Type-IB} : \mathbf{c} = (c_0 + (q - c_0)x) \cdot x^j, \, c_0 \in \{1, 2, \ldots, q - 1\}, \, j \in \{0, 10\}. \quad (9)$$

For each chosen c_0, we send a Type-IB ciphertext ($j = 0$) and a Type-IB ciphertext with shifted PoI ($j = 10$). We then replace $p(\text{PoI}_0)$ and $p(\text{PoI}_1)$ of ($j = 0$) with $p(\text{PoI}_{10})$ and $p(\text{PoI}_{11})$ of ($j = 10$), and perform clustering to recover δ. Although 4 more ciphertexts were used, the success rate can reach 100% without averaging, i.e., $4 + 4 = 8$ traces can recover δ, thus infer a correct $\mathbf{f}$.

In summary, our method is effective on the reference implementations of all the four NTRU instances. In the best situation (also the most common case), the secret key can be recovered with only 4 traces, while the worst case (one test in ntruhps4096821) requires at least 6 traces per ciphertext, meaning $4 \times 6 = 24$ traces in total. As the target function runs coefficient-wise, we can handle all n PoI together, and thus the complexity will not increase as n increases.

3.3   Attack 2: Recovery of the "Invisible" Secret Polynomial

In the previous subsection, we have shown that as few as four chosen ciphertexts could recover the secret key. The question arises of whether four is the lower bound for the chosen-ciphertext SCA. In this section, we propose a more efficient attack using an inherent property of NTRU that succeeds with fewer ciphertexts.

Analysis of NTRU-HPS. Firstly, we target the NTRU-HPS variant. We consider the polynomial $\mathbf{c}$ with the following structure:

$$\text{Type-II} : \mathbf{c} = r_0\mathbf{h}, \, r_0 \in \{1, 2, \ldots, q - 1\}. \quad (10)$$

Type-II-like ciphertext have been discussed in several CCA-based mismatch analysis schemes [10,28]: the idea of these techniques is to send a ciphertext $\mathbf{c} = \mathbf{r} \cdot \mathbf{h}$ with carefully chosen coefficients of $\mathbf{r}$ and utilize a decryption-failure oracle to recover the polynomial $\mathbf{g}$. Unfortunately this oracle is unavailable in CCA-secure NTRU KEM, as the chosen polynomial $\mathbf{r}$ in those schemes cannot pass the validation test (line 8 in Algorithm 3) in the decryption phase. Without depending on a decryption-failure oracle, we only leave the constant term of $\mathbf{r}$ and reveal the coefficients of the "invisible" polynomial $\mathbf{g}$ via the leakage of polynomial modular reduction. Inspired by the formula derivation in [10,28], we make the following proposition:

Proposition 1. *For NTRU-HPS, given the chosen ciphertext* $\mathbf{c} = r_0\mathbf{h}$, *each coefficient of the intermediate polynomial* $\mathbf{v}$ *only has three possible values:* $-3r_0$, 0 *and* $3r_0$.

Proof.

$$
\begin{aligned}
\mathbf{v} &= \mathbf{c} \cdot \mathbf{f} & \mathrm{mod}(q, \Phi_1 \cdot \Phi_n) \\
&= r_0\mathbf{h} \cdot \mathbf{f} & \mathrm{mod}(q, \Phi_1 \cdot \Phi_n) & \qquad (11) \\
&= 3r_0\mathbf{g} \cdot \mathbf{f}_q \cdot \mathbf{f} & \mathrm{mod}(q, \Phi_1 \cdot \Phi_n)
\end{aligned}
$$

Since $\mathbf{f}_q \cdot \mathbf{f} \equiv 1 \pmod{(q, \Phi_n)}$, $\mathbf{f}_q \cdot \mathbf{f} = 1 + \mathbf{k} \cdot \Phi_n \bmod q$, where $\mathbf{k} \in \mathbb{Z}_q[x]$ and $\deg \mathbf{k} = \deg \mathbf{f}_q + \deg \mathbf{f} - (n-1)$. As key generation in NTRU-HPS forces $\mathbf{g} \equiv 0 \pmod{(q, \Phi_1)}$, we can write it as $\mathbf{g} = \mathbf{g}^* \cdot \Phi_1 \bmod q$, where $\mathbf{g}^* \in \mathbb{Z}_q[x]$ and $\deg \mathbf{g}^* = \deg \mathbf{g} - 1$. Then we have:

$$
\begin{aligned}
\mathbf{v} &= 3r_0\mathbf{g} \cdot (1 + \mathbf{k} \cdot \Phi_n) & \mathrm{mod}(q, \Phi_1 \cdot \Phi_n) \\
&= 3r_0\mathbf{g} + 3r_0\mathbf{g} \cdot \mathbf{k} \cdot \Phi_n & \mathrm{mod}(q, \Phi_1 \cdot \Phi_n) \\
&= 3r_0\mathbf{g} + 3r_0\mathbf{k} \cdot \mathbf{g}^* \cdot \Phi_1 \cdot \Phi_n & \mathrm{mod}(q, \Phi_1 \cdot \Phi_n) & \quad (12) \\
&= 3r_0\mathbf{g} & \mathrm{mod}(q, \Phi_1 \cdot \Phi_n).
\end{aligned}
$$

□

From Proposition 1, we know v_i only has three possible values: $(-3r_0)\,\mathrm{mod}^+q$, 0 and $3r_0\,\mathrm{mod}^+q$. Utilizing the same HW leakage and strategy of choosing appropriate ciphertexts as in Sect. 3.2 (but on a different data set (h vs. (r_0, g_i)), we can recover $\mathbf{g}$ with only 2 chosen ciphertexts.

Next, we take the analysis of `ntruhps2048509` as an example to explain the process of recovery of secret polynomials $\mathbf{g}$ and $\mathbf{f}$. The chosen r_0 and corresponding partitions are shown below:

$$
\begin{aligned}
&\{\mathrm{H} : \{1\}, \mathrm{L} : \{-1, 0\}\}, \text{when } r_0 = 682; \\
&\{\mathrm{H} : \{-1\}, \mathrm{L} : \{0, 1\}\}, \text{when } r_0 = 1366.
\end{aligned} \qquad (13)
$$

With these two partitions, we can build a decision tree (see Fig. 6) to reveal each coefficient of $\mathbf{g}$.

The actual analysis process with the PoI in the traces is shown in Fig. 7. As $\mathbf{g}$ has a fixed type that $(q/16-1)$ coefficients are equal to 1 and another $(q/16-1)$ coefficients equal to -1, we can use a sorting algorithm instead of clustering to find the highest $(q/16 - 1)$ $|p\,(\mathrm{PoI}_i)|$, then the corresponding g_i will be assigned to 1 when $r_0 = 682$ (or -1 when $r_0 = 1366$).

Proposition 2. *For NTRU-HPS, with the knowledge of* $\mathbf{g}$ *and the relation* $3\mathbf{g} = \mathbf{f} \cdot \mathbf{h} \bmod (q, \Phi_1 \cdot \Phi_n)$, *one can recover* $\mathbf{f}$ *by computing* $3\mathbf{g} \cdot \mathbf{h}_q \bmod (q, \Phi_n)$.

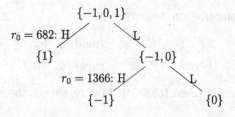

Fig. 6. Decision tree for recovery of **g**.

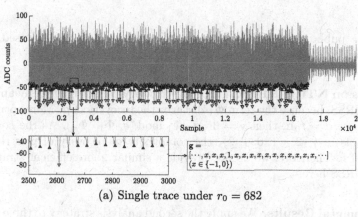

(a) Single trace under $r_0 = 682$

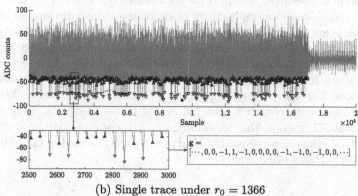

(b) Single trace under $r_0 = 1366$

Fig. 7. Traces used to recover the secret ternary polynomial **g**; yellow (or purple) triangle indicates the corresponding g_i is clustered to "H" (or "L"). (Color figure online)

Proof. From the above relation, we know that for some $t \in \mathbb{Z}_q[x]$, $\deg t = \deg f + \deg h - n$, the following equation holds:

$$3g - f \cdot h \equiv t \cdot \Phi_1 \cdot \Phi_n \pmod{q}. \tag{14}$$

So after modular computation, we have:

$$3\mathbf{g} - \mathbf{f} \cdot \mathbf{h} \equiv 0 \qquad (\mathrm{mod}\ (q, \boldsymbol{\Phi}_n)),$$
$$i.e.,\ \mathbf{f} \cdot \mathbf{h} \equiv 3\mathbf{g} \qquad (\mathrm{mod}\ (q, \boldsymbol{\Phi}_n)). \tag{15}$$

Furthermore, $\mathbf{h} \cdot \mathbf{h}_q \equiv 1 \ (\mathrm{mod}\ (q, \boldsymbol{\Phi}_n))$. Finally, we get the following formula:

$$\mathbf{f} \equiv 3\mathbf{g} \cdot \mathbf{h}_q \quad (\mathrm{mod}\ (q, \boldsymbol{\Phi}_n)). \tag{16}$$

$\square$

As $\mathbf{h}$ is a public polynomial, it is easy to compute its inverse. Based on the formula in Proposition 2, we can recover the secret polynomial $\mathbf{f}$.

Analysis on NTRU-HRSS. The above strategy cannot be directly applied to NTRU-HRSS, because the term $\boldsymbol{\Phi}_1$ disturbs the recovery of $\mathbf{g}$. More specifically, if we choose $\mathbf{c} = r_0\mathbf{h}$, then $\mathbf{v} = 3r_0\mathbf{g} \cdot \boldsymbol{\Phi}_1 \bmod (q, \boldsymbol{\Phi}_1 \cdot \boldsymbol{\Phi}_n)$. As the coefficient of $\mathbf{g} \cdot \boldsymbol{\Phi}_1$ is in $\{-2, -1, 0, 1, 2\}$, the complexity of recovery of $\mathbf{g}$ is the same as that of $\mathbf{f}$. Therefore, we do not provide a similar, more efficient strategy for NTRU-HRSS here.

Experimental Results. We apply the second analysis strategy to three NTRU-HPS instances. The chosen ciphertexts and corresponding partitions for the other two instances are provided in our data sets (please refer to the URL in Sect. 2.3). We again examine the success rate of recovery of $\mathbf{g}$ on 16 randomly generated key pairs. For ntruhps2048509, in a single-trace setting, we can recover all g_i correctly in all 16 cases. This means we can reveal the secret polynomial $\mathbf{f}$ with just 2 traces. The results for the remaining instances are shown in Table 3. In the worst case (with respect to the number of traces used), i.e., in one test in ntruhps2048677, we note that we could improve the success rate from 99.70% to 100% by averaging the corresponding PoI of three traces from the same challenge, i.e., $2 \times 3 = 6$ traces are needed.

Table 3. Keys with 100% success rate and lower success rate for all NTRU-HPS instances (cause $g_{n-1} = 0$, $\#g_i = n - 1$), the number after percentage indicates the needed number of traces in averaging for each ciphertext in order to improve the SR to 100%

Instance	$\#g_i$	#SR = 100%	#SR < 100%	SR in <100% cases
ntruhps2048509	508	16	0	–
ntruhps2048677	676	15	1	99.70% (3)
ntruhps4096821	820	14	2	99.51% (2), 99.76% (2)

4   Applicability to pqm4

We also checked if the same or similar vulnerability exists in pqm4 [16] implementations which have been optimized for ARM processors. In this library, the function that transfers an element in $\mathcal{R}_q$ to S_3[2] is slightly different from the one in the PQClean library (see Listing 1.2).

```
1 void poly_Rq_to_S3(poly *r, const poly *a) {
2     int i;
3     //1. Translation: integer in [0,q) ==> representative in [3q-q/2,3q+q
        /2)
4     for (i = 0; i < NTRU_N; i++) {
5         r->coeffs[i] = ((a->coeffs[i] >> (NTRU_LOGQ - 1)) ^ 3) << NTRU_LOGQ
        ;
6         r->coeffs[i] += a->coeffs[i];
7     }
8     //2. Reduce mod (3, Phi)
9     ...
10 }
```

Listing 1.2. Polynomial reduction from $\mathcal{R}_q$ to S_3 in pqm4 implementation

We found that the local upward peak (i.e., PoI) in the EM trace is related to the HW of the output in line 6 in Listing 1.2. From Fig. 8, we assume the following HW leakage model: ·

$$|p\,(\text{PoI}_i)| \propto h'(v_i) = \begin{cases} \text{HW}\,(3q + v_i), & \text{if } v_i < q/2; \\ \text{HW}\,(2q + v_i), & \text{otherwise.} \end{cases} \tag{17}$$

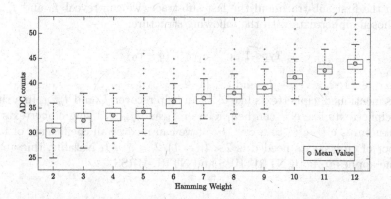

Fig. 8. $p\,(\text{PoI})$ vs. Hamming Weight of intermediate value in pqm4 (1000 PoI for each HW value, red line indicates median, "+" indicates outlier). (Color figure online)

---

[2] https://github.com/mupq/pqm4/blob/0b50e72/crypto_kem/ntruhps2048509/m4f/
owcpa.c#L150.

4.1   Direct Recovery of f

We first tried to apply the recovery of δ in Sect. 3.2 to pqm4. The same ciphertext (i.e., $\mathbf{c} = c_0 + (q - c_0)x$) was chosen. However, the two different implementations of polynomial reduction cause different effects on our analysis. In the reference implementation in PQClean, based on our leakage model, we could find c_0 to distinguish between positive numbers $\{1, 2\}$, negative numbers $\{-1, -2\}$ and zero. Furthermore, some c_0 can be used to distinguish 2 from 1, and -1 from -2. In the pqm4 implementation, it is possible to find c_0 to distinguish positive numbers, negative numbers, and zero through the leakage model. Yet, for arbitrary $c_0 \in \{1, 2, \ldots, q - 1\}$, we have:

$$\begin{aligned}
\left| h' \left((2c_0) \bmod{}^+ q \right) - h' \left((c_0) \bmod{}^+ q \right) \right| &\leq 1, \\
\left| h' \left((-2c_0) \bmod{}^+ q \right) - h' \left((-c_0) \bmod{}^+ q \right) \right| &\leq 1.
\end{aligned} \tag{18}$$

It will be impractical to distinguish 2 from 1, and -2 from -1 in actual analysis.

However, the ability to distinguish between positive, negative, and zero is sufficient to recover $\mathbf{f}$. According to Eq. (4) and $f_{n-1} = 0$, we can find the following pair of equations:

$$\begin{cases} f_0 = \delta_0 \\ f_{n-2} = -\delta_{n-1}. \end{cases} \tag{19}$$

There are just three possible values for δ_0 and δ_{n-1}. It is feasible to find two c_0 that form two partitions like Eq. (13). By observing the clustering result of the PoI in the first sub-trace and the last sub-trace, we can reveal f_0 and f_{n-2} by two chosen ciphertexts with the following structure:

$$\text{Type-I} : \mathbf{c}^{(i)} = c_0 + (q - c_0)x^i, \tag{20}$$

where $c_0 \in \{1, 2, \ldots, q - 1\}$ and $i \in \{1, 2, \ldots, (n - 1)/2\}$.

As mentioned, ciphertexts like $\mathbf{c}^{(1)}$ can help recover f_0 and f_{n-2}. In a similar way, ciphertexts like $\mathbf{c}^{(2)}$ can help recover f_1 and f_{n-3}. Using ciphertexts with the structures $\mathbf{c}^{(1)}, \mathbf{c}^{(2)}, \ldots, \mathbf{c}^{((n-1)/2)}$, we can recover all coefficients of $\mathbf{f}$. The number of ciphertexts needed is $2 \times (n - 1)/2 = n - 1$. Notably, this analysis scheme applies to both NTRU-HPS and NTRU-HRSS.

4.2   Recovery via g

As the leakage can still be used to distinguish between zero, positive, and negative numbers, an approach similar to the strategy in Sect. 3.3 can be applied to pqm4 implementation of NTRU-HPS. Only two ciphertexts with the structure of Eq. (10) are needed to launch the attack. However, note that the number of required traces may be much larger than 2, because the amplitude of variation of p (PoI) against HW of intermediate value is not as significant as that in PQClean implementations. We leave an experimental evaluation of this for future work.

5 Conclusion

In this paper, we proposed two chosen-ciphertext SCA schemes on the latest versions of NTRU. Targeting the HW leakage from the polynomial modular reduction—which is an essential component in NTRU—we first used four Type-I ciphertexts to exploit the differences between contiguous coefficients (i.e., δ) of the core secret polynomial $\mathbf{f}$. Combining the inherent property of NTRU-HPS, we put forward a more efficient second strategy—first revealing the "invisible" secret polynomial $\mathbf{g}$ via 2 Type-II ciphertexts, then recovering $\mathbf{f}$ from Eq. (16). In practical experiments, we found the required number of traces to be low, from two to 24. We also pinpointed similar leakage in the optimized pqm4 implementations. As the above two analysis strategies are succinct and efficient, an effective countermeasure is required to mitigate the side-channel leakage when one uses NTRU with a long-term secret key. We plan to investigate appropriate countermeasures like masking and shuffling in a future work.

Acknowledgements. This work is partially supported by the National Key Research and Development Program of China (2020YFB1005700) and by the Engineering and Physical Sciences Research Council (EPSRC) under grants EP/R012598/1 and EP/V000454/1. We thank the anonymous reviewers for the valuable comments and Sitong Zong for her helpful proofreading advice.

References

1. Alagic, G., et al.: Status report on the second round of the NIST post-quantum cryptography standardization process. Rep. NISTIR 8309, US Department of Commerce, NIST, July 2020. https://doi.org/10.6028/NIST.IR.8309
2. Alagic, G., et al.: Status report on the first round of the NIST post-quantum cryptography standardization process. Rep. NISTIR 8240, US Department of Commerce, NIST, January 2019. https://doi.org/10.6028/NIST.IR.8240
3. Alagic, G., et al.: Status report on the third round of the NIST post-quantum cryptography standardization process. Technical report NISTIR 8413, US Department of Commerce, NIST, July 2022. https://doi.org/10.6028/NIST.IR.8413
4. An, S., Kim, S., Jin, S., Kim, H., Kim, H.: Single trace side channel analysis on NTRU implementation. Appl. Sci. **8**(11) (2018). https://doi.org/10.3390/app8112014
5. Askeland, A., Rønjom, S.: A side-channel assisted attack on NTRU. Cryptology ePrint Archive, Paper 2021/790 (2021). https://eprint.iacr.org/2021/790
6. Atici, A.C., Batina, L., Gierlichs, B., Verbauwhede, I.: Power analysis on NTRU implementations for RFIDs: First results. In: RFIDSec 2008 (2008)
7. Bos, J., et al.: CRYSTALS–Kyber: a CCA-secure module-lattice-based KEM. In: 2018 IEEE European Symposium on Security and Privacy (EuroS&P), pp. 353–367 (2018). https://doi.org/10.1109/EuroSP.2018.00032
8. Chen, C., et al.: NTRU: algorithm specifications and supporting documentation. Technical report, NIST (2020). https://csrc.nist.gov/Projects/post-quantum-cryptography/post-quantum-cryptography-standardization/round-3-submissions

9. D'Anvers, J.-P., Karmakar, A., Sinha Roy, S., Vercauteren, F.: Saber: module-LWR based key exchange, CPA-secure encryption and CCA-secure KEM. In: Joux, A., Nitaj, A., Rachidi, T. (eds.) AFRICACRYPT 2018. LNCS, vol. 10831, pp. 282–305. Springer, Cham (2018). https://doi.org/10.1007/978-3-319-89339-6_16

10. Ding, J., Deaton, J., Schmidt, K., Vishakha, Zhang, Z.: A simple and efficient key reuse attack on NTRU cryptosystem. Cryptology ePrint Archive, Paper 2019/1022 (2019). https://eprint.iacr.org/2019/1022

11. Hoffstein, J., Pipher, J., Silverman, J.H.: NTRU: a new high speed public key cryptosystem. presented at the rump session of Crypto 96 (1996)

12. Hoffstein, J., Pipher, J., Silverman, J.H.: NTRU: a ring-based public key cryptosystem. In: Buhler, J.P. (ed.) ANTS 1998. LNCS, vol. 1423, pp. 267–288. Springer, Heidelberg (1998). https://doi.org/10.1007/BFb0054868

13. Huang, W.L., Chen, J.P., Yang, B.Y.: Power analysis on NTRU Prime. IACR Trans. Cryptogr. Hardw. Embed. Syst. **2020**(1), 123–151 (2019). https://doi.org/10.13154/tches.v2020.i1.123-151

14. Hülsing, A., Rijneveld, J., Schanck, J., Schwabe, P.: High-speed key encapsulation from NTRU. In: Fischer, W., Homma, N. (eds.) CHES 2017. LNCS, vol. 10529, pp. 232–252. Springer, Cham (2017). https://doi.org/10.1007/978-3-319-66787-4_12

15. Jaulmes, É., Joux, A.: A chosen-ciphertext attack against NTRU. In: Bellare, M. (ed.) CRYPTO 2000. LNCS, vol. 1880, pp. 20–35. Springer, Heidelberg (2000). https://doi.org/10.1007/3-540-44598-6_2

16. Kannwischer, M.J., Rijneveld, J., Schwabe, P., Stoffelen, K.: PQM4: post-quantum crypto library for the ARM Cortex-M4. https://github.com/mupq/pqm4

17. Karabulut, E., Alkim, E., Aysu, A.: Single-trace side-channel attacks on ω-small polynomial sampling: with applications to NTRU, NTRU Prime, and CRYSTALS-Dilithium. In: 2021 IEEE International Symposium on Hardware Oriented Security and Trust (HOST), pp. 35–45. IEEE (2021). https://doi.org/10.1109/HOST49136.2021.9702284

18. Kocher, P., Jaffe, J., Jun, B.: Differential power analysis. In: Wiener, M. (ed.) CRYPTO 1999. LNCS, vol. 1666, pp. 388–397. Springer, Heidelberg (1999). https://doi.org/10.1007/3-540-48405-1_25

19. Lee, M., Song, J.E., Choi, D., Han, D.: Countermeasures against power analysis attacks for the NTRU public key cryptosystem. IEICE Trans. Fundam. Electron. Commun. Comput. Sci. **E93.A**(1), 153–163 (2010). https://doi.org/10.1587/transfun.E93.A.153

20. Mujdei, C., Beckers, A., Mera, J.M.B., Karmakar, A., Wouters, L., Verbauwhede, I.: Side-channel analysis of lattice-based post-quantum cryptography: Exploiting polynomial multiplication. Cryptology ePrint Archive, Paper 2022/474 (2022). https://eprint.iacr.org/2022/474

21. Park, A., Han, D.G.: Chosen ciphertext simple power analysis on software 8-bit implementation of Ring-LWE encryption. In: 2016 IEEE Asian Hardware-Oriented Security and Trust (AsianHOST), pp. 1–6 (2016). https://doi.org/10.1109/AsianHOST.2016.7835555

22. Ravi, P., Ezerman, M.F., Bhasin, S., Chattopadhyay, A., Sinha Roy, S.: Will you cross the threshold for me? Generic side-channel assisted chosen-ciphertext attacks on NTRU-based KEMs. IACR Trans. Cryptogr. Hardw. Embed. Syst. **2022**(1), 722–761 (2022). https://doi.org/10.46586/tches.v2022.i1.722-761

23. Schanck, J.M.: A comparison of NTRU variants. Cryptology ePrint Archive, Paper 2018/1174 (2018). https://eprint.iacr.org/2018/1174

24. Sim, B., et al.: Single-trace attacks on message encoding in lattice-based KEMs. IEEE Access **8**, 183175–183191 (2020). https://doi.org/10.1109/ACCESS.2020. 3029521
25. Tizpaz-Niari, S., Cerný, P., Trivedi, A.: Data-driven debugging for functional side channels. In: 27th Annual Network and Distributed System Security (NDSS) Symposium, San Diego, California, USA, 23–26 February 2020. The Internet Society (2020). https://doi.org/10.14722/ndss.2020.24269
26. Ueno, R., Xagawa, K., Tanaka, Y., Ito, A., Takahashi, J., Homma, N.: Curse of re-encryption: a generic power/EM analysis on post-quantum KEMs. IACR Trans. Cryptogr. Hardw. Embed. Syst. **2022**(1), 296–322 (2021). https://doi.org/ 10.46586/tches.v2022.i1.296-322
27. Xu, Z., Pemberton, O., Sinha Roy, S., Oswald, D., Yao, W., Zheng, Z.: Magnifying side-channel leakage of lattice-based cryptosystems with chosen ciphertexts: the case study of Kyber. IEEE Trans. Comput. **71**(9), 2163–2176 (2022). https://doi. org/10.1109/TC.2021.3122997
28. Zhang, X., Cheng, C., Ding, R.: Small leaks sink a great ship: an evaluation of key reuse resilience of PQC third round finalist NTRU-HRSS. In: Gao, D., Li, Q., Guan, X., Liao, X. (eds.) ICICS 2021. LNCS, vol. 12919, pp. 283–300. Springer, Cham (2021). https://doi.org/10.1007/978-3-030-88052-1_17
29. Zheng, X., Wang, A., Wei, W.: First-order collision attack on protected NTRU cryptosystem. Microprocess. Microsyst. **37**(6), 601–609 (2013). https://doi.org/ 10.1016/j.micpro.2013.04.008

Security Assessment of NTRU Against Non-Profiled SCA

Luk Bettale[1], Julien Eynard[2]([✉]) [iD], Simon Montoya[1,3], Guénaël Renault[2,3] [iD], and Rémi Strullu[2]

[1] Crypto and Security Lab IDEMIA Courbevoie, Paris, France
luk.bettale@idemia.com
[2] ANSSI, Paris, France
{julien.eynard,guenael.renault,remi.strullu}@ssi.gouv.fr
[3] LIX, INRIA, CNRS, Ecole Polytechnique, Institut Polytechnique de Paris, Palaiseau, France
simon.montoya@lix.polytechnique.fr

Abstract. NTRU was first introduced by J. Hoffstein, J. Pipher and J.H Silverman in 1998. Its security, efficiency and compactness properties have been carefully studied for more than two decades. A key encapsulation mechanism (KEM) version was even submitted to the NIST standardization competition and made it to the final round. Even though it has not been chosen to be a new standard, NTRU remains a relevant, practical and trustful post-quantum cryptographic primitive.

In this paper, we investigate the side-channel resistance of the NTRU Decrypt procedure. In contrast with previous works about side-channel analysis on NTRU, we consider a weak attacker model and we focus on an implementation that incorporates some side-channel countermeasures. The attacker is assumed to be unable to mount powerful attacks by using templates or by forging malicious ciphertexts for instance. In this context, we show how a non-profiled side-channel analysis can be done against a core operation of NTRU decryption. Despite the considered countermeasures and the weak attacker model, our experiments show that the secret key can be fully retrieved with a few tens of traces.

Keywords: Non-profiled SCA · NTRU · Post-Quantum Cryptography

1 Introduction

The emergence of a quantum computer with the capacity to run Shor's algorithm [29] is a threat to classic cryptography. This threat has led the scientific community to investigate further the field of cryptographic primitives which are resistant against such computers: post-quantum cryptography. A quantum computer could break widely used asymmetric cryptosystems such as RSA or elliptic curve cryptography. In order to keep communications secure in the future, some national agencies have started to study proposals for new algorithms (*e.g.* [2,8]) and have conducted standardization processes for quantum safe algorithms (*e.g.*

[9,25]). One of them was started in 2016 by the National Institute of Standards and Technology (NIST) in the form of a competition [25]. It was seemingly the most attended and followed standardization process by the post-quantum cryptography community until the results were announced on July 5th, 2022.

Among the candidates of the final round, two key encapsulation mechanisms, named NTRU [10] (finalist) and NTRU Prime [4] (alternate finalist), are based on the same mathematical problem that was first introduced in the definition of the original NTRU cryptosystem [20]. In particular, these versions of NTRU are improvements of the original cryptosystem. Even though they were not selected to be part of the next standards, their efficiency, compactness, security make them relevant to be considered for quantum-resistant cryptographic applications. Moreover, the interest of NTRU-based cryptosystems is confirmed by real life experiments. In 2019, Google and Cloudflare jointly initiated an experiment, named CECPQ2 [11,16] that integrates the NTRU finalist key-exchange algorithm into TLS 1.3. Afterwards, Bernstein *et al.* optimized the NTRU Prime key-exchange algorithm in [5] and used it in TLS 1.3 in order to speed-up CECPQ2 protocol.

This kind of cryptographic primitive can be considered for embedded implementations within devices that are limited in terms of CPU frequency, RAM quantity. These kinds of device are naturally threatened by side-channels attacks. In order to protect the implementations against those kind of attacks, the NIST asked the competitors to provide constant-time implementations. However, this countermeasure only protects against timing attacks [22]. Therefore, this protection is not enough against other side-channel attacks (see *e.g.* [7,23]). The overhead required for additional security depends on the subroutines and parameters used within a cryptosystem. The present work provides new results about the security of NTRU like implementations against side-channel attacks.

Previous Works and Motivations

The previous works on the side-channel analysis of NTRU-based schemes were done in different contexts.

In [21], the authors apply several power analysis attacks on various implementations of NTRU Prime. The side-channel analysis focus on the polynomial multiplication during the decryption. These attacks are applied to the NTRU Prime reference implementation, an optimized implementation using SIMD instructions and some implementations using classic countermeasures such as masking. In all these cases, the attacker retrieves the whole secret key. To do so, a powerful attacker model is assumed as he is able to profile the targeted embedded device by using a fully controllable similar open device. In [30], the authors present a single-trace side-channel attack against several lattice-based KEMs. This attack aims to retrieve the ephemeral session key exchanged during the encapsulation routine on not secured implementations. Whereas the NTRU implementation is not directly attacked, the authors give a methodology to apply the attack to it. In [28], the authors combine malicious forged ciphertexts and side-channel analysis in order to retrieve the secret key in the NTRU decapsulation routine. The NTRU cryptosystem is IND-CCA secure which means that it is protected against chosen ciphertext attacks. However, this protection is ensured by a verification done after the whole computation. Therefore, an attacker using side-channel analysis can learn information about these computations and can deduce information about the secret key. Here again, the targeted implementation is not secured against side-channel attacks. Moreover, the attacker can choose the input ciphertext. Even when the secret is masked, side-channel analysis combined with forged ciphertexts are still devastating (see attack [18] for such an attack on Kyber, another finalist to the NIST PQC competition). In [3], the authors mount a side-channel attack against the NTRU round 3 implementation, which is not claimed to be secured against SCA. The targeted operation is the reduction modulo 3 applied to secret polynomials. By using information leakage during this modular reduction, the attacker can retrieve approximately 75% of the secret polynomial. Eventually, the whole key is found by using some lattice reduction techniques. To conclude their work, the authors suggest a more secure modular reduction algorithm that significantly reduces the side-channel leakage.

All these works have paved the way for good practice about secure implementations. Even if they are applied to different contexts, these attacks mainly focus on non-secure implementations and/or suppose that the attacker is powerful enough to profile the target power consumption/electromagnetic radiations. The purpose of the present work is to go further with the side-channel analysis of NTRU by studying the relevance of some classic side-channel countermeasures (tested on the NTRU reference implementation) against weak attackers.

Being able to assess the security provided by lightweight side-channel countermeasures is crucial for constrained devices. Indeed, implementing secure post-quantum algorithms on constrained devices such as smart cards is already a particularly arduous challenge [17]. Since the development of countermeasures can be resource consuming in many ways (development time, extra energy or entropy consumption within the device, etc.), the designer needs to fine-tune the set of side-channel countermeasures according to the considered attacker model. The present work aims to help with this by challenging the efficiency of some classic lightweight side-channel protections against a basic, weak attacker model. To do so, we make use of some classic non-profiled side-channel analysis techniques (clustering) that have already been shown useful to attack embedded implementations of classic asymmetric cryptography (*e.g.* see [19]).

Our Contributions

Our work explores how a weak attacker model can still be threatening for embedded implementations that incorporate low-cost SCA countermeasures. Finding a good balance between performance and effectiveness of SCA countermeasures is crucial for developers. For instance, the rotation-shuffle technique considered in this work is a typical example of a dedicated countermeasure that relies on the intrinsic algebraic structures of NTRU in order to limit the development time, performance and entropy costs.

Attacker Model and Device Under Test. As explained above, we assume that the attacker is weak in a sense that he cannot profile the device under test (DUT). In other words, he cannot perform any supervised attack. The attacker has access to the DUT that implements NTRU. Moreover, he knows the algorithms and the arithmetic that are implemented. He can only probe physical signals of a limited number of Decrypt executions. Also, he knows that the implementation integrates two classic countermeasures. The first one consists in the implementation of a blind decryption, by masking the ciphertext, in order to prevent the possibility of a chosen ciphertext attack. The second one is a rotation of the secret polynomial in order to randomize the operation order at each execution of the Decrypt algorithm. To simulate an experimental situation, the DUT is a STM32F407 discovery board with CPU running at the maximal frequency of 168 MHz. The side-channel leakage used by the attacker is electromagnetic radiations (EM).

Sketch of the Attack. Considering this attacker model, we perform a non-profiled side-channel attack on NTRU. The attack is performed during the first polynomial multiplication within the Decrypt algorithm and aims at retrieving a secret polynomial f. The secret coefficients belong to $\{0, 1, q - 1\}$, where q is a power of 2. The attack is performed in two steps: 1) finding the secret key rotations by locating the $q - 1$ coefficients; 2) distinguishing between the 0 and 1 coefficients. The first step is done by using a clustering algorithm in order to identify the coefficients with value $q - 1$. We use the property that these coefficients have a high Hamming weight. Afterwards, we perform dot products on the clustering results in order to retrieve the rotation indices. The second step uses another property in order to distinguish between 0's and 1's. We use the fact that a 0 coefficient does not change the value of the accumulator during the polynomial multiplication. In practice, it might mean that the CPU handles identical data and instructions during a part of two consecutive coefficient products (this phenomenon is called a collision).

Results. We provide a complete analysis of this side-channel attack against NTRU descapsulation. In particular, we present algorithms to make our approach fully reproducible. It is done in a realistic scenario with a protected implementation running on a non restricted (in terms of frequency) microcontroller. The attacker model is particularly weak (no profiling, no ciphertext forging). It is therefore representative of the kind of threat model that the developers must consider when designing secure embedded implementations. Despite lightweight SCA countermeasures that are wanted to be efficient (performance-wise) yet effective against weak attackers, the attack succeeds in 45 EM traces in average. To succeed, the attack exploits the particular structures of NTRU that we will exhibit later on (key distribution, polynomial arithmetic).

Organization

Section 2 introduces the NTRU Decrypt algorithm. The targeted polynomial multiplication algorithm is presented as well as the implemented countermeasures that are supposed to protect (to a certain extent) this polynomial multiplication against side-channel analysis. Finally, Sect. 3 contains the details of the attack and provides the results of the experiments.

2   NTRU Description and Implemented Countermeasures

In this section, we present the NTRU Decrypt algorithm that we evaluate against side-channel attacks in Sect. 3. We first introduce notations and then describe the reference implementation of NTRU Decrypt with a focus on the targeted operation. Finally, we describe the implemented countermeasures that we take into account in our analysis.

2.1   NTRU Algorithm and Notations

Notations. For any integer $q \geq 1$, we note $\mathbb{Z}_q = \mathbb{Z}/(q)$. For three integers $p, q, n \geq 1$, we define R_q and S_p the following two polynomial rings: $R_q = \mathbb{Z}_q[x]/(\phi_1\phi_n)$ (with $\phi_1\phi_n = x^n - 1$) and $S_p = \mathbb{Z}_p[x]/(\phi_n)$ (with $\phi_n = 1 + x + \ldots + x^{n-1}$). An element in R_q (resp. S_p) is a polynomial of degree at most $n - 1$ (resp. $n - 2$) with coefficients in $\mathbb{Z}_q$ (resp. $\mathbb{Z}_p$). For any polynomial f, we denote by $f[i]$ or f_i the coefficient associated with the monomial x^i.

NTRU Finalist of NIST Competition. NTRU [10] is a Key Encapsulation Mechanism (KEM) based on the NTRU problem which is often assimilated to a lattice-based one. The first version of NTRU was presented in 1996 in [20]. Over the years, several variants of the NTRU cryptosystem were proposed in order to improve the original one. The series of improvements finally led to the variants submitted to the NIST competition. The implementation supporting the application to the NIST competition is used as a reference for our work. The NTRU submission contains 3 routines: KeyGen, Encrypt and Decrypt. These routines can differ slightly depending on the security parameters and the chosen configuration (HRSS or HPS). However, these small differences do not change the overall structure of the algorithms. As the other ideal lattice KEM proposed in the NIST call, the NTRU submission performs polynomial arithmetic. The polynomials are defined over R_q or S_p with $n \in \{509, 677, 701, 821\}$, $q \in \{2048, 4096, 8192\}$ and $p = 3$ depending on the security level.

Decrypt Routine. In this work, we focus on retrieving some information about the long term secret polynomials. These secret data are used in the subroutines KeyGen and Decrypt. However, the KeyGen subroutine is generally run only once in the life cycle of an embedded device. That is not the case for the Decrypt subroutine. Moreover, KeyGen can be done during the production in factory, therefore preventing a possible side-channel attack during the secret key generation. For these reasons, the present attack focuses on a side-channel analysis on Decrypt. Algorithm 1 describes Decrypt as presented in the reference specification [10]. The secret key is made of (f, f_p, h_q) where f, f_p are secret polynomials

---

Algorithm 1. Decrypt $((f, f_p, h_q), c)$

Require: (secret) keys: f, f_p, h_q, ciphertext: c
1: if $c \not\equiv 0 \pmod{(q, \phi_1)}$ return $(0, 0, 1)$
2: $a \leftarrow (c \cdot f) \mod (q, \phi_1\phi_n)$
3: $m \leftarrow (a \cdot f_p) \mod (3, \phi_n)$
4: $m' \leftarrow \text{Lift}(m)$
5: $r \leftarrow ((c - m') \cdot h_q) \mod (q, \phi_n)$
6: if $(r, m) \in \mathcal{L}_r \times \mathcal{L}_m$ **return** $(r, m, 0)$
7: else **return** $(0, 0, 1)$

---

and h_q is the inverse of the public polynomial. The Lift operation depends on the NTRU variant (HPS or HRSS). As it is not important for the attack, we omit its definition.

2.2   Targeted Algorithmic Setting and Operation

Algorithm 2. poly_Rq_mul(c, f)

Require: c, f, where $\deg(c) = \deg(f) = n - 1$
Ensure: $a = c \times f \mod (x^n - 1)$
1: **for** $k = 0$ to $n - 1$ **do**
2:     $a[k] \leftarrow 0$
3:     **for** $i = 1$ to $n - k - 1$ **do** $a[k] \leftarrow a[k] + c[k + i] \times f[n - i]$
4:     **for** $i = 0$ to k **do** $a[k] \leftarrow a[k] + c[k - i] \times f[i]$
5: **return** a

In order to demonstrate the efficiency of our attack and without loss of generality, we choose the following setting: $n = 509$, $q = 2^{11}$ (ntruhps2048509 configuration). The results of the present attack can be straightforwardly generalized to any other setting. The side-channel analysis performed in this article targets the first polynomial multiplication at line 2 of Algorithm 1. This choice is motivated by the fact that this operation is performed on a secret polynomial. Therefore, this polynomial multiplication is highly sensitive and requires a thorough study of its potential sources of leakage. Algorithm 1 describes the decryption algorithm at a high level but our attack is performed against the reference implementation. Thus, we need to detail more precisely how the sensitive operations are defined in the source code. The same attack mounted against another implementation should therefore be adapted to its specificities.

The polynomial multiplication $a \leftarrow (c \cdot f) \mod (q, \phi_1\phi_n)$ is done using the function poly_Rq_mul(c, f) which is described in Algorithm 2. In the reference C implementation, the inputs are uint16_t arrays that represent polynomials and where each element corresponds to a coefficient. The modular reduction modulo q is not performed in this algorithm, but later on the final result.

2.3   Countermeasures

Let's recall that our attacker model implies that only non-profiled side-channel analysis is possible. Some countermeasures are integrated within the NTRU implementation as explained in this section. The reference implementation of NTRU does not claim to be secure against side-channel attacks. As mentioned previously, in order to investigate more thoroughly the vulnerability to side-channel analysis, our objective is to evaluate a more secure implementation, yet still efficient (a typical designer's constraint). In such a case, the considered attacker is not as powerful as in the previous works. Hereafter, we propose two countermeasures targeting such an attacker.

- The `poly_Rq_mul` algorithm is protected with a shuffling countermeasure which aims to randomize the order of the secret coefficients at every iteration.
- Blind decryption by masking the input ciphertext so as to prevent chosen ciphertext attacks.

These countermeasures do not modify the functions and the arithmetic of the reference implementation. All in all, the attack considers executions of the following operation:

$$\texttt{poly\_Rq\_mul}\left(c_1, \text{ROTATE}\left(f\left(x\right), r\right)\right)$$

where r (resp. c_1) is uniformly sampled in $[0, n-1]$ (resp. R_q) at each decryption, without access to their value. ROTATE and c_1 are defined below.

Polynomial Rotation as an Efficient Shuffling. Randomizing the secret coefficients' order is one way to secure an implementation against side-channel attacks. However, this countermeasure can be costly in terms of required entropy and may need to modify the high level polynomial arithmetic. To completely randomize a polynomial, it requires at least n random numbers ($n - 1$ being the polynomial degree) when using Knuth's algorithm. Moreover, some optimization of polynomial multiplication such as Karatsuba or Toom-Cook algorithms cannot be applied with a random order of the coefficients. One cheaper alternative consists in rotating the coefficients of the secret polynomials. The rotation ensures a kind of randomization at the cost of one random generation per polynomial and without modifying the polynomial arithmetic. More precisely, let $f(x) = f_0 + f_1 x + \ldots + f_{n-1} x^{n-1}$ be a secret polynomial, then:

$$\text{ROTATE}(f(x), r) = f_{n-r} + \ldots + f_{n-1} x^{r-1} + f_0 x^r + \ldots + f_{n-r-1} x^{n-1}$$

By considering that the underlying algebraic structure of NTRU is $\mathbb{Z}[x]/(x^n - 1)$, this countermeasure is just a multiplication of f by x^r. The rotation is done prior to the execution of `poly_Rq_mul` algorithm and it is refreshed at each call to `Decrypt`. The implementation of `poly_Rq_mul` does not require modification to be applied to rotated polynomials.

Blind Decryption. Chosen ciphertext attacks combined with side-channel attacks are devastating (*e.g.* [18,28]). The ciphertext shape is generally chosen in order to amplify the side-channel leakages during a computation with the secret key. In order to avoid these attacks, we mask the ciphertext at the beginning of the `Decrypt` algorithm. As a result, the ciphertext c is split into two random variables (aka shares) c_1 and c_2 such that $c = c_1 + c_2 \mod q$. Afterwards, we compute $a_1 \leftarrow (c_1 \cdot f) \mod (q, \phi_1 \phi_n)$, $a_2 \leftarrow (c_2 \cdot f) \mod (q, \phi_1 \phi_n)$ and we obtain $a = (c \cdot f) \mod (q, \phi_1 \phi_n) = a_1 + a_2 \mod q$. Even if the attacker forges a malicious ciphertext, the secret polynomial is multiplied to random/unknown shares. Let's also recall that, as shown in [18], lattice based cryptosystems could be attacked with chosen ciphertexts even if their secret key is masked. Thus, it is a necessity to mask the ciphertext. This masking operation could require as much entropy as masking the secret key f but it does not introduce new potential critical sources of leakage and it can be implemented with less effort.

3   Side-Channel Analysis

In the following section, we target NTRU HPS with parameters $n = 509$ and $q = 2^{11} = 2048$. This attack can be directly applied to any other parameter set. The dataset and scripts are available at `github.com/ANSSI-FR/scantru.git`.

3.1   Target

We focus our analysis on the blind decryption. In particular, the masked ciphertext countermeasure provides the attacker only with side-channel leakages of the multiplication of the secret f by a random value[1].

Thus, we target the rotation countermeasure applied to the product $c \times f$ within the decryption routine (see Algorithm 2). This polynomial product is implemented with a classic convolution product. Beside the rotation countermeasure which consists in shuffling the secret f through a multiplication with a random monomial x^r, the `for` loops at lines 3 and 5 of Algorithm 2 might also be starting at a random index without having to change the attack. In order to defeat these kinds of countermeasures, we focus on some specific operations. More precisely, we consider the following targeted instructions: the first iteration of the outermost `for` loop (line 1 of Algorithm 2).

For the sake of simplicity, we shall be referring to the coefficients $q-1 = 2^{11}-1$ of f as -1. For the experiment, we generated the following data:

– 200 random keys in $(\mathbb{Z}_3[x]) / (x^{509} - 1)$ with coefficients in $\{-1, 0, 1\}$;
– 100 random ciphertexts in $(\mathbb{Z}_{2^{11}}[x]) / (x^{509} - 1)$ for each key.

Then, we got 100 EM traces of the targeted instructions for each key. Let's note that the knowledge of the ciphertexts is not necessary to mount the attack, in accordance with our attacker model (impossible ciphertext forgery).

---

[1] If the computation of the multiplications by the two shares are done sequentially, the attacker might obtain twice more leakages per decryption, therefore reducing the required number of attack traces. Since we assume a weak attacker model, we consider only one multiplication per trace in our experiments.

3.2   Setup and Pre-processing

The original implementation comes from the reference package of NTRU [10].
We have added a rotation function of the secret polynomial. This feature does
not change the arithmetic used in the reference implementation. The attacked
operation is implemented within the KEM decapsulation. In order to ease the
acquisitions and post-processing, a trigger is placed at the beginning of the
product cf. In a more realistic attack setup, let's note that the patterns of the
polynomial product could be identified and used to start the acquisition. The
DUT is a STM32F407 discovery board with clock settings that are tuned for the
CPU to run at the maximal frequency, that is 168 MHz. The implementation was
compiled with -O3 optimization flag. We stress that this optimization flag does
not impact the countermeasures that are considered here (coefficient rotation
and random ciphertext). The acquisitions of electromagnetic radiations (EM)
are made with an EM probe Langer and a Lecroy Waverunner oscilloscope at
20 Gs/s and 2 GHz of bandwidth. Each trace contains approximately 700,000
samples and captures a period of time that lasts ∼35 μs (corresponding to the
first iteration of the for loop at line 1 of Algorithm 2).

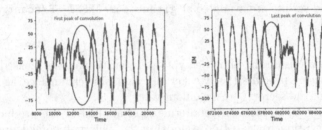

Fig. 1. Details of EM traces of targeted instructions. Red ellipsis point to first (left)
and last (right) peaks to be discarded. (Color figure online)

A code analysis shows that the targeted instructions are parted in two for
loops. Given f and c, we expect the EM peaks to be related to the following
products: (loop 1) $c_1 \cdot f_{(508+r)}, c_2 \cdot f_{(507+r)}, \ldots, c_{508} \cdot f_{(1+r)}$, (loop 2) $c_0 \cdot f_r$.

Figure 1 shows the beginning and the end of the attacked operations. We can
notice that the first and last peaks appear to be quite different than the others,
but similar to each other. Since these differences might affect the clustering
results, they are discarded. The rotation countermeasure allows the attacker to
remove some outliers (*e.g.* difformed peaks) because with several traces it is
possible to get EM peaks for each coefficient anyway.

To summarize, the recorded peaks correspond to the iterations $i = 2$ to
$i = n - 1$ of the for loop at line 3 of Algorithm 2. Even though there is almost
no jitter, a more accurate alignment is realized by using the Pearson correlation.

The assembly code corresponding to the acquired EM peaks is shown in
Fig. 2. This code can be obtained from the open source implementation. It can
help the attacker to mount the attack, even though it is not mandatory.

```
.L3:
    ldrh    fp, [r5], #2        @ loading ci
    ldrh    r4, [r9, #-2]!      @ loading fj
    smlabb  r3, r4, fp, r3      @ rk <- rk+ci*fj
    uxth    r3, r3              @ 16 to 32 bit copy
    cmp     r5, r10             @ loop counter check
    strh    r3, [r0]            @ store in r
    bne     .L3                 @ loop back if not done
```

Fig. 2. Assembly code of targeted first inner for loop of Algorithm 2

3.3 Defeating the Rotation Countermeasure

The rotation countermeasure may be defeated if we find a way to (partially) identify the EM peaks that correspond to the processing of coefficients -1 of the secret key f. The reason why these coefficients are expected to help us through our attack comes from the following remark and hypothesis.

- *Remark:* The coefficients $-1 = q - 1 \mod q$, with $q = 2^{11}$ in the attacked setup, are encoded with Hamming weight (HW) equal to 11, whereas coefficients 0 (resp. 1) are encoded with HW zero (resp. HW one).
- *Hypothesis:* The target device leaks in Hamming weight.

Henceforth, the strategy to defeat the countermeasure consists in identifying the time samples where the HW of the coefficients leaks in EM, allowing a partial recovery of the coefficients -1. These partial recoveries, for every trace, might be enough to shift them back in a synchronized position. When analyzing the code in Fig. 2, we expect this leakage to happen during the ldrh instruction that loads the coefficient of secret polynomial f for instance.

Non-profiled Leakage Assessment. From previous remarks we expect to be able to distinguish -1 from 0 and 1 more easily than 0 from 1. Let's note that, since q is a power of two, if the coefficients -1 are encoded in two's complement using the CPU register size (for instance $2^{32} - 1$ in a 32-bit architecture, or $2^{16} - 1$ when using a type such as uint16_t), then they are encoded with a maximal Hamming weight (HW). This would reinforce the assumed leakage property and would be therefore strongly in favor of the attack strategy we develop. In our experiments, we choose $q = 2^{11}$. Other possible settings correspond to $q = 2^{12}$ and $q = 2^{13}$. Thus, we stress that our attack might even be more efficient in such settings due to a higher Hamming weight for the -1 coefficients.

k-means Clustering. A clustering algorithm is an unsupervised machine learning technique which groups data into a given number of sets according to a given similarity metric. With a k-means clustering [24], the metric is the Euclidean distance to the centroids (means of the clusters), and the variance gives a measure of the spread of the clusters.

It is possible to explore different clustering techniques, such as fuzzy [15] or hierarchical clustering [26], and to investigate their impact on the efficiency of the attack. However, the k-means algorithm is sufficient for our purpose. An advantage of this technique is that it scales up well on the size of the dataset. Practically, a k-means algorithm, where k is the chosen number of clusters, is fed with (parts of) the EM peak traces of all the 507 coefficients of each rotated f at the same time. It gathers the traces in k different sets, each one of them being associated with a so-called *label* that lies in the set $\{0, 1, \ldots, k-1\}$. In order to identify the leakage, we perform a k-means clustering [27] at every time sample and gather the results. We expect that, where the leakage appears to be, the peaks corresponding to -1 will be gathered in one cluster and the peaks for 0 and 1 will be grouped in a second cluster. The credibility of this hypothesis can be first assessed by using the *elbow method*.

Elbow Method to Determine an Optimal Number of Clusters k. The elbow heuristic allows to determine an optimal number of clusters. The goal is to detect the number of clusters from which an overfitting phenomenon might start.

The metric used for this heuristic is called distortion which is the overall sum of the distances between the samples and the center of their own clusters. Thus, it is a measure of how spread the clusters are. A large decrease of the distortion means that adding an extra cluster allows to significantly decrease the overall variance (distance from the elements to their cluster centers). Whenever the distortion starts decreasing less, it means that adding an extra cluster becomes not as relevant when attempting to decrease the overall variance. This heuristic is tested for each time sample[2]. Since all the results suggest that 2 clusters is always an optimal choice, we carry on with the adopted strategy.

Non-profiled Leakage Assessment. In total, 15×507 peaks are used for this leakage assessment (*i.e.* the first 15 traces from one key). We use a classical metric in clustering which is the Davies-Bouldin index (DBI) [14]. This metric gives an insight of the compactness of the clusters respectively to their respective distance. For two clusters $\mathcal{C}_0$ and $\mathcal{C}_1$ that form a partition of the data, it can be expressed as follows:

$$\text{DBI} = \frac{\delta_0 + \delta_1}{2 \cdot d(\mu_0, \mu_1)} \in [0, +\infty)$$

where $\delta_j = |\mathcal{C}_j|^{-1} \cdot \sum_{x \in \mathcal{C}_j} d(x, \mu_j)$ is the average distance of the cluster points to the centroid μ_j (*e.g.* d is Euclidean distance), also known as cluster diameter. It measures the ratio between the cluster sizes and the distance between the cluster centroids. A lower DBI is preferred since it indicates more distinct clusters.

In order to gain higher confidence in the selection of points of interest, we combine DBI with another metric in order to gather more clues about the best area to select. Since the k-means clustering where $k = 2$ is assumed to be able to distinguish quite well between -1 and $0/1$ coefficients, we can expect that one cluster to be roughly twice bigger than the other one (since each coefficient

---

[2] www.scikit-yb.org/en/latest/api/cluster/elbow.html for the python module.

is chosen uniformly randomly in $[-1, 0, 1]$). We then use the following metric (cluster size ratio), that we expect to be as close to zero as possible:

$$\text{CSR} = \left| \frac{\text{size of smallest cluster}}{\text{size of biggest cluster}} - \frac{1}{2} \right| \in \left[0, \frac{1}{2} \right]$$

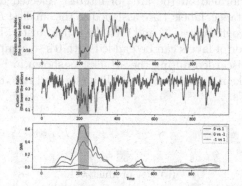

Fig. 3. Non-profiled leakage assessment, with area of interest (in red), performed on the collection of EM peaks of all the targeted coefficients. Top: DBI (the lower, the better). Middle: CSR (the lower, the better). Bottom: SNR with true labels (can only be performed when knowing the keys). (Color figure online)

To sum up, a combination of DBI and CSR is used in order to detect points of interest without the knowledge of the true labels (*i.e.* coefficient values). In practice, our heuristic is as follows. First of all, for each time sample, we perform one clustering per time sample (with 2 clusters as suggested by the *elbow heuristic*) on all the extracted EM peaks. From the clustering labels, we calculate the corresponding DBI and CSR metrics. Second of all, we select areas combining expected metric values (low DBI and CSR). The results are displayed in Fig. 3 (top and middle). One area is highlighted by the expected metrics behaviors. It combines both low DBI and low CSR. The area delimited by the red surface is chosen in order to mount the first phase of the attack. To demonstrate the validity of such approach, the Fig. 3 (bottom) shows the true Signal-to-Noise ratio SNR and the selected area of interest (between red dotted lines). The SNR is a classical metric in SCA. It requires to know the true labels to calculate it. In our case, it basically measures how bigger is the signal carried by the coefficient values (0, 1, -1) comparatively to the noise in the EM signal. Its formula is given by: $\text{SNR} = \mathbb{V}_{\mathbf{f}} \left(\mathbb{E}_{\mathbf{L}}(\mathbf{L} \mid \mathbf{f}) \right) / \mathbb{E}_{\mathbf{f}} \left(\mathbb{V}_{\mathbf{L}}(\mathbf{L} \mid \mathbf{f}) \right)$ where $\mathbf{L}$ is the random variable that represents the EM signal, and $\mathbf{f}$ is the variable that corresponds to the processed coefficients of f.

Let's notice that the Fig. 3 (bottom) also shows why a single trace SPA attack such as the one described in [1] is not possible. With a SNR lower than 1, it is difficult to distinguish the -1's with high confidence in a single trace.

Even though the average EM activity is clearly different when processing a −1 rather than 0 or 1, the impact of the signal variance (noise) must be decreased by averaging enough traces to reveal which coefficients are equal to −1.

Retrieving Random Rotations by Targeting the −1's

Clustering the Traces and Padding the Vectors of Labels. A *k*-means clustering with *k* = 2 is performed on the area of interest selected during the leakage assessment. For instance, labels 1 are assigned to the smallest cluster that should correspond to the coefficients −1, and 0 to the other cluster. For each trace, the corresponding vector of labels can be padded with 0's to compensate peaks that would have been discarded during the attack. As shown in Fig. 1, the first and last peaks are not kept for all the traces. Therefore, all the vectors of labels are padded with two 0's.

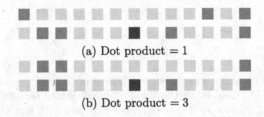

(a) Dot product = 1

(b) Dot product = 3

Fig. 4. Dot product of clustering labels for two traces (top and bottom). Bottom: labels for a reference trace (fixed). Top: labels of a trace to be re-aligned to the ref. trace; they are rotated to find the best candidate (*i.e.* largest dot product result). Red squares: coefficients −1, blue squares: coefficients in {0, 1}. More vivid colors: clustering labels 1, light tone colors: labels 0. (Color figure online)

Aligning Vectors of Labels and Retrieving Rotation Indices. For each rotated key, we use the partially/erroneously recovered patterns of −1 coefficients provided by the clustering in order to find the secret rotation indices.

The strategy is to use the labels to re-align the traces. As shown in Fig. 4, it is usually possible to find the rotation index between two traces. This index is expected to match with the rotation index of the vector of labels which provides the largest dot product with a reference vector of labels.

The tool used to re-align the shifted coefficients is a dot product (other tools like L1 distance were tried out without noticeable difference). This approach might be sensitive to the labels/trace used as a reference. The reference trace was chosen as one that minimizes the CSR metric (*i.e.* with a ratio (number of coeff. −1):(number of coeff. 0/1) being close to the expected ratio 1 : 2). If the results are not satisfying, *i.e.* too many traces are required to recover the key, we can simply choose another reference trace, randomly permute the data and try again. Each vector of labels is then compared to these reference vector. If they share of a big pattern, it can be detected with a peak when calculating the cyclical convolutions between the vectors of labels. In such case, we assume that it is likely that the two traces have been correctly aligned. Figures 5 display the differences between a trace for which we find the rotation index with high confidence, and one for which it is uncertain. The abscissa of the sharp peak points to the plausible rotation index to correctly re-align the two traces. When the scenario of Fig. 5 (right) comes up, we discard the corresponding labels/peak traces for the rest of the attack since it is most likely to add some noise in case the rotation index is not found.

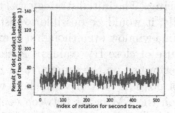 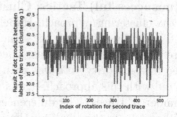

Fig. 5. Synchronizing two traces with (noisy) clustering labels. Left: a sharp peak = right re-alignment. Right: no sharp peak = uncertainty (trace discarded).

During the attack, this strategy allows to retrieve 100% of the rotation indices on the kept traces. Moreover, less than 10% of the traces were discarded during this process. Eventually, the kept traces are re-aligned with the reference trace which itself contains a random rotation. At the very end of the attack, this global rotation will be found with a simple brute force search.

Algorithm 3. HEURISTIC TO LOCATE THE -1'S AND RE-ALIGN THE TRACES

1: Leakage assessment to identify an area of interest
2: k-means with $k = 2$ on this area of each EM peak
3: Pick a reference vector of labels $\mathbf{v}_{ref}$
4: Pad vectors of labels with 0's to match size $n = 509$
5: **for** each vector of labels **do**
6:     Search for rotation providing a match with $\mathbf{v}_{ref}$
7:     If no good match, discard the labels/trace for the rest of the attack
8: Apply majority vote to the aligned labels
9: **return** the assumed indices $\mathcal{I}_{-1}$ of coefficients -1, and $\mathcal{I}_{01} := [0, n) \setminus \mathcal{I}_{-1}$ the assumed indices of coefficients 0 and 1

Majority Vote. Once most of the rotations are found, we can resynchronize the EM peaks and associate them to the right coefficient indices. Eventually, a majority vote is used on the synchronized labels to determine the right labels, *i.e.* which one of the following sets the key coefficients belong to: $\{-1\}$ or $\{0, 1\}$.

When the true keys are known, it is possible to check out the efficiency of this strategy. With around 20 traces, it is possible to retrieve almost 100% of the -1 coefficients. Eventually, we get a set of indices $\mathcal{I}_{-1}$ which is assumed to correspond with the locations of coefficients -1. By complementarity, the indices $\mathcal{I}_{01} = [0, n) \setminus \mathcal{I}_{-1}$ are those assumed to match with the locations of coefficients 0 and 1. The whole heuristic is summarized in Algorithm 3.

3.4   Last Phase of the Attack: Discriminating Between 0 and 1

Initial (unsuccessful) Tactics: Clustering. The initial strategy was to carry on with the clustering approach. Hopefully, it would be possible to distinguish between 0 and 1 by exploiting the same time window even though it is expected to require more traces in this case (because of close HW values). A clustering associated to a majority vote was then performed on the same area of interest. As expected with this approach, it was not possible to retrieve the keys with the given number of traces. The differences on the EM peaks between coefficients equal to 0 and 1 are too small for the attack to succeed in less than 100 traces. With known keys, this fact can be confirmed by analyzing the SNR between the true values. Figure 3 (bottom) shows that the SNR between classes 0 and 1 is around ten times smaller than the one between -1 and 0 or between -1 and 1. With several hundred of traces such approach might succeed. However, another approach is chosen in order to identify the coefficients 0 with less traces.

Changing Tactics: Finding Collisions. If we perform a clustering on the EM peaks then the discrimination between 0 and 1 coefficients is more difficult. Nevertheless, we use the following crucial remark in order to identify more quickly the coefficients equal to 0.

- *Remark:* when processing a coefficient $f_i = 0$, the accumulator (r3 reg. in Fig. 2) remains unchanged: $\mathtt{acc} \leftarrow \mathtt{acc} + 0 \cdot c_{509-i} = \mathtt{acc}$.
- *Hypothesis:* some parts of the EM peak (corresponding to the instructions uxth and strh in Fig. 2) should be similar to the preceding peak when corresponding to a coefficient 0, and differently when corresponding to a coefficient 1 or -1. This is called a collision between two consecutive peaks.

In other words, when processing a coefficient 0, we expect the difference in EM activity between the corresponding peak and the previous one to be quite low at a certain point in time. In such case, the difference is expected to be low since the same instructions (uxth and strh) and data (r3) are processed by the CPU.

Leakage Assessment. In order to use the previous remarks, we perform a non-profiled leakage assessment by using a k-means clustering with $k = 2$ at every time sample on the difference of a peak and its predecessor for the whole set of EM traces. The Davies-Bouldin metric was used unsuccessfully. This may be explained by the fact that the cluster corresponding to a coefficient $-1/1$ might be of bigger variance/diameter (the value contained in r3 register between two consecutive peaks varying randomly, while it is not the case for a 0 coefficient). Overall, we combine two metrics in order to identify an area of interest:

- CSR metric (the smaller, the better): we expect one cluster for the 0's (smallest cluster C_0), and another one for the $-1/1$'s (biggest cluster C_1).
- Distance between the centroids $\|\mathbb{E}(C_1) - \mathbb{E}(C_0)\|$ (the higher, the better): since we expect the EM difference to be noticeably higher in average when processing $-1/1$, we target areas with very distant centroids.

Figure 6 shows the result. We can notice that it is possible to target areas that combine low CSR and high distance between centroids. Moreover, these areas of interest are located in the second part of the peaks. It means that they should indeed correspond to the targeted instruction uxth and strh in Fig. 2.

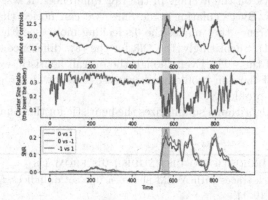

Fig. 6. Non-profiled Leakage assessment with area of interest (in red). Top: distance of centroids (the higher, the better). Middle: CSR (the lower, the better). Bottom: SNR with true labels (can only be performed when knowing the keys). (Color figure online)

Classification Heuristic based on L1-distance of Consecutive EM Peaks. The chosen heuristic (see Algorithm 4) that allows to label the targeted coefficients (*i.e.* the ones assumed to be either 0 or 1 at this point) is simple. All the couples of traces corresponding to an EM peak, supposedly corresponding to 0 or 1 thanks to previous clustering, and its predecessor are gathered. In order to combine several time samples, we consider the L1-distances (sum of absolute difference) between these couples of peaks over the area of interest. Indeed, we do not want the differences from each time sample to compensate each other. Then, for each targeted coefficient, we average all the obtained distances. Other distances were tested without giving better results. Eventually, the classification is based on the absolute difference in EM activity. The L1-distance (on the selected area of interest) between two consecutive peaks will be smaller, in average, when the processed coefficient is 0. Hence, the heuristic simply consists in comparing the distances to the global average of all the calculated distances. If the average for one coefficient is lower than this global average, then it is assumed to be 0, otherwise it is labeled as 1. Let's note that one might run a k-means clustering over the selected area of interest (the norm of the centroids would indicate the right labeling: 0 for the smallest centroid, 1 for the largest one). Even though it was tested, it did not improve the classification comparatively to the simple heuristic described in algorithm 4. Another method would consist in finding collisions through high correlation coefficients between consecutive peaks around the area of interest (see *e.g.* [31] for such kind of approach in a power analysis context). Given the low SNR, this approach did not allow us to retrieve the 0's.

One final remark about the attack phase is that we needed, for the attack to succeed, to consider the L1-distance on the area of interest whereas the leakage assessment was performing well on the raw difference at each time sample. When running the attack on the average of the raw differences, it was not possible to retrieve the keys. By calculating some SNRs, we can notice that it is actually the absolute difference that makes the 0's leaking more (see Fig. 6 (bottom) for the detailed SNR). Surprisingly, the SNR for the raw difference enables a better discrimination of the -1's. Even though the overall heuristic works, this shows the limitations of the non-profiled leakage assessment which remains a difficult task in such non-profiled SCA.

Finally, the Algorithm 4 summarizes the heuristic for this phase of the attack.

Algorithm 4. HEURISTIC TO DISTINGUISH 0'S FROM 1'S

Require: n_t attack traces with N EM peaks each, a set of indices $\mathcal{I}_{01}$ for coefficients supposedly in $\{0,1\}$

1: Leakage assessment to identify an area of interest $\mathcal{A}$
2: **for** each key coefficient f_i with $i \in \mathcal{I}_{01}$ **do**
3:    $\{(\mathbf{p}_{j_1-1}^{(i)}, \mathbf{p}_{j_1}^{(i)}), \ldots, (\mathbf{p}_{j_{n_i}-1}^{(i)}, \mathbf{p}_{j_{n_i}}^{(i)})\}$ collection of the EM peaks corresponding to f_i and their predecessors
4:    $m_i \leftarrow \mathbb{E}_{h \in [1,n_i]}\{\text{dist}_{|\mathcal{A}}(\mathbf{p}_{j_h-1}^{(i)}, \mathbf{p}_{j_h}^{(i)})\}$ average of the (L1-)distances on the area of interest $\mathcal{A}$
5: $m \leftarrow \mathbb{E}(\{m_i \mid i \in \mathcal{I}_{01}\})$
6: **for** each key coefficient f_i **do**
7:    Classification criteria: if $m_i \leq m$ then $f_i \leftarrow 0$ else $f_i \leftarrow 1$ # *based on hyp. that smaller difference in EM activity corresponds to coeff 0*
8: **return** the assumed locations $\mathcal{I}_0$ and $\mathcal{I}_1$ of coefficients 0 and 1 respectively

---

3.5 Summary of Attack and Results

Algorithm 5 sums up the attack. The traces are processed with sets of traces of increasing size. Among the 200 keys, the average and standard deviation of the number of traces to retrieve all the coefficients with no error are resp. $\mu \sim 45$ and $\sigma \sim 10$, with a min. of 23 and a max. of 69. Figure 7 (left) shows the average number of retrieved coefficients given the size of the attack trace set.

Residual Entropy. As we might find the rotation indices up to a global rotation index (the one from the first reference trace) thanks to the first clustering, $\log_2(n)$ bits of entropy remain. By assuming that for the final key we retrieve only $n - x$ coefficients, the residual entropy of the final brute-force correction is bounded by the following formula (searching from 1 to x errors; one erroneous coefficient can be replaced by 2 values): $\log_2(n) + \sum_{i=1}^{x} \binom{n}{i} 2^i$. If x becomes too large, using a lattice approach could be more efficient (see [13]). Figure 7 (right) shows that with less than 20 traces, the security of the implementation drops below 100 bits.

Algorithm 5. NON-PROFILED SIDE-CHANNEL ATTACK ON NTRU

Require: attack traces
1: Pre-processing traces (peak selection, alignment)
2: Apply Alg. 3 to get $\mathcal{I}_{-1}$ and $\mathcal{I}_{01}$, supposedly being the locations (up to a global rotation) of key coefficients -1 and $0/1$ respectively
3: Apply Alg. 4 with $\mathcal{I}_{01}$ and corresponding traces to get $\mathcal{I}_0$ and $\mathcal{I}_1$, supposedly being the locations (up to the same global rotation) of key coefficients 0 and 1 respectively
4: Build (rotated) candidate out of indices $\mathcal{I}_{-1}$, $\mathcal{I}_0$ and $\mathcal{I}_1$
5: Brute force search for last global rotation and potential remaining few incorrect coefficients

---

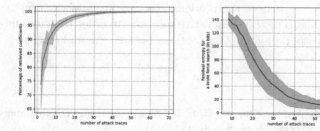

Fig. 7. Results on 200 keys: average ± 1 standard deviation. Left: percentage of coefficients retrieved. Right: residual entropy (in bits) for brute force correction.

Last Remarks. We stress that it should be possible to improve the results by tuning the heuristics better. For instance, shuffling the set of traces can lead to different, sometimes better, results. Moreover, the clustering outliers were not treated as such. An outlier is an element that lies unusually farther to the centroid of its own cluster than the majority of the elements. These outliers might potentially be mislabeled, therefore adding noise in the results of majority votes when trying to locate the -1. Let's also notice that, since the absolute difference in EM activity between consecutive peaks leaks less than the value of the coefficients themselves (see Fig. 3 (bottom) vs Fig. 6 (bottom)), it wouldn't be more advantageous to use the 0's to find the rotation indices.

Yet the heuristic is perfectible, it was possible to show that such a non profiled side-channel attack against a implementation of NTRU which embeds some low-cost countermeasures works well with pretty few traces against a realistic setup.

4   Conclusion

In this paper, we show how a non-profiled side-channel analysis can defeat a somewhat secure (secret key rotation) NTRU `Decrypt` algorithm. In particular, we assume a weak attacker model which cannot profile the targeted device nor use any chosen ciphertext. Despite this restrictive environment, we show that the number of required traces to recover the key is so low that it makes our attack practical. NTRU (and its alternate NTRUPrime) was a finalist of the NIST PQC competition, mostly because of its efficiency and well studied theoretical security. Our work shows that it might be challenging to design countermeasures, even against weak attackers, without noticeably impacting the efficiency of the primitive. Masking the secret key might be a solution but it would decrease the intrinsic efficiency and increase the RAM consumption by at least a factor 2 [6,12]. The security overhead and the requirement of additional security can be scaled down by reusing contemporary secure co-processors to improve the polynomial arithmetic, or by developing new PQC specific secure hardware accelerators.

More generally, we believe the academic community and the cryptography industry will be able to provide innovative solutions in order to make embedded post-quantum cryptographic implementations both secure and efficient.

References

1. An, S., Kim, S., Jin, S., Kim, H., Kim, H.: Single trace side channel analysis on NTRU implementation. Appl. Sci. **8**(11) (2018)
2. ANSSI: Technical position paper–ANSSI views on the Post-Quantum Cryptography transition (2022)
3. Askeland, A., Rønjom, S.: A Side-Channel Assisted Attack on NTRU. Presented at the Third PQC Standardization Conference and published in IACR Cryptology ePrint Archive (2021)
4. Bernstein, D.J., et al.: NTRU Prime (2021)
5. Bernstein, D.J., Brumley, B.B., Chen, M., Tuveri, N.: OpenSSLNTRU: faster post-quantum TLS key exchange. IACR Cryptol. ePrint Arch., p. 826 (2021). to appear in USENIX 2022
6. Bos, J.W., Gourjon, M., Renes, J., Schneider, T., van Vredendaal, C.: Masking kyber: first- and higher-order implementations. Cryptology ePrint Archive, Report 2021/483 (2021)
7. Brier, E., Clavier, C., Olivier, F.: Correlation power analysis with a leakage model. In: Joye, M., Quisquater, J.-J. (eds.) CHES 2004. LNCS, vol. 3156, pp. 16–29. Springer, Heidelberg (2004). https://doi.org/10.1007/978-3-540-28632-5_2

8. BSI: Migration zu Post-Quanten-Kryptografie - Handlungsempfehlungen des BSI (2020)
9. CACR: National Cryptographic Algorithm Design Competition (2018). https://www.cacrnet.org.cn/site/content/838.html
10. Chen, C., et al.: NTRU (2021)
11. Cloudflare: The TLS Post-Quantum Experiment (2019)
12. Coron, J.S., Gérard, F., Montoya, S., Zeitoun, R.: High-order polynomial comparison and masking lattice-based encryption. Cryptology ePrint Archive, Report 2021/1615 (2021)
13. Dachman-Soled, D., Ducas, L., Gong, H., Rossi, M.: LWE with side information: attacks and concrete security estimation. In: Micciancio, D., Ristenpart, T. (eds.) CRYPTO 2020. LNCS, vol. 12171, pp. 329–358. Springer, Cham (2020). https://doi.org/10.1007/978-3-030-56880-1_12
14. Davies, D.L., Bouldin, D.W.: A cluster separation measure. IEEE Trans. Pattern Anal. Mach. Intell. **PAMI-1**(2), 224–227 (1979)
15. Dunn, J.C.: A fuzzy relative of the ISODATA process and its use in detecting compact well-separated clusters. J. Cybern. **3**(3), 32–57 (1973)
16. Google: The Chromium Projects–The Chromium Projects (2019)
17. Greuet, A.: Smartcard and Post-Quantum Crypto (2021). https://csrc.nist.gov/Presentations/2021/smartcard-and-post-quantum-crypto
18. Hamburg, M., et al.: Chosen ciphertext k-trace attacks on masked CCA2 secure Kyber. IACR Trans. Cryptogr. Hardw. Embed. Syst. **2021**(4), 88–113 (2021)
19. Heyszl, J., Ibing, A., Mangard, S., De Santis, F., Sigl, G.: Clustering algorithms for non-profiled single-execution attacks on exponentiations. In: Francillon, A., Rohatgi, P. (eds.) CARDIS 2013. LNCS, vol. 8419, pp. 79–93. Springer, Cham (2014). https://doi.org/10.1007/978-3-319-08302-5_6
20. Hoffstein, J., Pipher, J., Silverman, J.H.: NTRU: a ring-based public key cryptosystem. In: Buhler, J.P. (ed.) ANTS 1998. LNCS, vol. 1423, pp. 267–288. Springer, Heidelberg (1998). https://doi.org/10.1007/BFb0054868
21. Huang, W., Chen, J., Yang, B.: Power analysis on NTRU prime. IACR Trans. Cryptogr. Hardw. Embed. Syst. **2020**(1), 123–151 (2020)
22. Kocher, P.C.: Timing attacks on implementations of Diffie-Hellman, RSA, DSS, and other systems. In: Koblitz, N. (ed.) CRYPTO 1996. LNCS, vol. 1109, pp. 104–113. Springer, Heidelberg (1996). https://doi.org/10.1007/3-540-68697-5_9
23. Kocher, P., Jaffe, J., Jun, B.: Differential power analysis. In: Wiener, M. (ed.) CRYPTO 1999. LNCS, vol. 1666, pp. 388–397. Springer, Heidelberg (1999). https://doi.org/10.1007/3-540-48405-1_25
24. MacQueen, J.: Some methods for classification and analysis of multivariate observations. In: Proceedings of the Fifth Berkeley Symposium on Mathematical Statistics and Probability Held at the Statistical Laboratory University of California 1965/66, no. 1, pp. 281–297 (1967)
25. Moody, D.: Post-Quantum Cryptography NIST's Plan for the Future (2016)
26. Nielsen, F.: Hierarchical clustering, pp. 195–211, February 2016
27. Pedregosa, F., et al.: Scikit-learn: machine learning in python. J. Mach. Learn. Res. **12**, 2825–2830 (2011)
28. Ravi, P., Ezerman, M.F., Bhasin, S., Chattopadhyay, A., Roy, S.S.: Will you cross the threshold for me? Generic side-channel assisted chosen-ciphertext attacks on NTRU-based KEMs. IACR Trans. Cryptogr. Hardw. Embed. Syst. **2022**(1), 722–761 (2022)
29. Shor, P.W.: Polynomial-time algorithms for prime factorization and discrete logarithms on a quantum computer. SIAM J. Comput. **26**(5), 1484–1509 (1997)

30. Sim, B.Y., et al.: Single-trace attacks on the message encoding of lattice-based KEMs. Cryptology ePrint Archive, Report 2020/992 (2020)
31. Zheng, X., Wang, A., Wei, W.: First-order collision attack on protected NTRU cryptosystem. Microprocess. Microsyst. **37**(6), 601–609 (2013)

Next-Generation Cryptography

Post-Quantum Protocols for Banking Applications

Luk Bettale[1], Marco De Oliveira[1], and Emmanuelle Dottax[2]([⊠])

[1] IDEMIA, Courbevoie, France
{luk.bettale,marco.deoliveira,}@idemia.com
[2] IDEMIA, Pessac, France
emmanuelle.dottax@idemia.com

Abstract. With the NIST competition for Post-Quantum (PQ) algorithms standardization coming to an end, it is urgent to address the question of integrating PQ cryptography in real-world protocols to anticipate difficulties and allow a smooth transition. This is especially true for banking applications where the ecosystem composed of a variety of cards and terminals is heterogeneous. Providing solutions to ensure efficiency and some kind of backward compatibility is mandatory. In this work, we provide the first analysis of card-based payments with respect to these questions. We integrate post-quantum algorithms in existing protocols, and propose hybrid versions. We implement them on banking smartcards and analyse the impacts on various aspects of the product, from production to actual transactions with terminals. Our work shows that such products are possible, but we identify several issues to overcome in the near feature in order to keep the same level of usability.

Keywords: Post-quantum cryptography · Card-based payments · EMV · Hybrid protocols · Smart-cards

1 Introduction

Though the path to a cryptographically relevant quantum computer is uncertain and may still be long, the time required to update our systems and the "record now, decrypt later" threat are requiring anticipation. Agencies and standardization bodies acknowledge this fact, and already launched actions towards a migration to Post-Quantum (PQ) cryptography. The NIST competition [26] is the main driver on this subject, and the announcement of the algorithms selected for standardization is going to accelerate the development, standardization and adoption of post-quantum protocols.

In parallel, industry is invited to start considering this migration, by looking for potential weaknesses in systems and protocols, by anticipating the transitioning phase, and by experimenting with the technology as of today. It is important to identify as soon as possible the difficulties that will be faced, in particular in terms of performances. Indeed, post-quantum algorithms have generally much

I. Buhan and T. Schneider (Eds.): CARDIS 2022, LNCS 13820, pp. 271–289, 2023.
https://doi.org/10.1007/978-3-031-25319-5_14

larger keys, ciphertext and signature sizes, and are more demanding in terms of computations and memory consumption. In addition to this "post-quantum overhead", agencies are asking to first implement *hybrid protocols*, that is to say protocols that combine "classical", well studied cryptography, and PQ cryptography (see [4] for instance). This is to cope with the immaturity of the novel PQ algorithms, and prevent any regression. Once the community will have sufficient confidence, PQ algorithms will be used alone.

Some experiments with real world protocols have already been conducted. Back in 2016, Google deployed a hybrid scheme for key establishment in TLS [12]: the PQ algorithm NewHope [2,3] was used in addition to a classical key exchange. On the same subject, [28] formalizes the security of PQ versions of TLS that rely on key encapsulation mechanism instead of signature for authentication. We can also mention the proposal [29] for a hybrid version of the Signal protocol, also relying on PQ key encapsulation. More industry-oriented works include [27] which focuses on M2M protocols, and [11] on secure-boot in the automotive context.

In this work, we consider protocols used for card-based payments, according to EMV specifications [20]. Those protocols make use of asymmetric cryptography (RSA) for card authentication by the terminal, and symmetric cryptography (usually TDES) for transaction certificates intended for the issuer. Contrary to protocols used to protect any kind of information, like TLS, the cryptographic data involved in a banking transaction are valuable only during a short period of time: after validation by the bank, they become useless. This makes banking transactions immune to the "record now, decrypt later" threat. On the other hand, *offline transactions* rely exclusively on card authentication. In this case, the validation by the bank is delayed: transactions certificates are stored some hours in the terminal before being transmitted to the bank. This is typically used to accelerate commercial exchanges in case of small amounts. Here, the authentication of the card is crucial for the security of the system. Though there is no short-term threat on RSA-based authentication, the industry should start today to explore PQ cryptography for this authentication step, so as to be ready for "Q-day". This is especially true because there are tight requirements on transactions timings, and more demanding computations could have a dramatic impact, as far as to require changes in the hardware of involved devices. This would be a very long process, that would add to the one of a cryptographic migration in the financial industry, itself long and complex. Early preparation and identification of obstacles can only facilitate this effort.

We here focus on the card part (including communications). As it is the most constrained device, this is where the repercussions are expected to be the greatest. To our knowledge, this is the first work that considers the integration of PQ cryptography into a banking protocol and that evaluates the corresponding impact on smart-cards in a real world context.

Contributions. We select two payment protocols: EMV CDA with offline PIN verification and offline transaction, and a variant providing protection against user tracking. These protocols offer different security properties and work with different (classical) cryptographic algorithms. It is thus relevant to see how they behave when "hybridized". We propose hybrid versions of these protocols. In both cases, we add the PQ resistance by careful addition of cryptographic operations. We do this with several objectives in mind. First, and obviously, we maintain the security features of the original protocols. Second, we respect as much as possible the general structure in terms of commands and exchanges. This aims at easing the migration, by limiting modifications and allowing retro-compatibility. Finally, we attempt to reduce the overhead on transaction timings, and to this end we carefully design the PQ version of protocols. We selected the PQ algorithms among the finalists of NIST's competition, and implemented the result on smart-cards. We then analyze the impacts in terms of size of data (code size, card personalization and communications) and computations timings, and expose some areas of work that should be considered if we want to reach good performances with hybrid protocols.

Outline. Section 2 describes the existing classic protocols we consider. Section 3 explains how we integrated PQ cryptography in the selected protocols and presents the hybrid versions we implemented. Section 4 describes our implementations and analyses the results. Section 5 concludes this paper.

2 Existing Protocols

A vast majority of banking cards and terminals use the Europay, Mastercard and Visa (EMV) standard. This standard has been created in 1996, and covers data formats and protocols. For our study, we consider two protocols that we think are the most interesting. Here, we will recall the main steps of a transaction, briefly list the different available options, and then focus on the description of the cryptographic parts of the protocol we selected. For interested readers, [13] provides a more thorough overview of EMV.

The EMV specifications consist in several documents. We here recall the main phases of an EMV session, the interested reader can look at [20,21] for more detailed information. (Note that in practice, these phases can be interleaved into several commands.)

1. **Initialization.** During this phase, the reader selects the right application on card, and some data are transmitted by the card, including card number and expiry date, and supported features.
2. **Data authentication.** This stage is meant to ensure that the card is a genuine one. Three different methods are supported:
   - *Static Data Authentication (SDA).* In this case, the card provides signed data to the terminal. The latter can then check the authenticity of the data. Note that this method is prone to cloning.
   - *Dynamic Data Authentication (DDA).* Here, the card is equipped with an asymmetric key pair, the corresponding certificate, and the ability to perform signatures. A "challenge-response" exchange where the card signs

a challenge sent by the terminal allows to authenticate the card – in a way that prevents cloning.

There is however a security issue with DDA: though it allows authentication of the card, the terminal has no assurance that the same card is involved in subsequent exchanges.

- *Combined Data Authentication (CDA).* This method is similar to DDA, with the only difference that the card signature is also used to authenticate some data, so as to repair the flaw of DDA.

3. **Cardholder verification.** This can be done by PIN verification or handwritten signature. When PIN is used, this latter is entered on the terminal, and the verification happens either online – the PIN is sent to and checked by the issuer, or offline – the PIN is sent to the card. The PIN can be transmitted in clear text, or encrypted. In this case, the terminal encrypts the PIN for the card using its public key.

4. **Transaction.** This is the final step, it can be online, or offline:
   - *Online transaction.* In this case, the card provides an *Authorisation Request Cryptogram*, generated thanks to a symmetric key shared with the issuer. The terminal forwards it to the issuer, who verifies it. In case of approval, the card provides a *Transaction Certificate (TC)* as a proof of completion.
   - *Offline transaction.* Here, the card provides directly a TC, which is transmitted to the issuer later. The commercial transaction is authorized only on the basis of the card authentication.

As can be seen, fully online transactions rely on symmetric cryptography, and are thus quite easily protected against quantum attacks: the already available 256-bit security symmetric cryptography provides quantum-resistance. However, those transactions cannot be executed everywhere (a good connection is required) and are less use-friendly, as they take more time to proceed. This is why offline transactions are often used. To protect these ones against quantum attacks, the card authentication needs to integrate new PQ algorithms. In the rest of this paper, we thus focus on offline transactions, with PIN transmitted encrypted and verified by the card, and delayed transmission of TC.

2.1 EMV CDA Protocol

For our analysis, we consider CDA, as it is the most secure variant. As explained above, we consider an offline transaction with PIN encryption. The resulting cryptographic protocol is depicted on Fig. 1. To ease reading and to focus on the parts relevant to our purpose, we adapted EMV notations and omitted some details (including some hash computations that are not relevant for our purpose).

This protocol relies on the RSA algorithm, used both for encryption and signature. The card is equipped with a secret key sk_C, and the corresponding certificate $cert_C$, emitted by the issuer. The certificate $cert_{IS}$ of the latter is also provided by the card, so that the terminal can check the chain up to the certification authority public key pk_{CA} (all certificates are signed with RSA). The function Verify is used to this aim: it takes as input a public key and a certificate, verifies the signature of the certificate and outputs the certified public key in case of success. The function RSA.Enc (resp. RSA.Dec) takes as input a public RSA key (resp. a private RSA key) and a message to encrypt (resp. to decrypt), and outputs the ciphertext (resp. the plaintext). Some nonces are used to ensure freshness of the encrypted PIN ($nonce_C$ sent by the card) and of the card signature $SDAD$ ($nonce_T$ sent by the terminal). The card is also equipped with a symmetric key MK_{AC}, shared with the issuer and used to compute the cryptograms. These ones are computed thanks to a MAC function – usually based on DES or Triple-DES – applied on required information related to the transaction, denoted by $transInfo$.

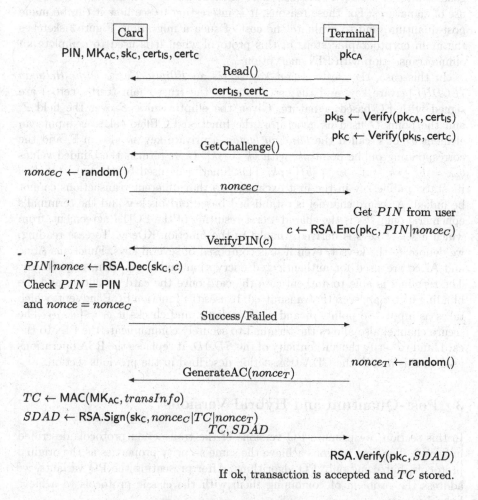

Fig. 1. EMV CDA protocol

2.2   BDH-Based Protocol

In 2012, EMVCo decided to replace RSA by ECC-based cryptography, and expressed the need for a protocol establishing a secure channel between the card and the terminal, while preventing card tracking – this one is indeed possible as soon as the card transmits its certificate in plain. Two protocols were identified and drafted in [22]: one called *Blinded Diffie-Hellman (BDH)* – where the card uses a static Diffie-Hellman key to authenticate itself –, and a 1-sided version of the Station-To-Station protocol [18]. BDH is the most efficient one and has been proven secure by [14], and further discussed in [24].

This protocol enjoys a provably sound design, substantially increasing the general level of security. It is also more efficient than CDA, and does not make use of signatures. For these reasons, it is interesting to see how it can be made post-quantum, and to evaluate the cost of such a migration. Figure 2 sketches the main cryptographic steps of this protocol when it is used to complete an offline transaction with PIN encryption.

In this case, the card's secret key sk_C is an *Elliptic Curve Diffie-Hellman (ECDH)* private key, and the certificates of the trust chain ($cert_C$, $cert_{IS}$) are signed with EC-based signature. Given the elliptic curve $\mathcal{E}$ over the field $\mathbb{F}$, and the point G used as generator, the function EC.Blind takes as inputs an element bf of $\mathbb{F}$ called the *blinding factor*, a private key sk also in $\mathbb{F}$, and the corresponding public point pk, with $pk = [sk] \cdot G$. It returns the blinded values $bsk = bf \cdot sk$ and $bpk = [bf] \cdot pk$. This function is used by the card to *blind* its static public key in the first exchange, so that different transactions cannot be linked. A secure channel is established based on this key and the terminal's ephemeral one: shs is the shared secret resulting of the ECDH agreement, from which a set of keys is derived, thanks to the function KDeriv. To ease reading, we denote K the keyset, even if it is composed of several keys. Functions AEnc and ADec are used for authenticated encryption and decryption, respectively. The terminal is able to authenticate the card once the card certificate and the blinding factor are secretly transmitted. It uses the function EC.BlindVerif, which takes as input two points p_1 and p_2, a scalar s, and checks if $p_1 = [s] \cdot p_2$. The secure channel also allows the terminal to securely communicate the PIN to the card, and to verify the authenticity of the $SDAD$. It replaces the RSA operations used to this aim in the CDA transaction described in the previous section.

3   Post-Quantum and Hybrid Versions

In this section, we propose PQ versions of the transaction protocols described in Sect. 2. The new versions achieve the same security properties as the original protocols, but using only PQ algorithms. After presenting the PQ variants, we address the problem of combining them with the classic protocols to achieve hybrid versions.

The NIST standardization process [26] aims at standardizing two types of cryptographic primitives: digital signature and key encapsulation mechanism.

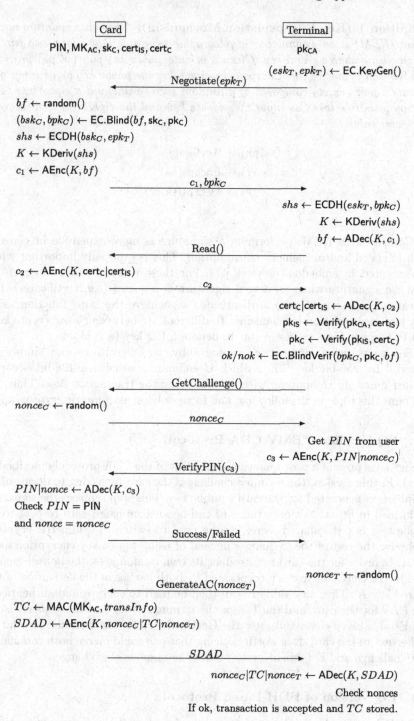

Fig. 2. BDH-based protocol

Definition 1 (Key Encapsulation Mechanism). *A key encapsulation mechanism (KEM) is an asymmetric cryptographic primitive that allows two parties to establish a shared secret key. Given a private/public key pair* (sk, pk) *provided by a* KeyGen *primitive, the* Encaps *primitive takes the public key* pk *as input and outputs a pair* (ss, ct) *composed of a random secret value and a ciphertext. The* Decaps *primitive takes as input the private key and the ciphertext and recovers the secret value.*

$$(\mathsf{sk}, \mathsf{pk}) \leftarrow \mathsf{KeyGen}()$$
$$(ss, ct) \leftarrow \mathsf{Encaps}(\mathsf{pk})$$
$$ss \leftarrow \mathsf{Decaps}(\mathsf{sk}, ct).$$

Compared to KEM, performing a signature is more expensive in general, both in speed and in memory consumption. This is especially important when implemented in embedded devices [25]. For these reasons, our strategy is to only use signature when an actual signature is required (e.g. certificates). For operations whose goal is to authenticate, we achieve the same function using only key encapsulation mechanisms. To differentiate between classic crypto keys and PQ crypto keys, we use a star to denote a PQ key (e.g. pk_C^*).

For parts using symmetric cryptography, the algorithms can simply be replaced by 256-bit key AES, which is enough to achieve a 128-bit security against quantum computers, without modifying the transaction flow. Thus, we will omit this topic in the following, and focus only on asymmetric cryptography.

3.1 PQ Version of EMV CDA Protocol

In Fig. 3, we present a post-quantum equivalent of the CDA protocol described in Fig. 1. In this version, the terminal challenges the card by asking to decapsulate a ciphertext generated for the card's public key. This step replaces the signature performed in Fig. 1. Not only the card can be authenticated with respect to its public key, but the shared secret can be used to switch to symmetric cryptography for the rest of the exchanges instead of using public key encryption as in Fig. 1. In order for the card to introduce its own randomness, the $nonce_C$ generated in the GetChallenge is incorporated by both parties in the derivation of the shared key K. This very value K can then be used to encrypt and authenticate the PIN for the card, and the TC for the terminal. Note that the sequence adds the EstablishKey command after the GetChallenge for the terminal to send its ciphertext to the card. It is worth noticing that one could merge both commands GetChallenge and EstablishKey to reduce the number of exchanges.

3.2 PQ Version of BDH-Based Protocol

The BDH protocol uses the mathematical properties of elliptic curves to derive a valid ephemeral public key from a static public key using the blinding factor.

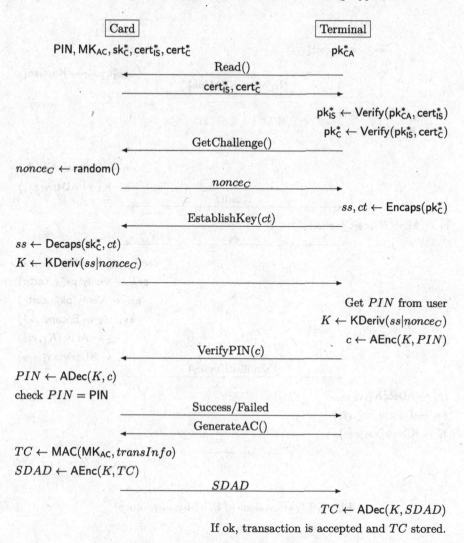

Fig. 3. PQ equivalent of CDA protocol

It is not possible to achieve this in general with PQ algorithms. To achieve authentication, we can use a protocol inspired by KEM-TLS [28]. In Fig. 4, the terminal generates an ephemeral key pair, and sends its public key to the card. This is equivalent to the previous Negotiate command. The card can encapsulate a secret and derive a first shared secret K_1 to securely communicate with the terminal. This is performed by the Encaps command. The shared secret K_1 is used by the card to send its static public key encrypted. The public key being encrypted, it cannot be linked to the actual user by an eavesdropper. The terminal can authenticate the card's public key. Now, the terminal can

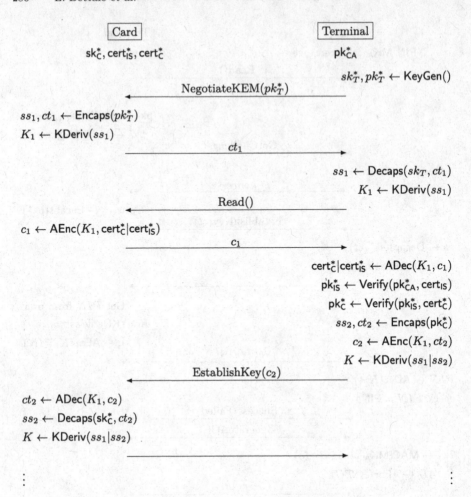

Fig. 4. PQ equivalent of BDH-based protocol

challenge the card by asking to decapsulate the ciphertext ct_2. To completely ensure privacy, this value is sent encrypted with K_1. Two passes are necessary to achieve the same privacy-preserving properties of classic BDH. Once the new derived secret K is shared between the card and the terminal, the PIN encryption and the rest of the transaction can be conducted in the same way as in Fig. 2, so this part is omitted in our description. Finally, please note that this protocol offers implicit authentication by sharing the common secret K. This can be made explicit by sending an acknowledgment encrypted with the shared key K to the terminal in response of the EstablishKey command if required.

3.3   Hybrid Versions

Though our PQ protocols do not make use of signatures, these ones are needed for certificates. Indeed, two certificates are sent by the card and verified by the terminal: the one of the issuer, and the one of the card. For hybrid versions, different approaches exist, one having two separated chains, and one using "hybrid certificates" (see [10] for instance). All options show no significant difference in terms of size: in all cases, the same number of public keys and signatures are present. As a consequence, we decided to implement separated chains for simplicity. Of course, considering certificate management and performances on the terminal side could lead to a different implementation.

For the CDA protocol, we can derive a hybrid equivalent by arranging together protocols presented by Fig. 1 and Fig. 3 as follows:

- The card stores both RSA and PQ KEM key pairs and certificates.
- In the first Read command, the certificates chain of the PQ static public key is sent to the terminal. The terminal then verifies the certificate chain.
- The terminal then asks the card to decapsulate ct, the encrypted shared secret ss.
- In a second Read command, the certificates chain of the classic static public key is sent to the terminal. The terminal then verifies the certificate chain.
- After the terminal receives $nonce_C$, a key K is derived from $nonce_C$ and the PQ shared secret ss. The PIN get from the user is first encrypted using K, then this value is encrypted using RSA: $c \leftarrow \mathsf{RSA.Enc}(\mathsf{pk}_C, \mathsf{AEnc}(K, PIN))$.
- The TC is encrypted/authenticated using the key K, it is also signed using the RSA private key as in Fig. 1: $\mathsf{RSA.Sign}(\mathsf{sk}_C, nonce_C|TC|nonce_T)$. Both values are sent as a response to the GenerateAC command.

The result is described in Fig. 5.

For the BDH-based protocol, we derive a hybrid version by arranging protocols presented by Fig. 2 and Fig. 4 as follows:

- The card stores both RSA and PQ KEM key pairs and certificates
- Also perform BDH key exchange in the Negotiate command: the terminal will send its ephemeral ECC key in addition to its ephemeral KEM key.
- the secret K must be derived not only from the PQ shared secrets ss_1 and ss_2, but also from the shared secret shs obtained from BDH: $K \leftarrow \mathsf{KDeriv}(shs|ss_1|ss_2)$

The result is described in Fig. 6.

4   Practical Implementation

In this section, we describe our implementations and give the results of our measurements. We analyse them so as to identify points that require specific attention in view of a post-quantum migration.

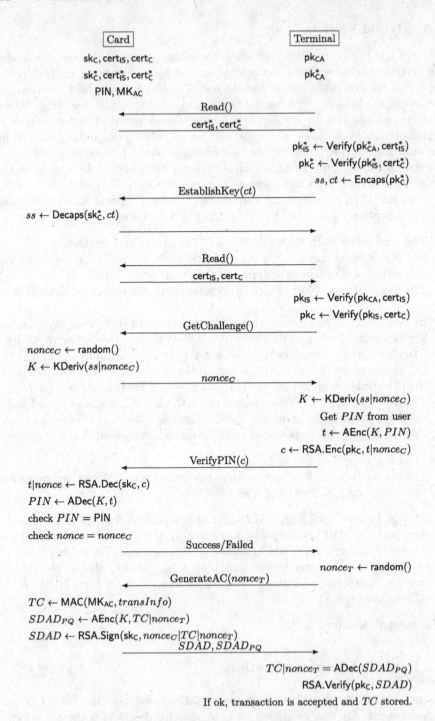

Fig. 5. Hybrid equivalent of CDA protocol

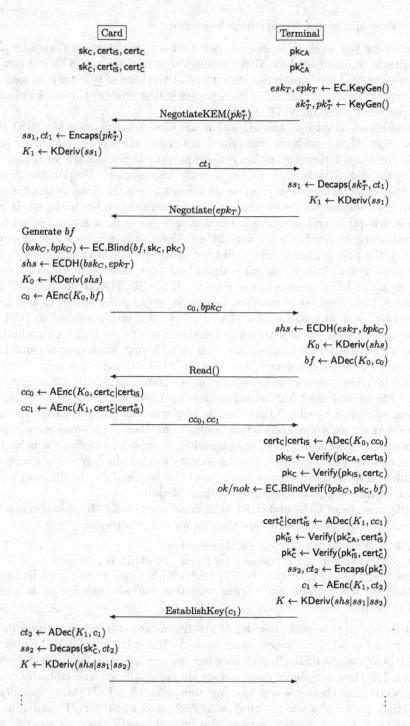

Fig. 6. Hybrid equivalent of BDH protocol

4.1   Post-quantum Algorithms Selection

In order for this work to be relevant for further experiments, we consider only algorithms recommended by NIST, or selected as finalists of NIST's PQ competition – the final selection is not announced at the time of this work. Indeed, we can expect that most standardization bodies will articulate their works towards PQ protocols based on NIST standards.

Some hash-based signature algorithms are already standardized by NIST [17]. They enjoy high confidence from the community and can be used alone (in a non-hybrid way). However, unlike general purpose signatures schemes, they are state-full, and as such, they require specific management. Despite the gain in size they could bring in our case, we decide not to use them, as we feel the gain is not large enough to justify putting more constraints on the back-end. It is of course debatable, and advances in the management of these signatures schemes (in particular in Hardware Security Modules) could justify their use. However, as said, the gain is limited and would not change drastically our figures. Thus, we considered the three signature algorithms that have been selected for the last round of NIST's competition: FALCON [23], CRYSTALS-Dilithium [6] and Rainbow [16]. We exclude Rainbow because of very large public keys and, independently, security issues that lead to a practical attack presented in [8,9]. As it is expected that one of FALCON or Dilithium will be selected for standardization, we considered both in our experiments. We first implemented candidates of security strength category 1, according to NIST's definition (categories range from 1 to 5, cat. 1 being the (somehow) equivalent of AES with 128-bit key, and cat. 5 the equivalent of AES with 256-bit key). However, as will be shown in our figures, we quickly realised that the size of certificates was the most impacting characteristic regarding transaction timings. We thus made some more experiments with signatures in categories 3 and 5, in order to measure the behaviour of timings with respect to certificates security level. Indeed, it seems reasonable to think that different security levels would be required for the different actors in the certificate chain, depending on their lifetime.

PQ versions of CDA and BDH both make use of a KEM. Several characteristics of the chosen algorithm are relevant for our experiment:

- the card needs to store the KEM secret key;
- the KEM public key is transmitted as part of card's certificate;
- one (for PQ CDA) or two (for PQ BDH) KEM ciphertexts are exchanged;
- the card will perform one Decaps operation, and also one Encaps in case of PQ BDH.

At the time of this work, four KEM algorithms are still in competition in the final round of NIST's competition: Classic McEliece [1], CRYSTALS-Kyber [5], NTRU [15] and SABER [7]. McEliece has key size incompatible with embedded usage. The three remaining have roughly similar public key and ciphertext sizes, but NTRU has shorter secret key. We thus selected NTRU. More specifically, we implemented the variant ntruhps2048509. It is a category 1 candidate. As smart-cards have a limited lifetime, and limited capabilities, this security level seems to offer the best tradeoff.

4.2   Implementation Description

The protocol has been implemented on an ARM Cortex-M3 derivative, which is a common architecture for banking smart-cards. Several applications support EMV protocols, we opted for a *Common Payment Application (CPA)*, compliant with [19], in which we add support of the hybrid protocols presented earlier. We use a basic, "toy" profile for the application. The version with the EMV-compliant CDA protocol described by Fig. 1 serves as a reference for timings and memory size. In practice however, very different figures can be observed depending on many parameters, including the chip and the profile of the target application.

The protocols presented in Sect. 2 are implemented through commands defined by EMV. We reused these commands as much as possible for the hybrid versions, however, some modifications have been necessary.

We modified two commands involved in the first phase, where the terminal selects the payment application and gets the data necessary to authenticate the card. First, the SELECT_APPLICATION now allows the card to indicate whether it supports a hybrid version of the protocol, and which signature is used in the certificates. Second, as EMV does not provide any command to read data of length superior to 256 bytes, we developed a new command GET_DATA_EXTENDED to this end. It is used to read $cert_{IS}$ and $cert_C$.

In addition, the PQ and hybrid versions require some operations related to the KEM scheme, which are not supported by EMV. We implemented a dedicated command PERFORM_PQ_OPERATION and use it to implement NegotiateKEM and EstablishKey, which invoque the Encaps and Decaps operations.

Finally, the original EMV command for PIN verification has been updated to decrypt the PIN using the key K. The command for transaction certificate generation has also been updated to generate $SDAD_{PQ}$ in addition to $SDAD$, as described in Fig. 5.

4.3   Performances Analysis

Memory Size. Regarding code, the implementation of ntruhps2048509 (Encaps and Decaps operations), of the two additional commands and other modifications described above amounts to 5.51 KB. Chips used for this kind of applications usually have several hundreds of kilobytes of Flash memory, so this overhead is manageable.

Regarding user-dependent data, which has to be loaded on the card during a specific operation called *personalization*, the PQ part of the protocols adds:

- $cert_{IS}^*$, which mainly includes the issuer's signature public key, and the signature by the authority,
- $cert_C^*$, which mainly includes the card's KEM public key, and the signature by the issuer,
- sk_C^*, the card's KEM secret key.

We consider the size of an instance with or without the PQ layer, and deduce the overhead. We perform the exercise for all proposed versions of FALCON and

Dilithium: FALCON-512 and FALCON-1024 (resp. cat. 1 and 5) and cat. 2, 3 and 5 versions of Dilithium, denoted in the rest of the document by Dilithium-2, Dilithium-3 and Dilithium-5 respectively. Results are shown in Table 1. We first give the *theoretical* overhead, i.e. the one obtained by considering only the size of involved data. We then give the overhead as measured in practice. We notice a difference between the two columns. This is due of course to some headers but most importantly, to the fact that writing in Flash memory is done *per page*, and that the considered chip's pages are 1024-byte long. This is why the actual numbers are multiples of 1024.

The increasing factor demonstrates how significant the impact of PQ cryptography is. In our case, it is mainly due to the size of certificates. For Dilithium-5, we even reached the maximum capacity of our application and were unable to load the data. This shows that in certain case, this migration will require modifications to allow more space for user data, and so modification to the hardware or optimizations to free some space. This figure has also an impact on the production. Indeed, as we said, these user-dependent data are introduced during the personalization, and any extension of the timing of this operation lowers the production capacity.

Transaction Timings. Table 2 presents the timings of transactions for hybrid versions of the CDA and BDH-based protocols, for the different signature algorithms we used in certificates. To ease the analysis, we measure separately the communication timings, for which we distinguish terminal-initiated (T->C) and card-initiated (C->T) transmissions, and the card processing timing. The total timing is indicated, as well as the ratio to a classic CDA transaction.

Overall, the global timing of a transaction stays in conceivable orders of magnitude. It might however be too high to be used in practice. The ratio column exhibits indeed the huge impact of hybridisation on the transaction timing. We can observe that most of the transaction timing is due to the transmission of the card certificate chain to the terminal. This is due to the large size of PQ certificates, and the relative slowness of the card-terminal interface. This latter has not been designed to support such exchanges and thus severely slows down the transaction.

The processing on the card is relatively reasonable, and only increases a little when the certificates get larger — this is due to the processing of more commands to send them. We should however keep in mind that our implementation of ntruhps2048509 is not secure against side-channel attacks. Implementing countermeasures against differential power analysis and fault attacks could bring significant overhead, up to doubling the execution time. Yet this would not change the conclusion: the communication part would still be overwhelming. Furthermore, future optimizations of algorithms and implementations, and dedicated hardware accelerators should improve the performances of the embedded part. We can also note that the hybrid BDH-based protocol is slower than the hybrid CDA one. This is mainly due to the additional Encaps operation required to provide tracking resistance.

Table 1. Personalization overhead

Algorithm	Cat.	Overhead (bytes)		Increase (%)
		Theoretical	Measured	
FALCON-512	1	3863	4096	34.29
FALCON-1024	5	5987	6144	43.91
Dilithium-2	2	7786	8192	51.07
Dilithium-3	3	10172	10240	56.61
Dilithium-5	5	13416	N/A	N/A

Table 2. Full transaction timings for hybrid protocols

Protocol	Certificate Algo.	Timings (ms)				Ratio
		Comm.		Card proc.	Total	
		T->C	C->T			
Hybrid CDA						
	FALCON-512	1075	4039	390	5504	3,28
	FALCON-1024	1111	6030	390	7531	4,49
	Dilithium-2	1155	7710	398	9263	5,53
	Dilithium-3	1197	9947	416	11560	6,90
Hybrid BDH						
	FALCON-512	1772	4753	450	6975	4,16
	FALCON-1024	1798	6746	479	9023	5,38
	Dilithium-2	1826	8436	501	10763	6,42
	Dilithium-3	1867	10696	530	13093	7,81

Using another KEM instead of NTRU, like Kyber or SABER, would not radically change the figures. Indeed, performances are of the same order of magnitude, and even if Kyber public keys are slightly larger (12%), the size of certificates — and hence the transaction timings — is dominated by the characteristics of the signature algorithm (size of public keys and signatures).

5   Conclusion

By carefully designing hybrid protocols that take into account the specificities of the current candidates to PQ standardization, we propose a solution that could be considered for future specifications of banking transaction protocols. In particular, we took care to limit the usage of digital signatures only when mandatory. We demonstrated the relevance and the feasibility of our propositions by implementing them on a real device. Our work allows to estimate the overhead of an hybrid version compared to the current protocols. It appears that a large part of the transaction time is due to the transmission of certificates. In this context, a PQ signature scheme with small signatures would be the best choice. Further work is needed to secure the embedded implementation against side-channel attacks, and to provide proofs of security for the protocols.

References

1. Albrecht, M.R., et al.: Classic McEliece: conservative code-based cryptography. Technical report (2020). https://classic.mceliece.org/
2. Alkim, E., Ducas, L., Pöppelmann, T., Schwabe, P.: NewHope without reconciliation. Cryptology ePrint Archive, Report 2016/1157 (2016). https://eprint.iacr.org/2016/1157
3. Alkim, E., Ducas, L., Pöppelmann, T., Schwabe, P.: Post-quantum key exchange - A new hope. In: Holz, T., Savage, S. (eds.) USENIX Security 2016, pp. 327–343. USENIX Association (2016)
4. ANSSI: ANSSI views on the Post-Quantum Cryptography transition (2022). https://www.ssi.gouv.fr/en/publication/anssi-views-on-the-post-quantum-cryptography-transition/
5. Avanzi, R., et al.: CRYSTALS-Kyber - Algorithm specifications and supporting documentation. Technical report (2021). https://pq-crystals.org/kyber/index.shtml, version 3.2
6. Bai, S., et al.: CRYSTALS-dilithium - algorithm specifications and supporting documentation. Technical report (2021). https://pq-crystals.org/dilithium/, version 3.1
7. Basso, A., et al.: SABER: Mod-LWR based KEM (Round 3 Submission). Technical report. https://www.esat.kuleuven.be/cosic/pqcrypto/saber/
8. Beullens, W.: Improved cryptanalysis of UOV and rainbow. In: Canteaut, A., Standaert, F.-X. (eds.) EUROCRYPT 2021. LNCS, vol. 12696, pp. 348–373. Springer, Cham (2021). https://doi.org/10.1007/978-3-030-77870-5_13
9. Beullens, W.: Breaking rainbow takes a weekend on a laptop. IACR Cryptol. ePrint Arch, p. 214 (2022). https://eprint.iacr.org/2022/214
10. Bindel, N., Herath, U., McKague, M., Stebila, D.: Transitioning to a quantum-resistant public key infrastructure. In: Lange, T., Takagi, T. (eds.) PQCrypto 2017. LNCS, vol. 10346, pp. 384–405. Springer, Cham (2017). https://doi.org/10.1007/978-3-319-59879-6_22
11. Bos, J.W., Carlson, B., Renes, J., Rotaru, M., Sprenkels, D., Waters, G.P.: Post-quantum secure boot on vehicle network processors. Cryptology ePrint Archive, Paper 2022/635 (2022). https://eprint.iacr.org/2022/635
12. Braithwaite, M.: (2016). https://security.googleblog.com/2016/07/experimenting-with-post-quantum.html
13. van den Breekel, J., Ortiz-Yepes, D.A., Poll, E., de Ruiter, J.: EMV in a nutshell (2016). https://www.cs.ru.nl/erikpoll/papers/EMVtechreport.pdf
14. Brzuska, C., Smart, N.P., Warinschi, B., Watson, G.J.: An analysis of the EMV channel establishment protocol. In: Sadeghi, A.R., Gligor, V.D., Yung, M. (eds.) ACM CCS 2013, pp. 373–386. ACM Press (2013). https://doi.org/10.1145/2508859.2516748
15. Chen, C., et al.: NTRU - algorithm specifications and supporting documentation. Technical report (2020). https://ntru.org/
16. Chen, M.S., et al.: Rainbow. Technical report (2020). https://www.pqcrainbow.org/

17. Cooper, D.A., Apon, D.C., Dang, Q.H., Miller, M.S.D.M.J.D.C.A.: Recommendation for stateful hash-based signature schemes. Technical report, NIST (2020). https://doi.org/10.6028/NIST.SP.800-208

18. Diffie, W., van Oorschot, P.C., Wiener, M.J.: Authentication and authenticated key exchanges. Des. Codes Crypt. **2**, 107–125 (1992)

19. EMVCo: EMV - Integrated Circuit Card Specifications for Payment Systems - Common Payment Application Specification (2005). version 1.0

20. EMVCo: EMV - Integrated Circuit Card Specifications for Payment Systems - Book 2 - Security and Key Management (2011). version 4.3

21. EMVCo: EMV - Integrated Circuit Card Specifications for Payment Systems - Book 3 - Application Specification (2011). version 4.3

22. EMVCo: EMV ECC Key Establishment Protocols (2012)

23. Fouque, P.A., et al.: Falcon: Fast-Fourier Lattice-based Compact Signatures over NTRU. Technical report (2020). https://falcon-sign.info/

24. Garrett, D., Ward, M.: Blinded Diffie-Hellman. In: Chen, L., Mitchell, C. (eds.) SSR 2014. LNCS, vol. 8893, pp. 79–92. Springer, Cham (2014). https://doi.org/10.1007/978-3-319-14054-4_6

25. Kannwischer, M.J., Rijneveld, J., Schwabe, P., Stoffelen, K.: pqm4: testing and benchmarking NIST PQC on ARM cortex-m4. IACR Cryptol. ePrint Arch, p. 844 (2019). https://eprint.iacr.org/2019/844

26. National Institute for Standards and Technology: Post-Quantum Cryptography Standardization. https://csrc.nist.gov/projects/post-quantum-cryptography/post-quantum-cryptography-standardization

27. Paul, S., Scheible, P.: Towards post-quantum security for cyber-physical systems: integrating PQC into industrial M2M communication. In: Chen, L., Li, N., Liang, K., Schneider, S. (eds.) ESORICS 2020. LNCS, vol. 12309, pp. 295–316. Springer, Cham (2020). https://doi.org/10.1007/978-3-030-59013-0_15

28. Schwabe, P., Stebila, D., Wiggers, T.: Post-quantum TLS without handshake signatures. In: Ligatti, J., Ou, X., Katz, J., Vigna, G. (eds.) ACM CCS 20, pp. 1461–1480. ACM Press (2020). https://doi.org/10.1145/3372297.3423350

29. Stadler, S., Sakaguti, V., Kaur, H., Fehlhaber, A.L.: Hybrid signal protocol for post-quantum email encryption. Cryptology ePrint Archive, Paper 2021/875 (2021). https://eprint.iacr.org/2021/875, https://eprint.iacr.org/2021/875

Analyzing the Leakage Resistance of the NIST's Lightweight Crypto Competition's Finalists

Corentin Verhamme[1]([✉]), Gaëtan Cassiers[1,2,3], and François-Xavier Standaert[1]

[1] ICTEAM, Université catholique de Louvain, Louvain-la-Neuve, Belgium
corentin.verhamme@uclouvain.be
[2] TU Graz, Graz, Austria
[3] Lamarr Security Research, Graz, Austria

Abstract. We investigate the security of the NIST Lightweight Crypto Competition's Finalists against side-channel attacks. We start with a mode-level analysis that allows us to put forward three candidates (Ascon, ISAP and Romulus-T) that stand out for their leakage properties and do not require a uniform protection of all their computations thanks to (expensive) implementation-level countermeasures. We then implement these finalists and evaluate their respective performances. Our results confirm the interest of so-called leveled implementations (where only the key derivation and tag generation require security against differential power analysis). They also suggest that these algorithms differ more by their qualitative features (e.g., two-pass designs to improve confidentiality with decryption leakage vs. one-pass designs, flexible overheads thanks to masking vs. fully mode-level, easier to implement, schemes) than by their quantitative features, which all improve over the AES and are quite sensitive to security margins against cryptanalysis.

1 Introduction

Security against side-channel attacks is explicitly mentioned by the NIST as a target in the ongoing standardization process for lightweight cryptography.[1] In this paper, we analyze the leakage resistance of 9 out of the 10 finalists of the competition. Our contributions in this respect are twofold.

First, we use a framework introduced by Bellizia et al. to evaluate the high-level leakage properties of the candidates' modes of operations [BBC+20]. (We exclude Grain-128AEAD from our study, which cannot be captured with such a mode vs. primitive granularity.) This high-level analysis allows us to observe that 6 candidates can mostly rely on (expensive) implementation-level countermeasures. By contrast, 3 candidates (namely Ascon [DEMS21], ISAP [DEM+20] and Romulus-T[2]) have leakage-resistant features enabling so-called leveled imple-

---

[1] https://csrc.nist.gov/Projects/lightweight-cryptography.
[2] https://romulusae.github.io/romulus/. Note that Romulus comes with different modes of operation. In particular, the (single-pass) N version does not provide mode-level leakage-resistance guarantees while the (two-pass) T version does.

© The Author(s), under exclusive license to Springer Nature Switzerland AG 2023
I. Buhan and T. Schneider (Eds.): CARDIS 2022, LNCS 13820, pp. 290–308, 2023.
https://doi.org/10.1007/978-3-031-25319-5_15

mentations, where different parts of the implementations require different (more or less expensive) implementation-level countermeasures.

Second, we investigate the hardware performances of these 3 leakage-resistant modes of operation and evaluate their leveled implementation. In leveled implementations, we distinguish between Differential Power Analysis (DPA), where the adversary is able to collect an adversarially chosen number of measurements corresponding to fixed secret inputs to the target primitive, and Simple Power Analysis (SPA), where the number of such traces is small and bounded by design. The goal of mode-level protections is to minimize the amount of computations that must be protected against DPA and SPA. For Ascon and Romulus-T, we protect the Key Derivation Function (KDF) and Tag Generation Function (TGF) against DPA with Hardware Private Circuits (HPC), a state-of-the-art masking scheme that jointly provides resistance against physical defaults and composability [CGLS21,CS21]. For ISAP, the KDF and TGF are based on a leakage-resilient PRF that embeds a fresh re-keying mechanism such that they only require security against SPA [MSGR10,BSH+14]. The latter is natively (and efficiently) obtained thanks to parallelism in hardware. For all 3 candidates, the bulk of the computation contains an internal re-keying mechanism. Hence, guarantees of confidentiality with leakage essentially require its SPA security (again achieved with hardware parallelism). This part of the implementation can even leak in an unbounded manner if only integrity with leakage is required.

The hardware design space of Ascon (and ISAP, that relies on the same permutation) has already been quite investigated in the literature, both regarding unprotected and masked implementations [GWDE15,GWDE17]. Our implementations heavily build on this state-of-the-art. By contrast, to the best of our knowledge such evaluations are a bit sparser for Romulus-T [KB22] and the Skinny block cipher it relies on [BJK+20], especially for higher-order masked implementations. Therefore, and as an additional technical contribution, we complete the study of masked Skinny implementations tailored for masking.

We conclude that more than the quantitative comparison of the finalists, the main criteria that should help the NIST in selecting a lightweight cryptography standard (if leakage is deemed important) are qualitative. The limited relevance of quantitative comparisons at this stage of the competition follows from two facts. For ciphers that rely on comparable countermeasures for their DPA security (like Ascon and Romulus-T, both leveraging masking), the performance gap is limited and quite sensitive to security margins against cryptanalysis. Both are nevertheless significantly easier to protect against leakage than the AES, as witnessed by simple proxies such as their number of AND gates or AND depth. For ciphers that rely on different countermeasures for their DPA security (like ISAP, that leverages re-keying), we currently lack (both theoretical and practical) tools that would allow a definitive comparison (e.g., with masking). By contrast, these three ciphers have different qualitative features, leading to at least two questions that could (and we think, should) guide the final selection:

- *Is confidentiality with decryption leakage wanted?* Ascon, ISAP and Romulus-T all reach the top of the hierarchy in [GPPS19] for integrity with leakage (coined CIML2). The leveled implementation of Ascon only provides confidentiality with encryption leakages and misuse-resilience (coined CCAmL1): decryption queries of a ciphertext leak the underlying plaintext via a straightforward DPA. The leveled implementations of ISAP and Romulus-T can additionally provide confidentiality with decryption leakages and misuse-resilience (coined CCAmL2) at the cost of being two-pass for decryption (they are only CCAmL1 if a single-pass decryption is performed).
- *Flexibility or simplicity for the KDF and TGF?* Ascon and Romulus-T require DPA countermeasures like masking to protect their KDF and TGF. Implementing masking securely is a sensitive process that requires expertise [MPG05,CGP+12]. But it comes with a lot of flexibility: countermeasures do not always have to be deployed, different security vs. performance tradeoffs can be considered and one can have different security levels in encryption and decryption. ISAP relies on a re-keying mechanism so that only SPA security is needed for the whole implementation, which is easy to obtain in hardware.[3] But it has no flexibility: the overheads of the leakage-resilient PRF have to be paid even if side-channel security is not a concern.

A slightly longer-term question relates to the choice between permutations and Tweakable Block Ciphers (TBCs). While the same leakage-resistant features can be obtained at somewhat similar costs from permutations and sponges, these two building blocks also come with some differences. On the one hand, TBC-based designs seem more amenable to security analyzes in the standard model [BGP+20,BGPS21], while permutations currently require idealized assumptions [DM19,GPPS20]. On the other hand, TBC-based schemes enable performing an inverse-based tag verification that can leak in full [BPPS17] while permutation-based schemes require masking [BMPS21] or additional computations [DM21] for securing this part of their design against leakage.

2   Mode-level Analysis

Our mode-level analysis follows the framework of Bellizia et al. [BBC+20]. Because of place constraints, we do not detail the specifications of the NIST's Lightweight Crypto Competition's Finalists and refer to the webpage https://csrc.nist.gov/Projects/lightweight-cryptography for this purpose. We rather focus on the features of these modes that are relevant for leakage.

The high-level decomposition of the modes we will rely on is depicted in Fig. 1. It includes a KDF that generates a fresh encryption key K^*, the bulk of the scheme that processes the message blocks, the TGF that generates the authentication tag T and the verification (that checks whether T is correct). Some parts may naturally be empty for some candidates.

---

[3] Security in low-end embedded software implementations is unclear both for masking and re-keying, which can be the target of strong attacks in low-noise contexts: see [BS21] for masking and [KPP20,BBC+20] for re-keying.

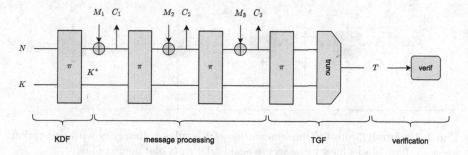

Fig. 1. Leakage-resistant modes of operation decomposition.

The goal of this decomposition is to identify the parts of the modes that must be implemented in a DPA-resistant manner and the parts of the modes that can be implemented with weaker guarantees. When analyzing confidentiality, these weaker guarantees correspond to SPA security. When analyzing integrity, it is even possible to implement those parts without any guarantee (which is referred to as the unbounded leakage model in formal analyzes). We next classify the designs based on the amount of mode-level protections they embed. At high-level (details are given in [BBC+20]), Grade-0 designs do not provide mode-level leakage-resistance; Mode-1 designs can be leveled to preserve confidentiality and integrity as long as only encryption leakages are given to the adversary (i.e., CCAmL1 and CIML1); Mode-2 designs can be leveled even if integrity with decryption leakage is required (i.e., CIML2); Mode-3 designs complete the picture by allowing leveled implementations that preserve both confidentiality and integrity with decryption leakages (i.e., CCAmL2 and CIML2).

Grade-0 Designs (no Mode-Level Protections). A first way to design modes of operation for lightweight cryptography is to focus exclusively on performance and to ignore leakage. This is the case of modes where the long-term secret key is used by most of the underlying primitives. In the NIST lightweight crypto competition, it is for example what happens for Elephant, GIFT-COFB, Romulus-N, Romulus-M and TinyJambu. A protected implementation of Romulus-N targeting integrity with encryption leakage is illustrated in Fig. 2, where the blue color is used to reflect that the corresponding computations must be protected against DPA. This requirement essentially holds for any security target (i.e., for confidentiality and integrity, with or without nonce-misuse and leakage available in encryption or decryption). We insist that being Grade-0 does not imply that these modes cannot be protected against leakage. It rather implies that this protection will be expensive because uniformly applied to all the components of the modes. The following (higher-level) designs gradually increase the mode-level protections, leading to different trade-offs between the efficiency of their unprotected implementations (that mildly decreases) and the efficiency of their protected implementations (that significantly increases for long messages).

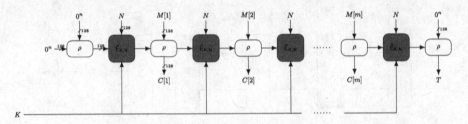

Fig. 2. Uniformly protected implementation of Romulus-N (integrity with encryption leakage). Blue blocks have to be secure against DPA. (Color figure online)

Grade-1 Designs (internal Re-Keying). A first step towards building modes of operation that cope better with leakage is to embed an internal re-keying mechanism. In this case, the mode first generates a fresh key K^* from the long-term key and the nonce, which is then updated after the processing of each message block. As a result, and as long as the adversary can only observe encryption leakage without nonce misuse, only the KDF needs security against DPA (as there is a DPA using the nonce) and all the other computations must only be protected against SPA. Such a leveled implementation is illustrated in Fig. 3 for PHOTON-Beetle. Unfortunately, this guarantee vanishes as soon as nonce misuse or decryption leakage are granted to the adversary. In this case the adversary can target the processing of one message block with many different messages (while keeping the nonce and all the the other message blocks constant) and perform a DPA to recover the corresponding intermediate state. In the case of a P-sponge construction [BDPA07], it is then possible to invert the permutation and get back to the long-term key. In the NIST lightweight crypto competition, it is the case of PHOTON-Beetle, Sparkle and Xoodyak.

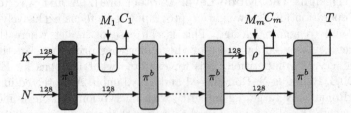

Fig. 3. Leveled implementation of PHOTON-Beetle (integrity with encryption leakage). Blue (resp., green) blocks have to be secure against DPA (resp., SPA). (Color figure online)

Grade-2 (Grade-1 + Strengthened KDF/TGF). The second step towards building modes of operation that cope better with leakage is to strengthen the KDF/TGF so that the recovery of an internal state of the mode cannot lead to

long-term secrets. This is easily (and efficiently) done by making the KDF and
the TGF non-invertible. In the case of sponges, it can be achieved by XORing
the long-term key before and after the permutation used to generate the fresh
key K^* and the tag T. For TBCs, it is a direct consequence of their PRP
security. In the NIST lightweight crypto competition, it is for example the case of Ascon.
For illustration, its leveled implementation is illustrated in Fig. 4.

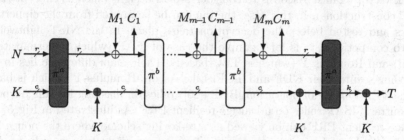

(a) Integrity requirements (with decryption leakage).

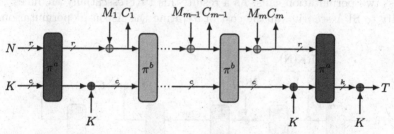

(b) Confidentiality requirements (without decryption leakage).

Fig. 4. Leveled implementation of Ascon. The blue blocks have to be protected against
DPA and the green blocks have to be protected against SPA, while the white ones do
not require protection against side-channel leakage. (Color figure online)

The top of the figure depicts the integrity requirements. In this case, only the
KDF and the TGF (in blue) must be protected against DPA and the rest of the
computations (in white) can leak in full. This guarantee holds even when nonce
misuse and leakage in decryption are granted to the adversary. It intuitively
derives from the fact that the ephemeral secrets cannot be used to infer long-
term ones, and corresponds to the top of the hierarchy introduced in [GPPS19].
The bottom of the figure depicts the confidentiality requirements. In this case,
it is naturally not possible to tolerate unbounded leakage. Yet, as long as the
adversary is not granted with decryption leakage, only SPA security (in green)
is required for this part of the computation. (The orange color for the plaintexts is used to reflect that even their very manipulation may leak sensitive
information). The main attack vector that remains against this construction
happens with decryption leakage. Since the message is decrypted before verify-
ing the tag, an adversary can then target the processing of one message block

with many different messages (keeping the nonce and all the the other message blocks constant) and perform a DPA to recover the corresponding intermediate state. This reveals the ephemeral keystream, hence the message, but does not affect the confidentiality of messages encrypted with a different nonce.

Grade-3 (Grade-2 + Two Passes). The natural way to get rid of the last attack vector against Ascon is to consider 2-pass designs (such as encrypt-then-MAC constructions). In this case, the tag can be computed from the ciphertext blocks and tested before the decryption takes place. In the NIST lightweight crypto competition, it is for example the case of ISAP (which is permutation-based) and Romulus-T (which is TBC-based). Their main difference lies in the way they secure their KDF and TGF. Like Ascon, Romulus-T (which is based on the TEDT mode of operation [BGP+20]) relies on masking for this purpose. By contrast, ISAP relies on a leakage-resilient PRF. As illustrated in Fig. 5, the leakage-resilient PRF can be viewed as a re-keying scheme where the nonce bits are absorbed one by one so that each of its intermediate keys is only used to process two permutation calls. As a result, this PRF essentially "reduces" DPA security to SPA security, at the cost of iterating the Ascon-p^1 permutation.[4]

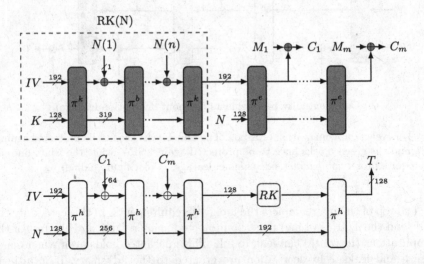

Fig. 5. Leveled implementation of ISAP (confidentiality with decryption leakage). The green blocks have to be protected against SPA (with averaging), while the white ones do not require any protection against side-channel leakage. (Color figure online)

Overall, ISAP's confidentiality with decryption leakage requires two calls to the leakage-resilient PRF and a plaintext processing that is secure against SPA. We provide a similar picture for the leveled implementation of Romulus-T in

---

[4] Increasing the rate to absorb more bits and get a more efficient design is possible but it then opens a DPA attack vector (so we do not consider this option here).

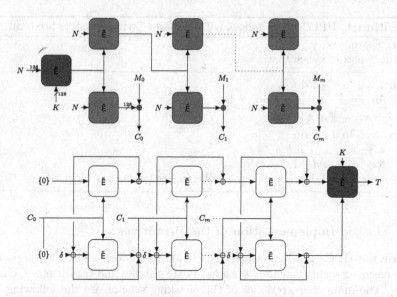

Fig. 6. Leveled implementation of Romulus-T (confidentiality with decryption leakage). The blue blocks have to be protected against DPA and the green blocks have to be protected against SPA (with averaging), while the white ones do not require any protection against side-channel leakage. (Color figure online)

Fig. 6, where the KDF and TGF are directly instantiated with a masked TBC. Note that the SPA-secure blocks of these two figures are in dark green to reflect the possibility that the adversary repeats the same measurements to average the noise. And as in the case of Ascon, integrity with decryption leakage only requires the two protected calls (to the leakage-resilient PRF or masked TBC) and can let all the other computations leak in full.

3   Hardware Implementations

Given the previous analysis, it appears that Ascon, ISAP and Romulus-T are the most promising candidates of the NIST lightweight crypto competition for leakage-resistant implementations. In this section, we therefore investigate their hardware implementations. For this purpose, we focus on the security guarantees that they enable without a uniformly protected implementation. We first investigate their primitives, with a special focus on Romulus-T and its underlying TBC Skinny-384+ (for which, as mentioned in introduction, the literature is a bit scarcer). We then detail the implementation of the modes and report their performances (with ASIC synthesis). We use these results to confirm the relevance of leveled implementations and to discuss the respective interest of the three implemented ciphers in the context of the NIST competition.

Algorithm 1. HPC2 AND gadget with d shares (sync. registers are omitted).

Input: Sharings $\mathbf{x}$, $\mathbf{y}$
Output: Sharing $\mathbf{z}$ such that $z = x \cdot y$.

1: **for** $i = 0$ to $d - 1$ **do**
2:     **for** $j = i + 1$ to $d - 1$ **do**
3:         $r_{ij} \xleftarrow{\$} \mathbb{F}_2$; $r_{ji} \leftarrow r_{ij}$
4: **for** $i = 0$ to $d - 1$ **do**
5:     $\mathbf{z}_i \leftarrow \mathbf{x}_i \mathbf{y}_i$
6:     **for** $j = 0$ to $d - 1, j \neq i$ **do**
7:         $\mathbf{z}_i \leftarrow \mathbf{z}_i \oplus \mathsf{Reg}\left((\mathbf{x}_i \oplus 1)\, r_{ij}\right) \oplus \mathsf{Reg}\left(\mathbf{x}_i \mathsf{Reg}\left(\mathbf{y}_j \oplus r_{ij}\right)\right)$

3.1   Masked Implementation of the Primitives

We use the HPC2 masking scheme, as it allows almost arbitrary composition while ensuring security against both hardware glitches and transitions [CGLS21, CS21]. The main characteristics of this masking scheme are the following. The linear operations are very efficient, since they are made of purely combinational logic and have a linear overhead in the masking order. On the other hand, the non-linear operation, which is the 2-input AND gate (see Algorithm 1), has quadratic overhead and asymmetric latency: 2 cycles with respect to one input, and only 1 cycle with respect to the other input.

Skinny Sbox. The Skinny Sbox (depicted in Fig. 7) is made of XOR gates (that are linear, and therefore easy to implement) and of NOR gates that we implement with a AND gate whose inputs are inverted.

We next propose two area-optimized architectures for this Sbox, that both instantiate two masked AND and two masked XOR gates. First, the high-throughput one is based on the observation that the Sbox can be decomposed into 4 applications of a simpler function (as visible in Fig. 7), followed by a wire shuffling. This function has a latency of at least two cycles with our masking scheme, due to the AND gate. We therefore build the high-throughput Sbox by looping 4 times on a two-stage pipeline. The core of this pipeline is a simple function block B shown in Fig. 8a, which is then used twice and connected to combinational logic to form the Sbox (Fig. 8b). The two-stage pipeline can be used to perform simultaneously two Sbox operations. We finally add input and output synchronization registers such that the logic performs two Sbox evaluations in 9 clock cycles, without the need for any external synchronization mechanism. Second, we design a low-latency architecture with an ALU-style design: the Sbox inputs are stored in a register, as well as the outputs, and the data fed to the two AND-XOR blocks are selected from these states (and from the input wires) when needed (see Fig. 10). This flexible architecture allows to benefit from the asymmetric latency of the HPC2 AND gadgets, leading to a latency of 6 cycles for one Sbox evaluation (see Fig. 9).

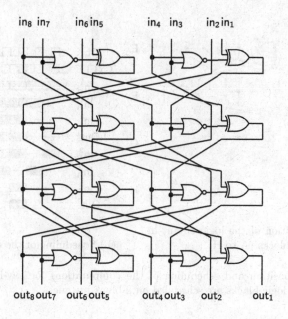

Fig. 7. Skinny Sbox circuit representation with NOR and XOR gates.

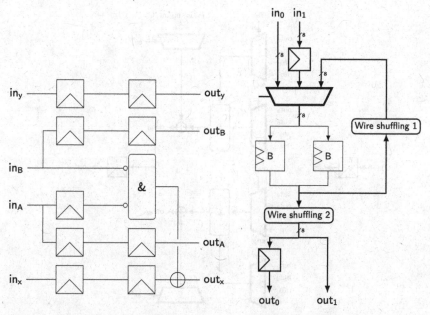

(a) Inner logic block "B" for Sbox implementation, with one AND and one XOR.

(b) High-throughput Sbox architecture with input and output sync. registers.

Fig. 8. High-throughput masked Skinny Sbox.

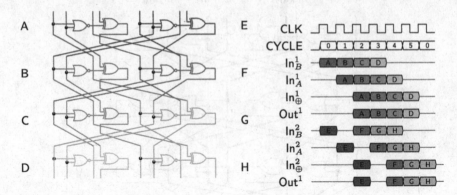

(a) Decomposition of the logic circuit in iterated logic blocks (A to H).

(b) Scheduling of the computations.

Fig. 9. Decomposition and scheduling of the computations for low-latency masked Skinny Sbox: 8 logic blocks are scheduled on 2 block instances.

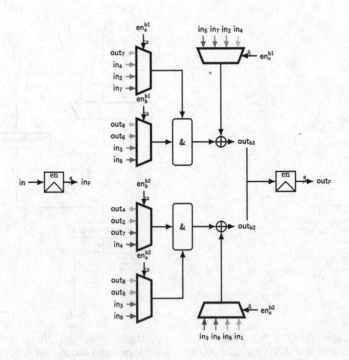

Fig. 10. Architecture of low-latency masked Skinny Sbox.

Lastly, we discuss fully pipeline architectures. While such architectures can achieve very high throughput that compensate for their large area, they are difficult to use in our case. Indeed, we are not interested in parallel Skinny evaluations (since this is not useful for encrypting a single message with a Romulus-T leveled implementation). Therefore, the latency overhead of filling the pipeline (of at least 6 cycles) is significant when it is used for only 16 Sbox evaluations (a Skinny round). We however note that the only other HPC Skinny implementation we know of uses that strategy, and uses a depth-12 pipeline [KB22].[5]

Skinny. Based on these Sbox architectures, we design three Skinny implementations with various area vs. latency trade-offs. The two first ones are simple round-based architectures, as shown in Fig. 11, where either 16 low-latency Sboxes ("low-latency Skinny", S_{LL}) or 8 high-throughput ("balanced Skinny", S_B) Sboxes are used. The third implementation ("small Skinny", S_S) targets lower area: it is a serialized architecture that instantiates only one high-throughput Sbox (that is used 8 times per round), as shown in Fig. 12.

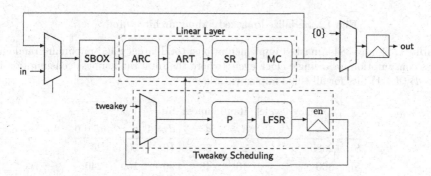

Fig. 11. Round-based masked Skinny architecture (S_{LL} and S_B).

Let us now discuss the performance of these implementations. We consider latency, randomness requirements and area as performance metrics, since the critical path will be similar in all cases (it lies in the linear layer). The latency and maximum randomness requirements per cycle of the implementations are shown in Table 1. We can see that the S_S implementation has 8 times the latency of S_B (due to 8x serialization), while S_{LL} reduces latency by 33 % compared to S_B. Regarding randomness, the maximum randomness throughput of S_S is 8 times lower than S_B, and the one of S_{LL} is twice the one of S_B.

Next, we look at area requirements in Fig. 13. The Sbox logic area clearly reflects the architectural choices: 2 AND and XOR gadgets for S_S, 16 of each for S_B, and 32 of each for S_{LL}. Next, the remaining Sbox area is fairly high for S_{LL}

---

[5] Which can be improved to depth-6 thanks to the asymmetric latency of HPC2 ANDs.

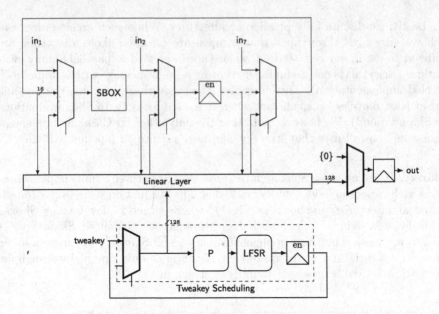

Fig. 12. Serialized masked Skinny architecture (S_S).

Table 1. Skinny-384+ masked implementations: total latency and maximum randomness consumption for a single clock cycle (where the total randomness consumption is $64 \cdot d \cdot (d-1)$ bits for all three implementations).

	Latency [cycle]	Randomness [bit]				
		$d = 2$	$d = 3$	$d = 4$	$d = 5$	$d = 6$
S_S	2880	2	6	12	20	30
S_B	360	16	48	96	160	240
S_{LL}	240	32	96	192	320	480

due to the large number of Sbox instances and due to their large MUXes and registers. For S_B and S_S, the larger number MUXes and registers in the Sboxes of the former compensate for the more complex datapath of the latter, resulting in a similar "routing" area for both of them. The remaining parts of Skinny are the same for all three architectures. Overall, the difference in area between the architectures is small for low number of shares, and increases as the latter grows. For all considered number of shares ($d \leq 6$), the Sboxes do not dominate the area of neither S_S nor S_B, hence S_S brings a limited area gain at a large latency cost compared to S_B. On the other hand, S_{LL} has an area overhead of up to 39 % (for $d \leq 6$), and a latency gain of 33 % over S_B.

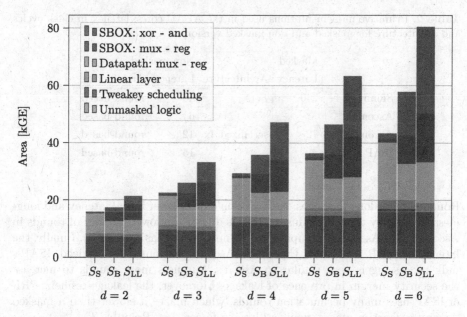

Fig. 13. Area requirements for the three masked Skinny hardware implementations in a 65 nm ASIC technology using the HPC2 masking scheme.

3.2   Implementation of the Modes

We implemented side-channel protected hardware accelerators for Romulus-N, Romulus-T, Ascon and ISAP, using the primitives described in Table 2. The Romulus-N implementation is fully masked and uses one S_{LL} instance. Next, the implementation of Romulus-T is leveled with one masked instance of Skinny (we also used S_{LL}) and four non-masked Skinny instances (with a round-based architecture). Similarly, the Ascon implementation is also leveled. The masked Ascon-p primitive is based on the HPC2 masking scheme and is serialized with 16 Sbox instances (each Sbox is a 2-stage pipeline performing 4 Sbox evaluations per round)[6], while the non-masked permutation is round-based (1 cycle per round). Finally, ISAP uses two instances of the non-masked Ascon-p primitive.

Let us first discuss the latency of these implementations with Fig. 14. The encryption time of Romulus-N grows very quickly with the message size due to the need of masking all Skinny calls, which are slow compared to non-masked calls (as shown in Table 2). However, for very short messages, Romulus-N is fairly competitive thanks to its low number of Skinny calls it that case. On the other hand, Romulus-T has a larger upfront cost, due to the larger number of Skinny calls even for short messages, however the mode-level leakage resistance allows to use non-masked calls for the bulk processing, resulting in lower latency than

---

[6] This choice is somewhat arbitrary: we took a serialization factor that gives a good latency versus area trade-off. It also happens to lead to a latency of 6 clock cycles per round, which is the same latency as a round of S_{LL}.

Table 2. Primitive implementations used in the AEAD cores: latency in clock cycles and architecture for masked and non-masked versions.

	Masked		Non-masked	
	Latency	Architecture	Latency	Architecture
Skinny-384+	240	S_{LL}	40	round-based
Ascon-p^6			6	round-based
Ascon-p^{12}	72	serialized 4x	12	round-based
ISAP RK			152	round-based

Romulus-N for long messages. Next, Ascon enjoys lower initial latency and long-message latency than Romulus-T. This is due to the lower number of rounds in Ascon-p^6 and Ascon-p^{12} compared to Skinny (which has 40 rounds). Finally, the latency of ISAP is between the one of Ascon and Romulus-T. Indeed, ISAP's bulk processing is very similar to Ascon's, but uses more rounds to increase the security margin in presence of leakage. Moreover, the leakage-resilient PRF of ISAP uses many permutation rounds, which makes it slower than a masked Ascon for short messages, while still being faster than Romulus-T.

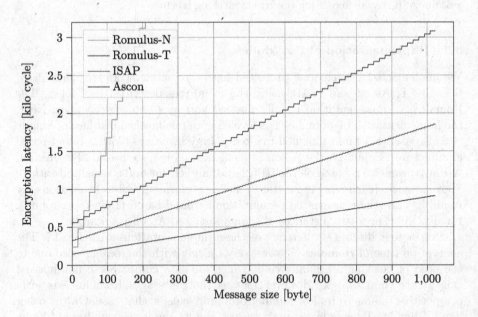

Fig. 14. Encryption latency as a function of the message size.

Let us now discuss the area usage with Fig. 15. As a general trend, for implementations with a masked primitive, the area for that primitive dominates the overall area, with the exception of Romulus-T with $d = 2$ shares (where the area

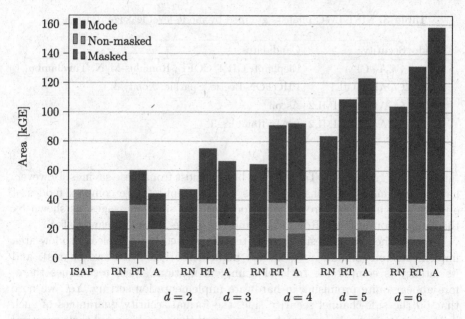

Fig. 15. Area requirements of leakage-resistant hardware AEAD cores in 65 nm ASIC technology. For leveled implementations, the area is split in three part: DPA-protected (i.e., masked) primitives, SPA-protected primitives (implemented in parallel), and mode (i.e., logic not in a primitive). RN/RT respectively stand for Romulus-N/Romulus-T and A stands for Ascon.

of the four non-masked Skinny instances dominates). These results therefore confirms the interest of leveled implementations. Next, the areas for all these modes is fairly similar, with a slight advantage to Ascon and Romulus-N at $d = 2, 3$ shares thanks to their lower unmasked area, while Romulus-T and Romulus-N are a bit better than Ascon for larger numbers of shares, thanks to using less masked AND gadgets. Lastly, the area for ISAP is similar to the area of the leveled implementations with $d = 2$ shares.

4   Conclusion

Even though the previous quantitative results should be interpreted with care, since they explore only a few points in the design space (e.g., we considered only round-based architectures for non-masked primitives and did not optimize the masking randomness usage), their comparison highlights a few general trends.

First, for long messages, leveled implementations bring large latency improvements while their area overheads remain small over non-leveled implementations. This is because unmasked primitives are small compared to the masked ones, especially when the number of shares is large. This smaller unmasked area as well as lower latency naturally translate into large energy savings. The candidates that can be implemented in such a way for both encryption and decryption

Table 3. NIST LWC finalists grouped by mode-level leakage resistance.

Grade	Security	Candidates
0	CCA+CI	Elephant, GIFT-COFB, Romulus-M/N, TinyJambu
1	CCAL1+CIL1	PHOTON-Beetle, Sparkle, Xoodyak
2	CCAmL1+CIML2	Ascon
3	CCAmL2+CIML2	ISAP, Romulus-T

(grade-2 and grade-3, see Table 3) will benefit most from these savings. However, having such mode-level characteristics usually implies more complex modes of operations, which leads to worse performance for small messages, as shown by the comparison between Romulus-N and the grade-2/grade-3 candidates.

Second, the leakage-resilient PRF technique used by ISAP leads to low area implementation (similar to a leveled implementations with two shares), and its latency is comparable to leveled implementations. Such techniques therefore appear quite promising in hardware implementation setting. Yet, we note that on the side-channel security side, the formal security guarantees of such implementations have been much less analyzed than masking and their practical security evaluation can be more challenging as well (see, e.g., [UBS21]).

Finally, we observe that the different security margins of the algorithms we implemented can explain some of the observed performance differences. For example, the differences in latency between the leveled implementations of Ascon and ISAP are explained by their number of rounds: the hashing part of ISAP uses Ascon-p^{12} while the Ascon inner sponge part uses only Ascon-p^6. When considering CIML2, both should however withstand the same attacks.

Overall, these results backup our suggestion that from a side-channel security viewpoint, the finalists of the NIST's Lightweight Crypto competition differ more by their qualitative features than by their quantitative performances.

Acknowledgments. Gaëtan Cassiers and François-Xavier Standaert are respectively Research Fellow and Senior Associate Researcher of the Belgian Fund for Scientific Research (FNRS-F.R.S.). This work has been funded in part by the ERC consolidator grant number 724725 (acronym SWORD), by the Walloon region CyberExcellence project number 2110186 (acronym Cyberwal) and by the Horizon Europe project 1010706275 (acronym REWIRE).

References

[BBC+20] Bellizia, D., et al.: Mode-level vs. implementation-level physical security in symmetric cryptography. In: Micciancio, D., Ristenpart, T. (eds.) CRYPTO 2020. LNCS, vol. 12170, pp. 369–400. Springer, Cham (2020). https://doi.org/10.1007/978-3-030-56784-2_13

[BDPA07] Bertoni, G., Daemen, J., Peeters, M., Van Assche, G.: Sponge functions. In: ECRYPT Hash Workshop (2007)

[BGP+20] Berti, F., Guo, C., Pereira, O., Peters, T., Standaert, F.-X.: TEDT, a leakage-resist AEAD mode for high physical security applications. IACR Trans. Cryptogr. Hardw. Embed. Syst. **2020**(1), 256–320 (2020)

[BGPS21] Berti, F., Guo, C., Peters, T., Standaert, F.-X.: Efficient leakage-resilient MACs without idealized assumptions. In: Tibouchi, M., Wang, H. (eds.) ASIACRYPT 2021. LNCS, vol. 13091, pp. 95–123. Springer, Cham (2021). https://doi.org/10.1007/978-3-030-92075-3_4

[BJK+20] Beierle, C., et al.: SKINNY-AEAD and skinny-hash. IACR Trans. Symmetric Cryptol. **2020**(S1), 88–131 (2020)

[BMPS21] Bronchain, O., Momin, C., Peters, T., Standaert, F.-X.: Improved leakage-resistant authenticated encryption based on hardware AES coprocessors. IACR Trans. Cryptogr. Hardw. Embed. Syst. **2021**(3), 641–676 (2021)

[BPPS17] Berti, F., Pereira, O., Peters, T., Standaert, F.-X.: On leakage-resilient authenticated encryption with decryption leakages. IACR Trans. Symmetric Cryptol. **2017**(3), 271–293 (2017)

[BS21] Bronchain, O., Standaert, F.-X.: Breaking masked implementations with many shares on 32-bit software platforms or when the security order does not matter. IACR Trans. Cryptogr. Hardw. Embed. Syst. **2021**(3), 202–234 (2021)

[BSH+14] Belaïd, S., et al.: Towards fresh re-keying with leakage-resilient PRFs: cipher design principles and analysis. J. Cryptogr. Eng. **4**(3), 157–171 (2014). https://doi.org/10.1007/s13389-014-0079-5

[CGLS21] Cassiers, G., Grégoire, B., Levi, I., Standaert, F.-X.: Hardware private circuits: from trivial composition to full verification. IEEE Trans. Comput. **70**(10), 1677–1690 (2021)

[CGP+12] Coron, J.-S., Giraud, C., Prouff, E., Renner, S., Rivain, M., Vadnala, P.K.: Conversion of security proofs from one leakage model to another: a new issue. In: Schindler, W., Huss, S.A. (eds.) COSADE 2012. LNCS, vol. 7275, pp. 69–81. Springer, Heidelberg (2012). https://doi.org/10.1007/978-3-642-29912-4_6

[CS21] Cassiers, G., Standaert, F.-X.: Provably secure hardware masking in the transition- and glitch-robust probing model: better safe than sorry. IACR Trans. Cryptogr. Hardw. Embed. Syst. **2021**(2), 136–158 (2021)

[DEM+20] Dobraunig, C., et al.: ISAP v2.0. IACR Trans. Symmetric Cryptol. **2020**(S1), 390–416 (2020)

[DEMS21] Dobraunig, C., Eichlseder, M., Mendel, F., Schläffer, M.: Ascon v1.2: lightweight authenticated encryption and hashing. J. Cryptol. **34**(3), 1–42 (2021). https://doi.org/10.1007/s00145-021-09398-9

[DM19] Dobraunig, C., Mennink, B.: Leakage resilience of the duplex construction. In: Galbraith, S.D., Moriai, S. (eds.) ASIACRYPT 2019. LNCS, vol. 11923, pp. 225–255. Springer, Cham (2019). https://doi.org/10.1007/978-3-030-34618-8_8

[DM21] Dobraunig, C., Mennink, B.: Leakage resilient value comparison with application to message authentication. In: Canteaut, A., Standaert, F.-X. (eds.) EUROCRYPT 2021. LNCS, vol. 12697, pp. 377–407. Springer, Cham (2021). https://doi.org/10.1007/978-3-030-77886-6_13

[GPPS19] Guo, C., Pereira, O., Peters, T., Standaert, F.-X.: Authenticated encryption with nonce misuse and physical leakage: definitions, separation results and first construction. In: Schwabe, P., Thériault, N. (eds.) LATINCRYPT 2019. LNCS, vol. 11774, pp. 150–172. Springer, Cham (2019). https://doi.org/10.1007/978-3-030-30530-7_8

[GPPS20] Guo, C., Pereira, O., Peters, T., Standaert, F.-X.: Towards low-energy leakage-resistant authenticated encryption from the duplex sponge construction. IACR Trans. Symmetric Cryptol. **2020**(1), 6–42 (2020)

[GWDE15] Gross, H., Wenger, E., Dobraunig, C., Ehrenhöfer, C.: Suit up! - made-to-measure hardware implementations of ASCON. In: DSD, pp. 645–652. IEEE Computer Society (2015)

[GWDE17] Groß, H., Wenger, E., Dobraunig, C., Ehrenhöfer, C.: ASCON hardware implementations and side-channel evaluation. Microprocess. Microsyst. **52**, 470–479 (2017)

[KB22] Khairallah, M., Bhasin, S.: Hardware implementations of Romulus: exploring nonce-misuse resistance and Boolean masking. In: NIST Lightweight Cryptography Workshop (2022)

[KPP20] Kannwischer, M.J., Pessl, P., Primas, R.: Single-trace attacks on Keccak. IACR Trans. Cryptogr. Hardw. Embed. Syst. **2020**(3), 243–268 (2020)

[MPG05] Mangard, S., Popp, T., Gammel, B.M.: Side-channel leakage of masked CMOS gates. In: Menezes, A. (ed.) CT-RSA 2005. LNCS, vol. 3376, pp. 351–365. Springer, Heidelberg (2005). https://doi.org/10.1007/978-3-540-30574-3_24

[MSGR10] Medwed, M., Standaert, F.-X., Großschädl, J., Regazzoni, F.: Fresh re-keying: security against side-channel and fault attacks for low-cost devices. In: Bernstein, D.J., Lange, T. (eds.) AFRICACRYPT 2010. LNCS, vol. 6055, pp. 279–296. Springer, Heidelberg (2010). https://doi.org/10.1007/978-3-642-12678-9_17

[UBS21] Udvarhelyi, B., Bronchain, O., Standaert, F.-X.: Security analysis of deterministic re-keying with masking and shuffling: application to ISAP. In: Bhasin, S., De Santis, F. (eds.) COSADE 2021. LNCS, vol. 12910, pp. 168–183. Springer, Cham (2021). https://doi.org/10.1007/978-3-030-89915-8_8

Author Index

Printed in the United States
by Baker & Taylor Publisher Services